OPTICAL DETECTORS FOR ASTRONOMY

ASTROPHYSICS AND SPACE SCIENCE LIBRARY

VOLUME 228

OPTICAL DETECTORS FOR ASTRONOMY

Proceedings of an ESO CCD Workshop
held in Garching, Germany,
October 8–10, 1996

Edited by

JAMES W. BELETIC

and

PAOLA AMICO

European Southern Observatory,
Garching bei München, Germany

KLUWER ACADEMIC PUBLISHERS
DORDRECHT / BOSTON / LONDON

Library of Congress Cataloging-in-Publication Data

ESO Workshop (1996 : Garching, Germany)
Optical detectors for astronomy : October 8-10, 1996, Garching, Germany / edited by James W. Beletic and Paola Amico.
p. cm. -- (Astrophysics and space science library ; v. 228)
ISBN 0-7923-4925-3
1. Astronomical instruments--Congresses. 2. Optical detectors--Congresses. 3. Charge coupled devices--Congresses. I. Beletic, James W. II. D'Amico, Paolo, M.D. III. Title. IV. Series.
QB84.5.E76 1998
522--dc21 97-49807

ISBN 0-7923-4925-3

Published by Kluwer Academic Publishers,
P.O. Box 17, 3300 AA Dordrecht, The Netherlands.

Sold and distributed in North, Central and South America
by Kluwer Academic Publishers,
101 Philip Drive, Norwell, MA 02061, U.S.A.

In all other countries, sold and distributed
by Kluwer Academic Publishers,
P.O. Box 322, 3300 AH Dordrecht, The Netherlands.

Printed on acid-free paper

Printed in the Netherlands

TABLE OF CONTENTS

SECTION IV:
MOSAIC PROJECTS, SPECIAL TECHNIQUES **185**

These Proceedings are dedicated to the memory of

Walter F. Kosonocky

«Walter's remarkable career spanned the entire quarter century of CCD history. His contributions in the early years include the ubiquitous floating-diffusion output amplifier, as well as pioneering work in key issues for imagers such as noise, trapping, and blooming control. Later he made major contributions to the development of PtSi Schottky-barrier infra-red imagers. In this volume, you will find the talk he presented at this conference on a novel high-frame-rate burst-mode imager that demonstrated his continued high level of creativity.

Until this conference, I knew Walter primarily through his numerous publications on CCDs, but here in Munich I came to appreciate another side of him. Though we both arrived tired from an all-night flight he coaxed me to go with him to Oktoberfest. It was a wonderful opportunity to get to know this energetic and affable person. I thought this photo I took of Walter enjoying a beer and wurst would be a nice memento of our last opportunity to see him. »

Barry E. Burke

A gaggle of CCD "nerds" at ESO - October 1996

LIST OF PARTICIPANTS

ABBOTT, Tim	Canada-France-Hawaii Telescope	USA
AMICO, Paola	European Southern Observatory	Germany
ANDERSEN, Johannes	Niels Bohr Institute for Astronomy Physics & Geophysics Copenhagen	Denmark
ANDERSEN, Michael	Niels Bohr Institute for Astronomy Physics & Geophysics Copenhagen	Denmark
ANGUITA, Jaime	European Southern Observatory	Chile
ASPIN, Colin	Nordic Optical Telescope	Spain
AVILA, Gerardo	European Southern Observatory	Germany
BAADE, Dietrich	European Southern Observatory	Germany
BAUER, Florian	Services de la Physique des Particules Dapnia	France
BEHNKE, Thomas	Institute of Planetary Exploration Berlin	Germany
BEIGBEDER, Francis	Observatoire Midi-Pyrenees	France
BELETIC, James	European Southern Observatory	Germany
BERGER, Jean-Pierre	CNRS - Observatoire de Haute Provence	France
BLECHA, Andre	Observatoire de Geneve	Switzerland
BLOUKE, Morley	Scientific Imaging Technologies Inc.	USA
BÖHM, Torsten	European Southern Observatory	Germany
BONACCINI, Domenico	European Southern Observatory	Germany
BORGEAUD, Pierre	CEA Saclay DAPNIA/SEI	France
BOULADE, Oliver	CEA Saclay DAPNIA/SPP	France
BREDTHAUER, Richard	Lockheed Martin Fairchild Systems	USA
BURKE, Barry	M.I.T. Lincoln Laboratory	USA
BURNAGE, Robert	CNRS - Observatoire de Haute Provence	France
CANNON, Russell	Anglo-Australian Observatory	Australia
COMARI, Maurizio	Osservatorio Astronomico di Trieste	Italy
CORRIE, Brian	Scientific Imaging Technologies Inc.	USA
COSENTINO, Rosario	Osservatorio Astrofisico di Catania	Italy
COWLEY, David	University of California at Santa Cruz Lick Observatory	USA
CROCKER, James	European Southern Observatory	Germany

CUMANI, Claudio	European Southern Observatory	Germany
D'ODORICO, Sandro	European Southern Observatory	Germany
DEIRIES, Sebastian	European Southern Observatory	Germany
DE KAT, Jean	CEA Saclay DAPNIA/SEI	France
DEKKER, Hans	European Southern Observatory	Germany
DENNEFELD, Michel	Institute d'Astrophysique de Paris	France
DONALDSON, Robert	European Southern Observatory	Germany
DUDZIAK, Gregory	European Southern Observatory	Germany
DUPIN, Jean-Pierre	Observatoire Midi-Pyrenees	France
DUVARNEY, Ray	Emory University	USA
EPSTEIN, Gerard	Observatoire de Paris-Meudon	France
EVANS, Geoffrey	South African Astronomical Observatory	South Africa
GAI, Mario	Osservatorio Astronomico di Torino	Italy
GEARY, John	Smithsonian Astrophysical Observatory	USA
GEIMER, Christoph	European Southern Observatory	Germany
GERDES, Rolf	European Southern Observatory	Germany
GILMORE, David	University of California at Santa Cruz Lick Observatory	USA
GILMOZZI, Roberto	European Southern Observatory	Germany
GOILLANDEAU, Michel	CNRS - Observatoire de Haute Provence	France
GRIMM, Bernhard	MPI für Astronomie Heidelberg	Germany
GRUNEISEN, Rachel	Observatoire de Paris-Meudon	France
HADDAD, Nicolas	European Southern Observatory	Chile
HARDY, Tim	Dominion Astrophysical Observatory	Canada
HARRIS, Frederick	U.S. Naval Observatory	USA
HOLTOM, Ralph	EEV	UK
HUBIN, Norbert	European Southern Observatory	Germany
IVES, Derek	Royal Greenwich Observatory	UK
IWERT, Olaf	European Southern Observatory	Germany
JORDEN, Paul	Royal Greenwich Observatory	UK
KELLER, Uwe	Max Planck Institut für Aeronomie Katlenburg-Lindau	Germany
KLEIN, Peter	University of Dortmund	Germany

KNISPEL, Gerard	CNRS - Observatoire de Haute Provence	France
KOHLEY, Ralf	Sternwarte - Universität Bonn	Germany
KOSONOCKY, Walter	New Jersey Institute of Technology	USA
KRAMM, Rainer	Max Planck Institut für Aeronomie Katlenburg-Lindau	Germany
KRAUSE, Norbert	MPE Halbleiterlabor Pasing	Germany
LEIBUNDGUT, Bruno	European Southern Observatory	Germany
LOCHMANN, Christian	Laser Component GMBH Olching bei München	Germany
LONGINOTTI, Antonio	European Southern Observatory	Germany
LUCIANO, Alessio	CNR - Experimental Medicine Roma	Italy
LUPPINO, Gerard	University of Hawaii	USA
MACKAY, Craig	University of Cambridge	UK
MACQUEEN, Phillip	The University of Texas at Austin	USA
MAIRE, Charles	Observatoire de Geneve	Switzerland
MANDEL, Holger	Landessternwarte Heidelberg	Germany
MARIEN, Karl-Heinz	Max Planck Institut für Astronomie Heidelberg	Germany
MAURY, Alain	Observatoire de la Cote d'Azur	France
MEIDINGER, Norbert	MPE MPI-Halbleiterlabor München	Germany
MENZIES, John	South African Astronomical Observatory	South Africa
MICHAELIS, Harald	Institute of Planetary Exploration Berlin	Germany
MIYAZAKI, Satoshi	National Astronomical Observatory.	Japan
MOHAMMADZADEH, Ali	European Southern Observatory	Germany
MONNET, Guy	European Southern Observatory	Germany
MORRIS, David	EEV	UK
MUELLER, Philipp	Radioastr. Inst. Bonn University	Germany
MUROWINSKI, Richard	Dominion Astrophysical Observatory	Canada
NØRREGAARD, Preben	Copenhagen University Observatory	Denmark
OATES, Anthony	Royal Greenwich Observatory	UK
PEACOCK, Tone	ESA/ESTEC Nordwijk	The Netherlands
PIEPER, Holger	European Southern Observatory	Germany
POOL, Peter	EEV	UK

RAGEY, Jean Pierre	Observatoire de la Cote d'Azur	France
RAHMER, Gustavo	European Southern Observatory	Chile
REED, Richard	National Optical Astronomy Observatory	USA
REIF, Klaus	Sternwarte - Universität Bonn	Germany
REISS, Roland	European Southern Observatory	Germany
ROTH, Martin	Astrophysical Institute Potsdam	Germany
RUPPRECHT, Gero	European Southern Observatory	Germany
SCHILDKNECHT, Thomas	University of Bern	Switzerland
SCHMIDT, Ricardo	Cerro Tololo Inter-American Observatory	Chile
SEIFERT, Walter	Landessternwarte Heidelberg	Germany
SEKIGUCHI, Maki	National Astronomical Observatory	Japan
SINCLAIRE, Peter	European Southern Observatory	Chile
SMITH, Roger	Cerro Tololo Inter-American Observatory	Chile
SØRENSEN, Anton	Niels Bohr Institute for Astronomy Physics & Geophysics Copenhagen	Denmark
STARR, Barry	Canada France Hawaii Telescope	USA
STOVER, Richard	UCO/Lick Observatory	USA
STUBBS, Christopher	University of Washington	USA
TARENGHI, Massimo	European Southern Observatory	Germany
TAYLOR, Keith	Anglo-Australian Observatory	Australia
THEUWISSEN, Albert	Philips Imaging Technology	The Netherlands
TIPHENE, Didier	Paris-Meudon Observatory	France
TONRY, John	Institute for Astronomy	USA
TROTTIER, Claude	EG&G Canada Ltd	Canada
TULLOCH, Simon	Royal Greenwich Observatory	UK
VIOT, Herve	Observatoire de la Cote d'Azur	France
VON ZANTHIER, Christoph	MPI Halbleiterlabor and KETEK München	Germany
WEI, Mingzhi	UCO/Lick observatory	USA
ZELLNER, Burkard	VISION 2000 Intl.	USA

FOREWORD

The Workshop "*Optical Detectors for Astronomy*" was held during October 8-10, 1996 at the headquarters of the European Southern Observatory in Garching, Germany. This was the third meeting of its kind, previous meetings being held in 1991 and 1993, but this is the first ESO "CCD Workshop" that has published proceedings.

Most of the leading manufacturers and major astronomical observatories were represented, with the 117 attendees coming together from 14 different countries that spanned every continent on Earth.

The motivation for the ESO CCD Workshop series is the creation of informal and open venue of information exchange about astronomical CCD detectors and systems. Judging from the reaction and feedback of the participants, the 1996 workshop was as successful as the previous editions, which is a credit to all who attended.

The Workshop was organized as a mixture of invited talks, oral presentations, poster sessions and roundtable discussions, the latter used to foster a free exchange of ideas among participants. These technical sessions were complemented by an opening reception and a congenial evening in downtown Munich, which included a walking tour of the historic area followed by dinner at the famous Franziskaner brewery and an after dinner talk by Walter Kosonocky, who reviewed the history of CCD technology.

These proceedings are divided in four main parts, following the structure of the meeting:

1. Manufacturers, Products and R&D Efforts
2. Observatory Plans
3. CCD Devices, Controllers, Testbenches and Instruments
4. Mosaic Projects and Special Techniques

The papers in each section are presented in alphabetical order of the first author's name.

We wish to thank everyone who contributed to the success of this Workshop, which is basically the list of the participants, since the workshop depends on the active participation and open discussion of all who are present. We give special thanks to the invited speakers and to the session chairmen, who promoted stimulating discussions and kept the programme on schedule.

We thank our colleagues who contributed a lot to this endeavor: Elena Zuffanelli, our secretary, who organized the logistical aspects of the workshop with efficiency and grace, Ed Janssen, ESO's graphic designer, whose creativity enhanced the look of these proceedings and Hans Heyer, ESO's photographer, who took the group photo.

A special note of gratitude goes to Jacqueline Bergeron, head of ESO's Science Division, who gave financial support to this workshop.

An inescapable rule of editorship is that the editors usually underestimate the amount of work required to bring proceedings to completion. We had some delays in publication due to bureaucratic issues, but we also deliberately indulged in taking care of the small details that make a book nicer to read. We hope you will enjoy reading these proceedings and that you will join us at the next ESO CCD Workshop, planned for August/September of 1999.

James W. Beletic & Paola Amico

SECTION I: MANUFACTURERS PRODUCTS AND R&D EFFORTS

LARGE AREA, BUTTABLE IMAGE SENSORS FOR MOSAIC APPLICATIONS

M.M. BLOUKE[1], *T. WOODY*[1], *T. DOSLUOGLU*[1],

S.T. ELLIOTT[2], *J.R. JANESICK*[3], *R. REED*[4], *R. STOVER*[5]

ABSTRACT

This paper will describe in some detail two new large area CCD image sensors designed specifically to be used either as a single imager or assembled in mosaics of CCDs. The devices have 2048×4096, 15 μm pixel. Performance data are presented on both front- and back-illuminated parts.

1. INTRODUCTION

There are a number of applications for which a large focal plane is a requirement: fluoroscopy, x-ray crystallography, astronomy, etc. Naturally, it is desirable to cover this focal plane with detecting material. This is especially important for astronomy where it costs many thousands of dollars/night to operate an instrument. Photographic film is one medium that can be manufactured in very large sizes. Amorphous silicon arrays in the 60×60 cm range and potentially larger are now being manufactured.

For CCDs to cover a large area there are several strategies. The most straightforward method would be to make a single device on a single very large wafer. Although progress in this arena is being made, this strategy has some obvious limits today in the sense that the largest wafers commercially available are 400 mm (8") wafers. No one is currently fabricating scientific quality CCDs on these wafers. Several IC fabs that do make CCDs are utilizing 125, and 150 mm wafers (e.g., Philips [1], EEV [2], and Lincoln Labs [3]). Consequently, the largest chip that can be manufactured now is limited to something on the order of 125 mm on the diagonal. While this qualifies as a very large device, it is not large enough for some focal planes currently under consideration or being developed.

An alternative is to pave the focal plane with a number of smaller chips. This has a two advantages. First, the smaller chips are less expensive. Secondly, one achieves a natural multiplex advantage during the readout in that the data for each chip is read out separately. A disadvantage is that there are non-imaging gaps between the chips, although with fiber optics these gaps can be minimized. Another is that all the chips must be carefully aligned to ensure that the imaging surface is in the focal plane.

At SITe we have developed two chips that are designed to be used in mosaics: these are the ST002A and the ST008A. The ST002A is a new CCD that is designed to be three side buttable in order to assemble large focal plane arrays. The ST008A is designed to be buttable along only one edge making 2×1 mosaics possible. In section 2

[1] Scientific Imaging Technologies, Inc., Beaverton, OR97075-0569, USA. Email: **Morley@site-inc.com**
[2] Jet Propulsion Laboratory, Pasadena, CA 91109, USA
[3] PixelVision, Inc., Huntington Beach, CA 92649, USA
[4] National Optical Astronomy Observatories, Tucson, AZ 85726-6732, USA
[5] UCO/Lick Observatory, Santa Cruz, CA 95064, USA

J. W. Beletic and P. Amico (eds.), Optical Detectors for Astronomy, 3–11.

we describe the layout and fabrication of these chips. Section 3 discusses measurements taken on a front- and back-illuminated versions of these parts. Finally, we summarize the results in section 4.

2. ARCHITECTURE AND FABRICATION

2.1 Layout and Design

As indicated above the medical, nondestructive testing and astronomical communities all have need for high resolution, low noise, back-illuminated imagers. The ST002A was designed to fill those requirements and be able to form large mosaics with minimum gaps between chips. The chip 2048×4096, 15 µm square pixels and is buttable along both of the 4096 pixel edges and one of the shorter edges. The ST008A has similar performance goals and has 2048×4096, 15 µm pixels. In this case the die is designed to be butted only along one of the 4096 pixel edges. This latter device was designed specifically for the third generation Hubble Space Telescope instrument, the Advanced Camera for Surveys (ACS).

ST002A

A functional diagram of the 2k×4k device is presented in Fig. 1. As noted above, the chip has 2048×4096, 15 µm square pixels arranged as two 2048×2048 pixel imagers, one on top of the other in a frame transfer configuration. An overflow drain is located at the top of the array to prevent dark current generated in the field region from accumulating in the top row. In addition, there are single dummy columns on each side to prevent dark current from the field region from accumulating in the first and last columns.

There are four parallel clocks; common phase one and phase two clocks and each section has a separate phase three clock. Clocking both the parallel phase three clocks together allows full frame operation. The device is designed for MPP operation with the phase three (i.e., the third level of polysilicon gates) being the MPP phase. Readout of the memory section is effected by holding the phase 3 clocks low in the upper, imaging section while clocking all the gates in the memory section. In this manner, signal charge in the imaging section is transferred between phase one to phase two gates but remains confined to its respective pixel by the barrier phase three.

The serial register is located at the bottom along one of the 2048 pixel edges of the device. The serial register contains 25 overscan pixels at each end and is bent 90° away from the parallel section. The bend is necessary to provide room for the overscan pixels, and at the same time to conserve space so that the imaging array is as close as possible to the edge of the chip. The register is split in the middle so that the entire device can be read out either one or both of the amplifiers that terminate the serial register. The amplifiers are single stage MOSFET source followers based on a design that has proven to give between 2-4 electrons rms read noise at 50 kpixel/s data rates. The serial register is designed with twice the well capacity of a parallel pixel and is terminated in a last gate and a summing well. The summing well is the same size and charge capacity as a normal serial gate.

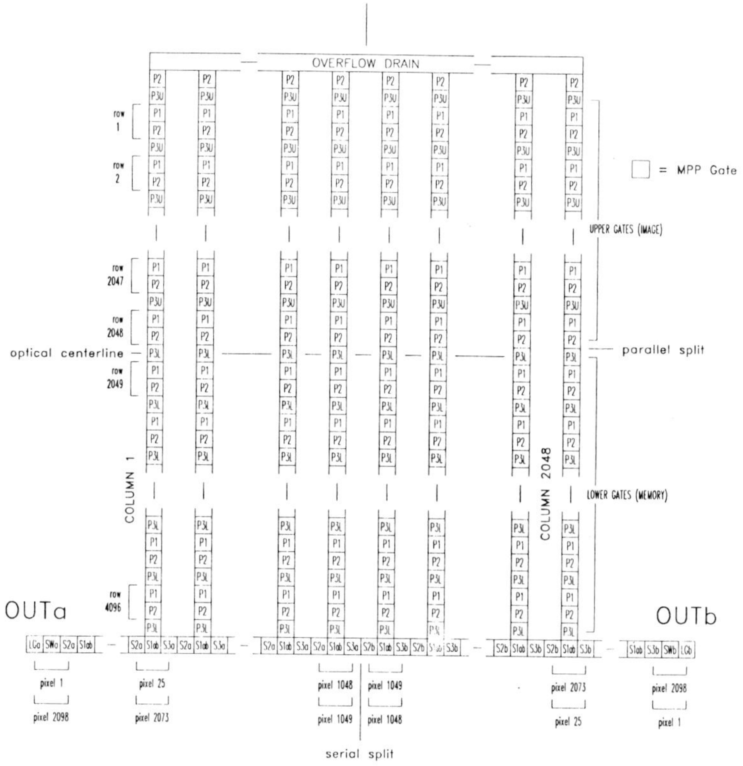

***Figure 1**: Functional diagram of the ST002A*

The bond pads are all brought out on the end of the chip next to the serial register. The chip size is approximately 6.34×3.15 cm. Two such devices fit on a 100 mm wafer.

ST008A

The major difference between the ST002A and the ST008A is in the orientation and layout of the serial register. In the case of the ST008A, the serial register is along the 4096 pixel edge of the device. This orientation was preferred for the ACS application as the devices will operate in a radiation environment. A damaging event can affect only a maximum of 2048 pixels in any given column. A plot of the devices on a wafer is presented in Fig. 2. As noted above, the chip has 4096×2048, 15 μm square pixels. It is designed as a full frame imager with no provision for frame transfer operation. As with the ST002A there is an overflow drain located at the top of the array, and there are single dummy columns on each side, to prevent dark current from the field region from accumulating in the imaging area of the sensor.

The serial register is located at the bottom along one of the 4096 pixel edges of the device. The register contains 25 overscan pixels at each end and are brought straight out in the normal configuration. The register is split in the middle so that the entire device can be read out either one or both of the amplifiers. The amplifier structure is identical to that used on the ST002A. The serial register is designed with twice the well capacity of a parallel pixel and is terminated in a last gate and a summing well. The summing well is the same size and charge capacity as a normal serial gate.

Bond pads are brought out on three sides of the device since only one side was planned for butting. The chip size is approximately 6.88 × 3.46 cm^2. The chip is larger than the ST002A because of the unique method of packaging the die. Two such devices will fit on a 100 mm wafer.

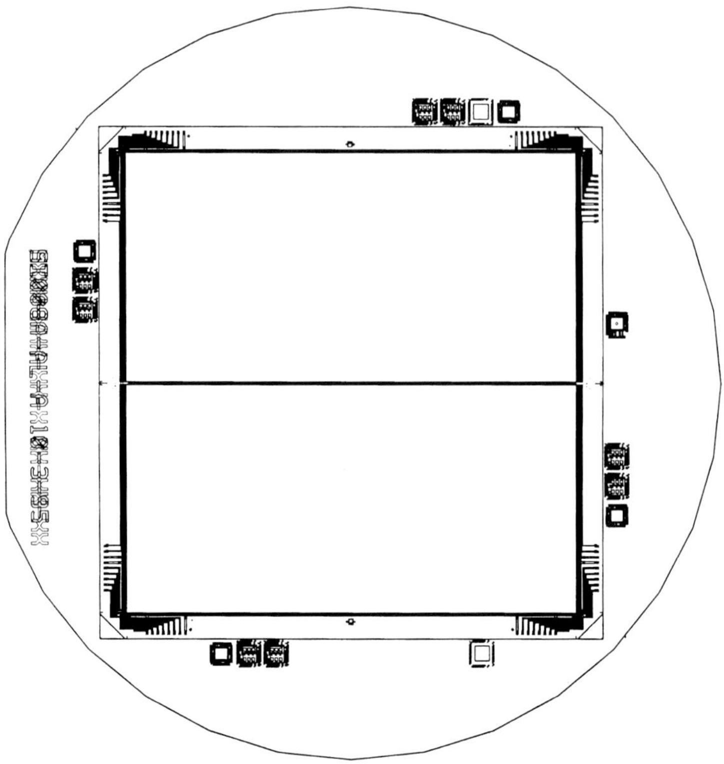

***Figure 2**: Plot of ST008A showing the position of the two die on a wafer.*

2.2 Device Fabrication

These devices are fabricated using the three level polysilicon gate technology with which all standard devices at SITe are fabricated. The CCDs are built on 20-40 Ω-cm p-type silicon epitaxial wafers. In the fabrication process, the CCD channels are first defined by a p+ channel stop diffusion and field oxide. A buried channel is then formed by an n-type implant. Next a gate dielectric is grown, a layer of polysilicon deposited and the first level electrodes defined and etched. The exposed gate dielectric is etched and regrown and a second polysilicon layer deposited, patterned and etched to form the second level gates. This process is repeated a third time to produce the third and final level of gate electrodes. The output diodes and MOSFETs are formed by an n+

diffusion. An insulating layer of reflow glass is deposited, followed by metallization for the interconnections. This completes the processing for a front-illuminated device.

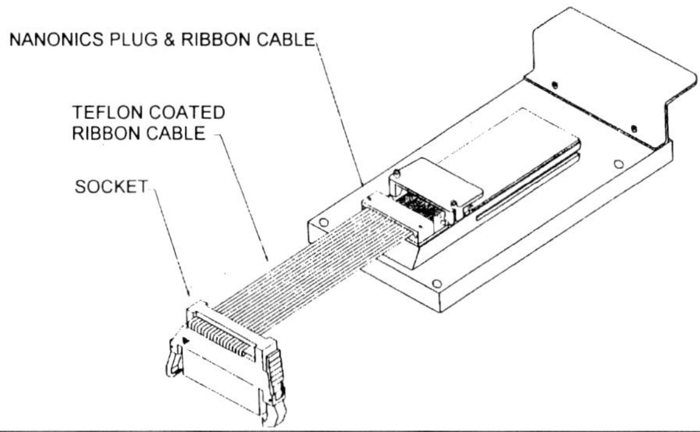

Figure 3: Drawing of the packaged ST002A including handling substrate.

In order to fabricate a back-illuminated sensor, additional processing is required. The surface of the 525 μm thick silicon wafer containing the CCD structure is attached to a glass/ceramic substrate material. The resulting sandwich is turned over and all but about 15 μm of the original wafer is removed.

The device regions are then photo-lithographically patterned and etched to isolate the individual chips. The back surface receives an enhancement treatment to reduce recombination of the photo-generated carriers near the exposed surface. An optional antireflection coating is applied. This may be either a coating designed for enhanced visible response or a coating which extends response to $<$ 200 nm in the ultraviolet. The support of the device using the glass/ceramic substrate has several advantages. First, the thinned device is completely supported and very rugged. Secondly, by mounting the device in this manner it is possible to use standard semiconductor processing techniques and equipment to process the devices through the thinning process. Since multiple devices are handled in parallel, this process will produce less expensive chips.

Processing options for the devices include MPP operation, minichannel implant and front- or back-illumination with visible or UV optimized AR coatings.

2.3 Packaging

ST002A

The packages for the ST002A and the ST008A are substantially different. The package for the ST002A device is based on a design developed by Lesser at the Steward Observatory, University of Arizona [4]. Figure 3 shows an overview of the package design with a mounted chip. One of the goals of the package is to minimize the spacing between chips when used in a mosaic. It will be noted that the edge of the die is offset from the edge of the package. This was done deliberately as it necessary to protect the edges of the chip. Although silicon is a very strong material, it is fairly easy to inadvertently nick the edge of a die. Since the active imaging area of these devices is so close to the edge of the die, some protection is desirable. Lastly, the package must be easily attached to some larger focal plane.

The chip is mounted on an Invar 36 chip carrier that is about 50 μm wider than the chip on the long (4096) side. In addition, the carrier extends 25 μm beyond the top of the chip. The chip is soldered to the carrier with an Indium alloy. Assuming the chip is centered on the chip carrier, the separation between active imaging areas on two

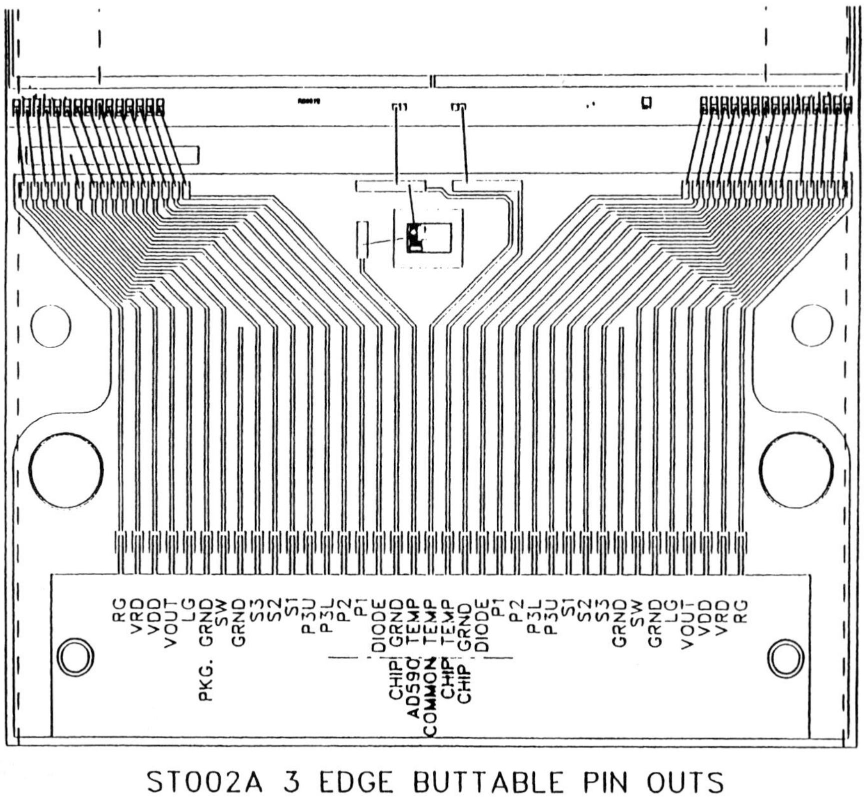

Figure 4: *Detailed view of the pinout of the 2k × 4k CCD.*

adjacent chips along the 4096 side will be less than 650 μm and between imaging areas on the short side, less than about 250 μm.

The lower end of the chip carrier extends beyond the soldered die. This shelf provides a place on which to mount a flexible circuit that connects the die to a Nannonics connector and a 37 lead ribbon cable. The flex circuit is illustrated in Fig. 4. Note that the bonding pattern is symmetric. This means that the pinout for the front- and back-illuminated versions of the device are identical. It requires 32 pins to operate the device including 2 pins used for an on-chip temperature sensor. In addition, provision is made for mounting an optional AD590 chip or other sensor on the flexible circuit to measure temperature. The chip ground is connected to the Invar 36 chip carrier for both front- and back-illuminated parts.

For handling purposes, the chip carrier is mounted on a large Al substrate with a handle. A Robinson-Nugent connector is permanently mounted on the free end of the 37 lead cable and forms the connection to the outside world. This is also shown in Figure 3.

ST008A

The package for the ST008A is an AlN substrate with two flex circuits. The package was designed by R. Wootten at Ball Aerospace and is designed to operate at

−80°C with minimum thermal load on the chip other than the radiation load. The package will be described in greater detail in a future article [5].

3. TEST DATA

Table 1 presents a summary of the performance expected from these devices.

Parameter	ST002A	ST008A
CTE		
Parallel	> 0.99999	> 0.99999
Serial	> 0.99999	> 0.99999
Noise	2-4 e rms.	2-4 e rms.
Conversion gain	~1.8 μV/e	~1.8 μV/e
Full Well	~100 ke	~100 ke
Dark Current -MPP	< 50 pA/cm^2	< 50 pA/cm^2
non-MPP	< 250	< 250
Quantum efficiency	see Fig. 8	see Fig. 8
Flatness	± 5 μm	~200 μm

***Table 1**. Expected CCD performance.*

A number of front- and back-illuminated devices have been fabricated. Several have been characterized thoroughly. We report detailed measurements on two ST002A devices here. Similar results in terms of noise, conversion gain, full well and CTE have

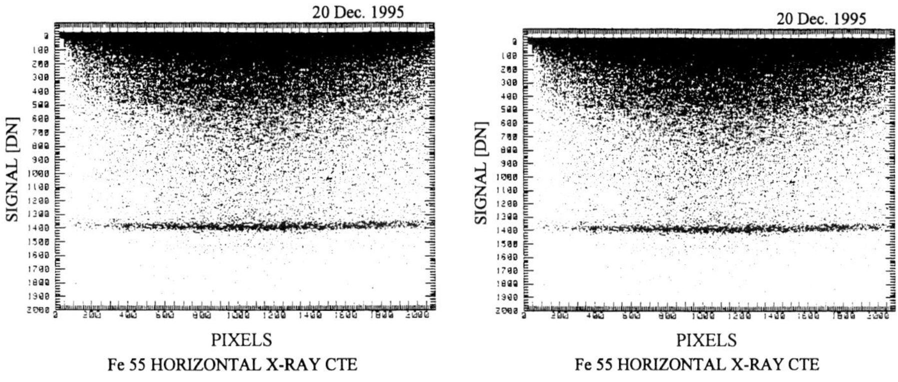

***Figure 5**: Horizontal and vertical stacked line traces using Fe^{55} x-rays. CTE is >0.99999.*

been obtained on ST008A devices.

Figure 5 presents Fe^{55} x-ray data for both the vertical and the horizontal charge transfer directions for a front-illuminated ST002A device. In both cases, the CTE is greater than 0.99999. Similar measurements on a back-illuminated part also exhibit equally high transfer efficiency.

Full well measurements on the front-illuminated part indicated an optimum well capacity of 90000 electrons. Measurements on other parts have given values between 70 and 100 ke$^-$

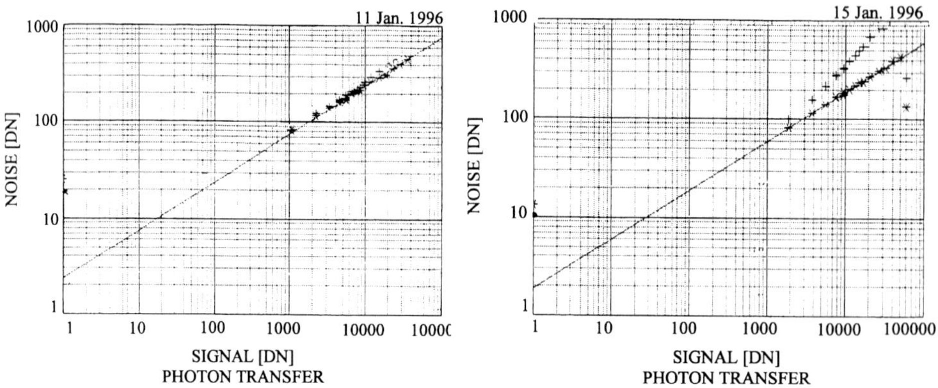

***Figure 6**: Photon Transfer curves for two amplifiers. (left). Front-illuminated part, 0.17 e$^-$/DN, (right). back-illuminated part, 0.28 e$^-$/DN.*

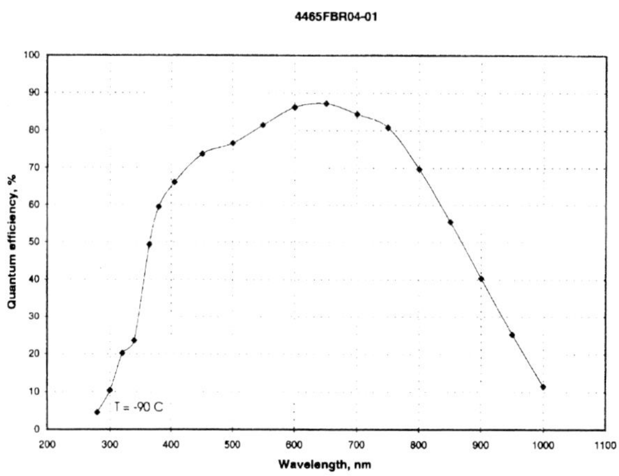

***Figure 7:** Quantum efficiency as a function of wavelength for a back-illuminated ST0012A.*

Photon transfer curves for amplifiers on both parts are shown in Fig. 6. The noise for one amplifier of the front-illuminated part is 3.2 e$^-$ rms for a 4 μsec clamp-to-sample time (the other amplifier was non-functional). For the back-illuminated part the noise was 2.9 e$^-$ using an 8 μsec clamp-to-sample time on one amplifier and 4.7 e$^-$ on the other amplifier. The conversion gain for this amplifier was 1.84 μV/e. In all cases the data rate was 50 kpixel/s.

Dark current is well behaved. For back-illuminated part the dark current is 27 pA/cm^2 in the MPP mode at 293 °K. Measurements on other MPP parts operated in the non-MPP mode yield values as low as 3-5 e$^-$/pixel/hr at -80°C. This translates to approximately 150 pA/cm^2 at 293 K.

Figure 7 presents quantum efficiency data for a back-illuminated part. The measurements were taken at -90C. The part had the SITe standard visible AR coating and a peak QE of 85%.

The bow of a number of thinned devices has been evaluated. The bow on unmounted parts has been measured from between 80 to 180 μm and is largely spherical in nature. With no effort expended to flatten the die, the bow remains at the presoldered level. It is expected, however, that flatness of ±5 μm can be achieved with this package and with improved packaging techniques. Efforts are currently in progress to achieve these results.

4. Summary

The ST002A and ST008A are large area scientific quality, CCD image sensors intended for use in the medical, nondestructive testing, and astronomical communities. The devices are configured with 2048x4096 pixels in imaging area but the serial register oriented along the short dimension in the case of the ST002A and along the long dimension in the case of the ST008A. In both cases the pixel size is 15 μm. The ST002A is designed to be three side buttable with leads or electrical connection to the chip occupying only one side of the die. A number of front and back illuminated parts have been fabricated and tested. In general, the device is well behaved with low noise, high CTE and low dark current.

Acknowledgments: The authors would especially like to thank P. Marriott and T. Yamaguchi of SITe for technical assistance in the fabrication of these devices. They would also like to thank Dr. Michael Lesser of the University of Arizona for many useful discussions regarding the packaging of these devices. The authors would also like to thank Dr. C. Claver of NOAO for assistance in the evaluation of certain devices.

5. References

[1] A. Theuwissen, these proceedings, p. 37

[2] P. Pool, these proceedings, p. 29

[3] B. Burke, these proceedings, p. 19

[4] Dr. M. Lesser, private communication.

[5] Woody, et al., *Proc SPIE* **3091**, Feb 97.

THE STATE OF CCD TECHNOLOGY AT LOCKHEED MARTIN FAIRCHILD SYSTEMS

R. BREDTHAUER[1]

GHOSTWRITTEN BY J. W. BELETIC AND P. AMICO

ABSTRACT

This paper presents an overview of the latest CCD devices manufactured at Lockheed Martin Fairchild Systems. Astronomical oriented applications are also discussed, with particular emphasis on space based and large area imagers.

1. CHARGED COUPLED DEVICE TECHNOLOGY

Three main features characterize CCD technology development at Lockheed Martin:

- It is tailored to meet Specific System Requirements
- It provides Custom Design and Process Modifications to optimize:
  - spectral performance - backside thinned and thin Polysilicon technologies
  - signal charge capacity
  - speed
- it makes use of Multiple Foundries, including Milpitas (4″) CCDs (commercial and platinum Silicide IR CCDs), Manassas (5″) CMOS & CCDs (High reliability and radiation hardened for military applications)and China Huang Guang (5″/6″) CMOS & CCDs (volume commercial production, on line early 1997)

2. MANUFACTURED DEVICES

2.1. Integrated CCDs

Two devices of this kind are being manufactured; namely, a high resolution TDI array, built from 7 chips butted end-to-end for a total of 15232×64 11μm pixels, with drive circuitry and CDS on-chip (Fig. 1) and a 2.9 mm diameter Endoscope Imager, 190×165 rectangular pixels (8×10 μm) with 4 pins integrated drivers (see Fig. 2.).

2.2. 2k Still Frame Imagers

These CCDs (see Fig. 3) have evolved from the scientific imager family which included 512×512, 1024×1024, 2048×2048 15 μm pixels devices. They are

[1] Lockheed Martin Fairchild Systems,14251A Chambers Rd,Tustin, California, CA 92680, USA
Email: **richard.bredthauer@lmco.com**

J. W. Beletic and P. Amico (eds.), Optical Detectors for Astronomy, 13–18.

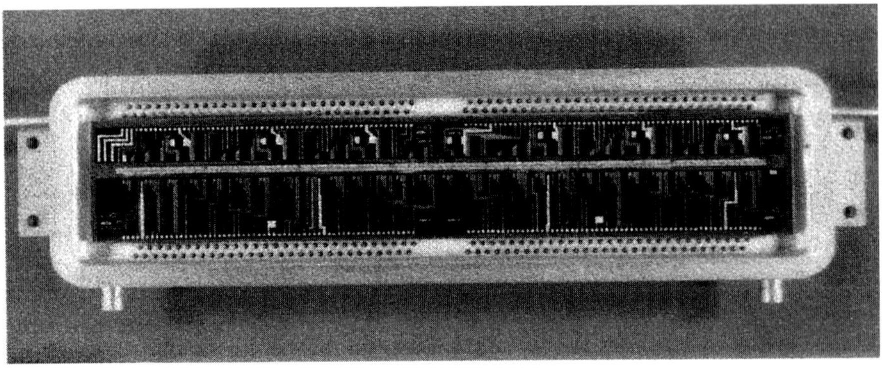

Figure 1: High Resolution TDI Array

built in MPP configuration to provide low dark current and high charge capacity. The commercial production is foreseen for digital electronic photography applications for Leaf and Megavision. The evolved 4096×4096 version will be produced for Dicomed and Hasselblad Cameras.

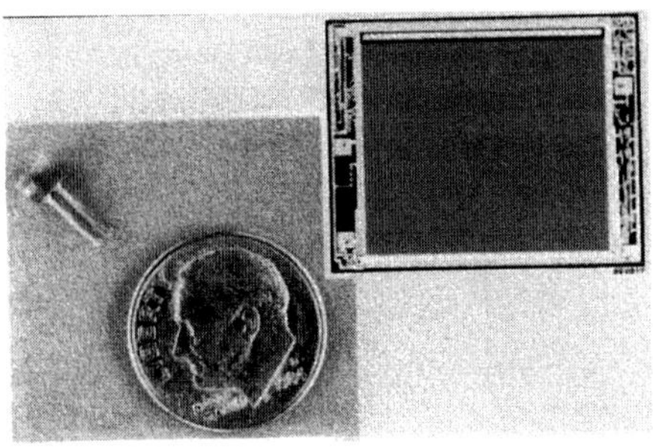

Figure 2: Endoscope Imager

Figure 3: 2K Still Frame Imager.

2.3 X-ray Imagers

Fig. 4 shows the schematic of a digital system for x-ray imaging. Another popular application is dental x-rays; for this, Lockheed Martin has a 2.6 × 3.3 cm intra-oral chip in volume production. In addition, extra aural devices are in being manufactured for Siemens.

3. SPACE BASED IMAGER APPLICATIONS

Lockheed Martin staff have a long history in space based imagers, including:

- SOHO Michelson Doppler Imager (NASA) developed in collaboration with Palo Alto Research, it is a 1024×1024 21 μm pixel device with custom flight package. The mission has been launched in 1995 to provide observations of the solar magnetosphere.

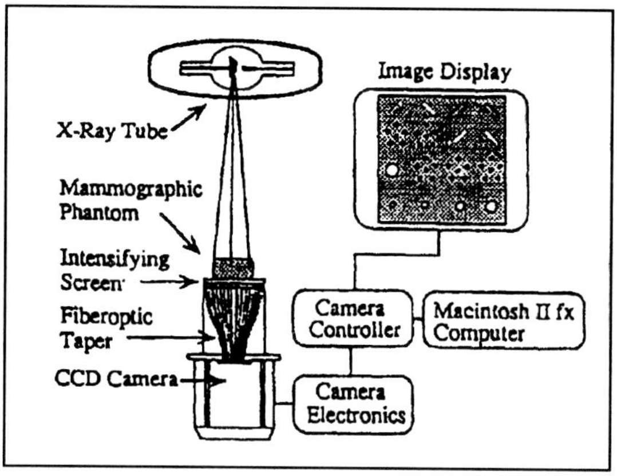

***Figure 4**: Schematic of the X-Ray Imager*

- Lightning image sensor (NASA): a 256×256 split frame storage device with 30 μm pixels, QE > 55% at 777.4 nm, and read out at 1000 frame per sec.
- Daylight StarTracker operated in MPP mode (Northrop).
- Nikon Still Camera (NASA).
- Mars observer mission (Caltech). In addition there are two new Mars missions: 1) the Mars Global Surveyor Mission launched in November 1996, which includes the backup CCDs from Mars Observer, 2) the Mars Pathfinder, launched in December 1996, which includes the Max Planck chip from the Cassini mission.
- CMWS – Common Missile Warning System equipped with a 512 × 512 15 μm pixels split frame store, integrated with fiber optics and read out at rates greater than 500 frame per sec.
- CUBIC, MISR (JPL).

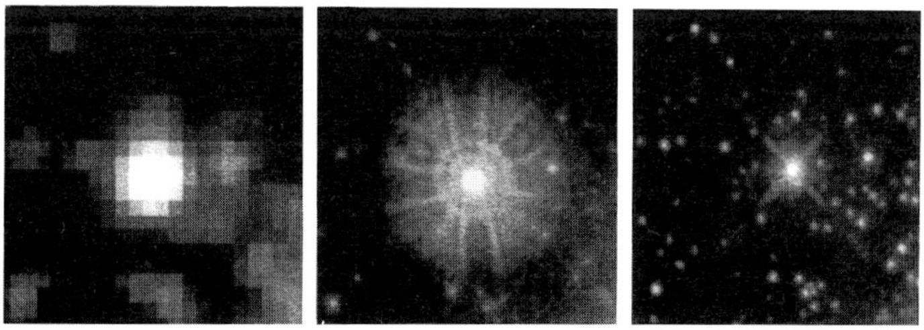

***Figure 5**: Images from HST*

- Cassini Mission (JPL): the main camera is equipped with a 1024×1024 12 μm pixel CCD selected from space qualified lots. In addition, a device operated in MPP mode for low dark current for the JPL Star Tracker, one with special anti-blooming requirements for the Max-Planck Institute Huygens Probe experiment and a frame transfer 24 μm pixel CCD for the Officine Galileo VIMS.

- Hubble Space Telescope WF/PC II (JPL): 800×800 15 μm pixel chip with three-level polysilicon layers and nitride-oxide gate dielectric. The new process technology developed for this device yields, in MPP mode, a dark current lower than 10 pA/cm ^{2} and an LDD output noise less than 4 e$^-$rms. (see Fig. 5)
- SPOT satellites (CHINA/INDIA)
- Brilliant Eyes (Rockwell)
- Brilliant Pebbles (Hughes)

4. LARGE AREA IMAGER DEVELOPMENT

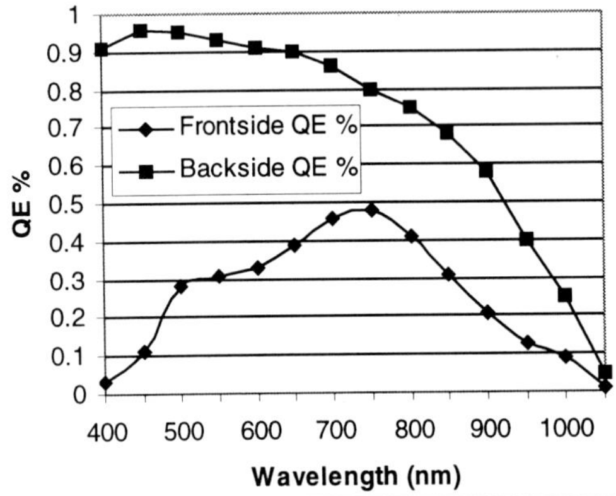

Figure 6: Typical QE curves for frontside and backside illuminated chips for astronomical applications

Manufacturing has to deal with a variety of requirements coming from different applications: high end digital camera backs, 70 mm movie camera replacement, dental X-ray imagery, large format X-ray imagery, astronomical and surveillance applications. The two near term requirements at Lockheed for large area imaging come from x-ray applications and for high resolution reconnaissance applications. The design approach for large devices is constrained by the 3 phase, 3-levels polysilicon, 1.5 micron design rules and manufacturability requirements.

The Dicomed professional camera is one example. Its essential features include a full format of 6 × 6 cm and 4k × 4k resolution (4096 × 4096 pixels). Two models, the BigShot 4000 for instant color capture and the BigShot 3000 for 3-shot sequential color and black and white, will be

Figure 7: CCDs for astronomical applications

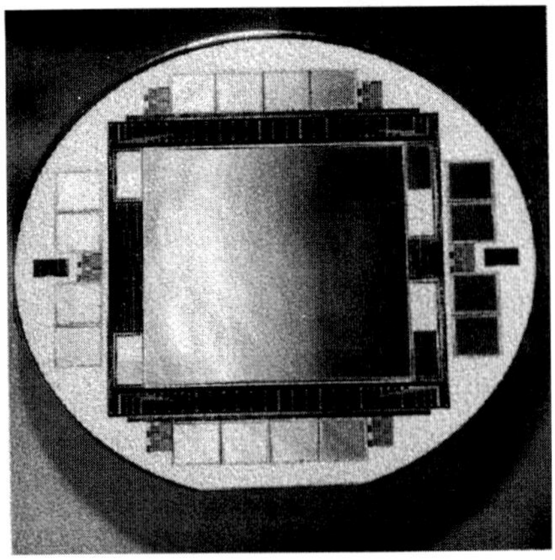

Figure 8: CCD for ESO VLT Test Camera

mounted on a Hasselblad 553 ELX camera body and will be equipped with a CCD array of the above mentioned size, giving a 48MB file capture and pixel color depth, with 12 bits per color channel or 12 bit monochrome.

***Figure 9**: Image from satellite*

Astronomy requirements include noise less than 3 e^- rms, maximum quantum efficiency (see Fig. 6) and very large imaging area for devices to be commissioned at the several new 8 meter telescopes, Keck, JNLT, NOAO ESO VLT, Carnegie, Hawaii, Cambridge and Smithsonian. Figures 7 and 8 show some examples of astronomical type CCDs.

One of these is foreseen for an asteroid tracking system, a project whose PI is Dr. Eleanor Helin. The chip is a 4k×4k CCD mounted onto the 1m GEODSS telescope located at Haleakala, Maui, Hawaii. The program requires the detection of moving celestial objects, which have to be observed 3 times over 1 hour. The camera electronic has been made by Bob Leach and gives ~10 e^- rms noise at –5 °C with 50Hz readout rate.

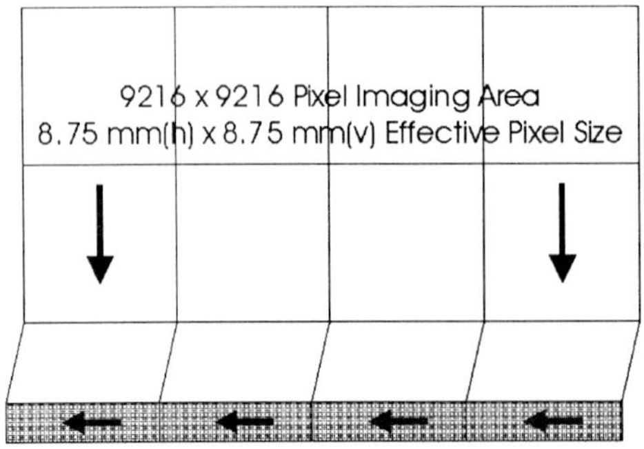

Figure 10: Imager Block diagram (not in scale)

The 4k large imager are used for very high resolution digital photography, and are especially meant for medical imaging and reconnaissance (see Fig. 9) applications. Each chip has one dual stage output at each corner. The 4k×4k device comes in two sizes:

1. 15μm pixels, 61.4 × 61.4 mm on 100 mm wafer

2. 19.5 μm pixel, 79.9 × 79.9 mm on 125 mm wafer.

The pixel full well capacities are 100ke$^-$ and 250ke$^-$ respectively.

The largest CCD is the 9k Ultra – Imager fabricated at Manassas, from a full 125 mm wafer, with a 80.6 × 80.6 mm image area whose block diagram is shown in Fig. 10. The 9216 × 9216 8.75 μm pixels are read out of 4 unidirectional output amplifiers, and has a 50 ke$^-$ full well capacity. It is the world's highest resolution array, with a total of 85 million pixels and 170 millions bytes per frame. The CCD and a test image are shown in Fig. 11.

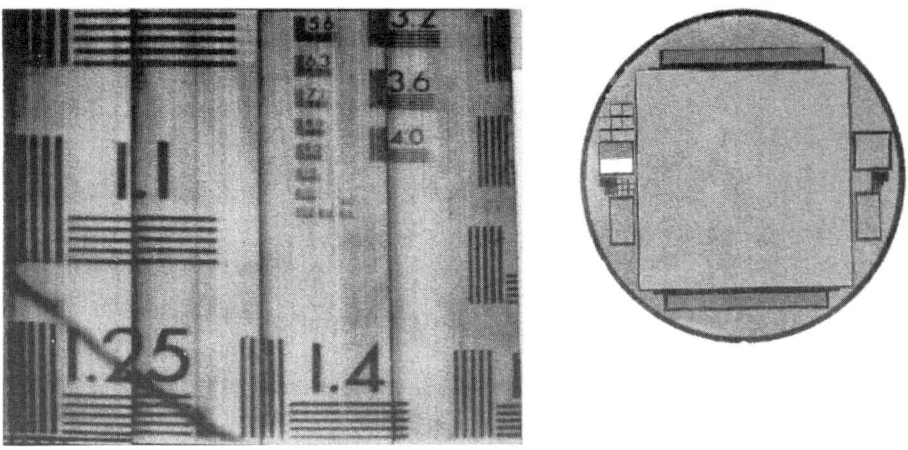

***Figure.11**: the 9k Ultra-Imager (right) and a test image (left)*

5. SUMMARY

Lockheed Martin's CCD technology is at the cutting edge of the digital image revolution. The requirements imposed to its essential components by the need of optimum performance military systems will grant benefits to all CCD users. Lockheed Martin aims to deploy its vast technical resources to capture a large portion of the commercial imager markets. The continued technology evolution pushed by these two major users will positively affect the smaller astronomical market.

LARGE-AREA BACK-ILLUMINATED CCD IMAGER DEVELOPMENT

B. E. BURKE[1], J. A. GREGORY[1], R.W. MOUNTAIN[1], B. B. KOSICKI[1], E. D. SAVOYE[1], P. J. DANIELS[1], V. S. DOLAT[1], T. L. LIND[1], A. H. LOOMIS[1], AND D. J. YOUNG[1]
G. A. LUPPINO[2] AND J. L. TONRY[2]

ABSTRACT

We describe recent work in the area of large, back-illuminated CCD imagers at M.I.T. Lincoln Laboratory as well as new technology applicable to astronomy. We completed in 1995 the development of a 2560 × 1960-pixel frame-transfer imager that filled a 100-mm wafer and several back-illuminated versions of this device were completed. More recently we have begun the development, in collaboration with the U. of Hawaii, of a three-side abuttable 2k × 4k CCD for a multi-chip focal plane. In the unused chord area of the wafer layout, we added test imagers as development vehicles for blooming control and for the demonstration of a CCD that is capable of charge transfer in all four directions. We expect the latter to find application as an electronic means of performing tip-tilt correction to compensate for atmospheric turbulence.

1. INTRODUCTION

Large-area, back-illuminated CCD imager technology has been developed at Lincoln Laboratory over the past several years to support various space surveillance programs where high search rates, high sensitivity, and low noise are essential. These same features plus deep depletion depths are required in soft-x-ray astronomy, and we have been involved in multi-chip focal-plane development for x-ray space missions that include the ASCA and AXAF satellites [1,2]. All this technology is applicable to the needs of the astronomers as well, and we are now seeking ways to bring our advanced capabilities to the service of this community as well.

Among the large-area sensors described here are wafer-scale imagers for space surveillance. The development of this device was completed approximately a year ago, and we delivered several back-illuminated versions of this chip to the sponsor. We will briefly summarize the design and performance of this device. The second large-area device to be discussed is a 2k × 4k three-side abuttable imager with 15-μm pixels, and is the result of a collaboration between Lincoln Laboratory and the U. of Hawaii Consortium. The goal in this case is to push new imager technology for the next generation of very large focal-plane arrays. We currently have completed the first wafer lots and as well as back-illumination processing of selected wafers. Some test results from low-temperature wafer probing will be described.

On the same mask set with the 2k × 4k devices are two smaller, experimental devices that represent new technology of interest to the astronomy community. One of

[1] Lincoln Laboratory, Massachusetts Institute of Technology, 244 Wood Street, Lexington, MA 02173-9108. Email: **bburke@ll.mit.edu**
[2] Institute for Astronomy, U. of Hawaii, 2680 Woodlawn Drive, Honolulu, HI 96822

J. W. Beletic and P. Amico (eds.), Optical Detectors for Astronomy, 19–28.

these is a vehicle for developing blooming control, and the other is an imager that features charge transfer in all four directions. The latter offers an interesting capability of performing tip-tilt correction electronically, and results of field tests with this device are described in another paper in these proceedings [3].

All of the devices described here are fabricated at the Microelectronics Laboratory at Lincoln Laboratory and features 750 m^2 of class 10 and 900 m^2 of class 100 clean-room space. A significant recent development in this facility was the conversion from 100-mm to 150-mm wafers in October 1996. This conversion included the acquisition of a new full-wafer scanner with 1-μm resolution specifically for large-area CCDs on the larger wafers.

2. Wafer-Scale Imager

Figure 1 illustrates a schematic of a large imager that fills virtually all the useable area of a 100-mm wafer and was developed for space surveillance. The device was required to operate at frame rates of about 1 Hz, and this dictated a frame-transfer format and eight output ports. The central imaging section consisting of 1960 × 2560 pixels with the frame stores situated above and below, so that half the image is clocked to the upper frame stores and half to the lower. The imaging pixels measure 24 × 24 μm, but because of severe space constraints on the wafer the frame-store pixels are reduced to a height of 12 μm. Moreover, the outer portions of the frame store are slanted in order to keep the corners of the device a sufficient distance from the wafer edge, and this results in the pixel width tapering down to 18 μm at the interface with the serial registers. The output circuits are capable of a 2 MHz data rate.

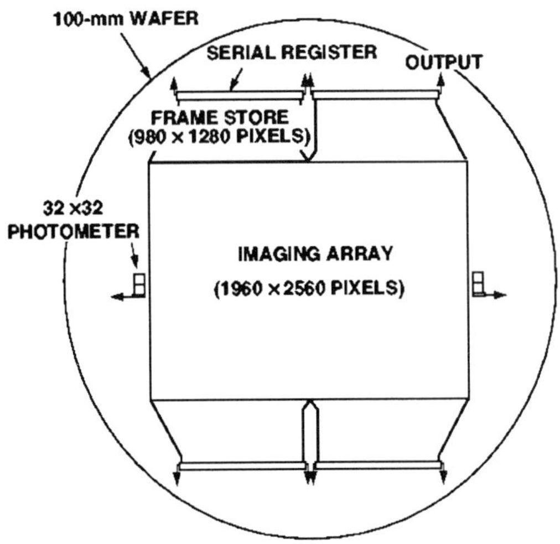

Figure 1. Schematic of 2560 × 1960-pixel wafer-scale CCD imager. Imaging pixels are 24 × 24 μm. Half the image is clocked to frame-store regions on the top and half to the bottom.

Two small (32 × 32 pixels) imagers used as tracking photometers are located on both edges of the device. These are operated at 1-kHz frame rates and are used to analyze the temporal brightness variations of tumbling space objects.

Figure 2 shows front and back photos of a packaged device. Since conventional integrated-circuit packages are not available for a device of this size, we developed a novel custom package consisting of a molybdenum (chosen for its good CTE match to silicon) plate to which the chip is bonded and a multi-layer printed circuit board which surrounds the chip and is bolted to the plate. All electrical leads are brought onto the board via standard Positronics multipin connectors, and the CCD is wire-bonded to gold-plated traces on the printed circuit (PC) board. The PC board allows a number of useful features to be incorporated such as ESD protection circuitry, video buffer circuits, and temperature sensors.

Several back-illuminated devices were completed under this program. In our

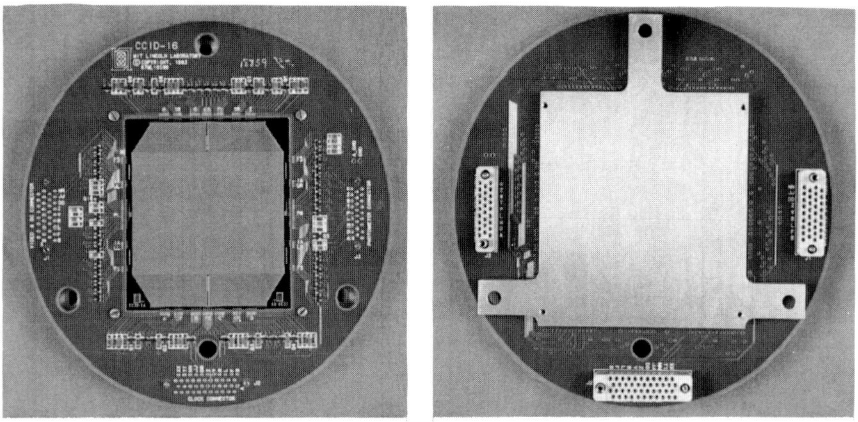

Figure 2: *Photographs of a packaged device showing a) front and b) rear views.*

process the thinned wafer is bonded with epoxy to another silicon wafer for mechanical support, and we then etch away the areas of silicon and field oxide covering the bond pads in order to attach wire bonds to the device. Typically the final thickness of the thinned device is 20 μm. The silicon resistivity is 100-300 Ω-cm, so the depletion depth can extend through most but probably not all of the membrane thickness. The critical process step in back illumination is the treatment of the back surface, and for this device we used a shallow ion implantation of boron followed by a pulsed laser anneal. This produces a p^+ layer of about 100 nm depth at the back surface.

Figure 3 shows quantum efficiency data taken on a typical back-illuminated device. This application did not require sensitivity in the UV, and therefore we used a simple single-layer anti-reflection coating of SiO that gives poor performance below 400 nm. Above this wavelength, the QE is routinely as high as 90% around 550-700 nm before falling off in the near IR.

A physical feature of the device, which is important to its optical performance, is the device flatness. Our back-illuminated devices are inherently flat because they are mounted on flat silicon wafers with a thin, uniform glue line. The Molybdenum plates on which they are mounted are ground flat to within about 4 μm. When mounted on a plate and held flat with a small weight (~150 gm) during the cure of the epoxy the device is flat to within ±4 μm. This is probably sufficient for fairly fast optical systems, but could likely be further improved with flatter Molybdenum plates and larger weights.

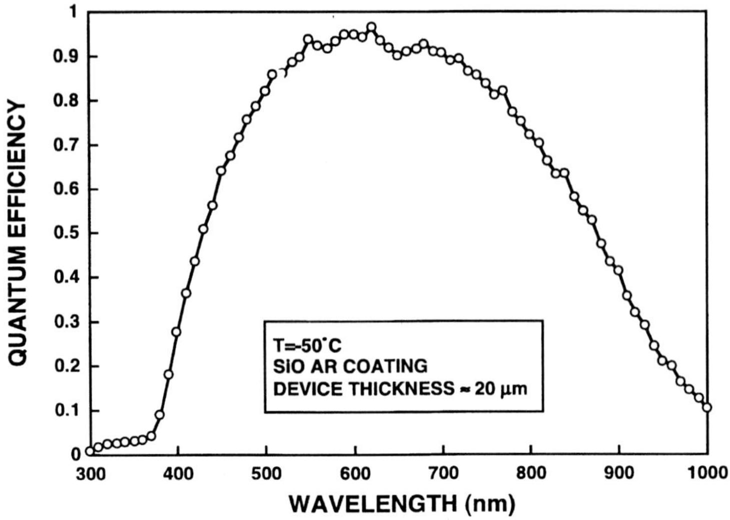

Figure 3: *Quantum efficiency of a back-illuminated imager using an ion-implant/laser anneal process. A single-layer anti-reflection coating of SiO was applied to the device.*

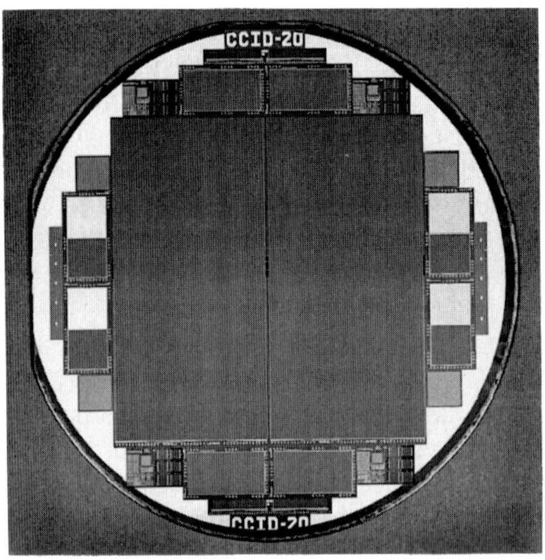

Figure 4: *Wafer photo showing two 2k ×4k imagers with additional 512 ×512 test imagers in the surrounding chords of the wafer.*

Because of the high-frame-rate requirement, the output circuit was designed primarily for wide bandwidth and not low noise at slow-scan rates. At a 2-MHz data rate, the noise level in a Photometrics camera was of order 10-12 e^-. Miyazaki, et al. have measured the noise to be as low as 3 e^- under slow-scan conditions on a sample part [4].

3. 2K× 4K DEVICE DEVELOPMENT

This and the following two sections discuss results from the collaboration between Lincoln Laboratory and the U. of Hawaii Consortium on the development of new imager technology for astronomy. The principal aim of this effort is a 2k × 4k imager with 15-μm pixels and three-side abuttability, and a photograph of a completed wafer in Figure 4 shows two of these imagers just fitting on a 100-mm wafer. Additional smaller devices are located around the periphery of the wafer and are described in the next two sections. The four chips above and below the 2k × 4k devices are test imager for blooming control development (described in the next section), while those on the left and right are the so-called orthogonal-transfer CCDs described in section 5.

The large imager is illustrated schematically in Figure 5. The imaging section is partitioned into upper and lower 2k×2k blocks of pixels with independent three-phase clocking, thus enabling the imager to be operated in a frame-transfer mode. The bi-directional serial register has output ports at each end, and the device can be operated in either a 1- or 2-port mode. This device employs an output circuit, which is used on the x-ray CCDs for the NASA AXAF program and routinely delivers less than 2 e^- noise at 100-kHz data rate.

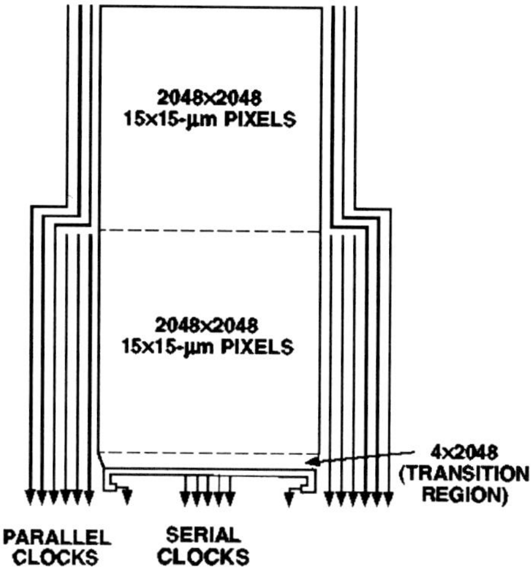

Figure 5.Schematic of the 2k × 4k imager

Because the device is to be closely abutted on three sides the clock lines along the left and right edges must be kept close to the pixel array. This created a design problem at the lower corners where the lines pass the output circuit. Rather than bend these lines outward and around the output and thereby increase the gap to the adjacent chips, we pushed the output circuit inward from the edges. This was done by adding an additional four rows of pixels at the bottom that are tapered inward slightly at the outer edges. In addition, we added ten stages to the end of the serial register and bent them around in a loop so that the output circuit could be comfortably placed away from the corner without crowding. These features are shown in the photograph of Figure 6. As a result of this design the imaging clock lines do not cross or otherwise interfere with the serial register or output circuit.

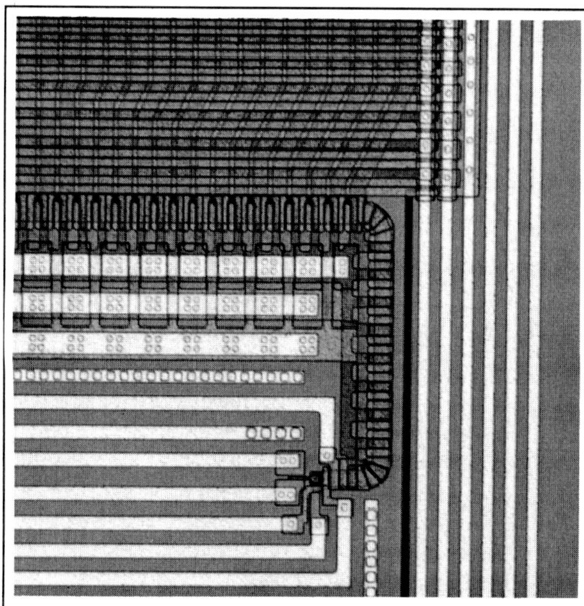

Figure 6. Photo of the output circuit area of the device.

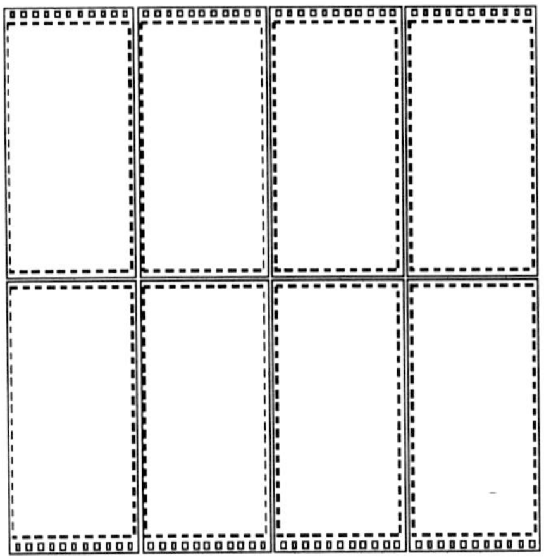

Figure 7: Depiction of an 8-chip array of 2k × 4k devices

We have fabricated these devices on both 300 Ω–cm, 20-μm epi and 7000 Ω–cm bulk wafers. The high-resistivity material will be thinned to 40-45 μm (in contrast to 20 μm for the epi), and will be fully depleted under normal clocking conditions. This should result in enhanced near-IR quantum efficiency, reduced Fabry-Perot fringing effects, and minimal resolution degradation arising from carrier diffusion.

The ultimate application of this device is in an 8-chip array (64 Mpixels) illustrated in Figure 7. The device dead space (sawn die edge to outside pixel edge) along the top of the device will be 70 μm, while along the sides this figure must be increased to 180 μm because of the six clock lines. These numbers derive from experience in several programs involving abutted arrays of chips. The dead space between the chips in the mosaic will depend additionally on the mechanical tolerances of the packages and their placement relative to adjacent devices. Our goal is a gap of less than 500 μm.

The package and the focal-plane hardware have been developed by the U. of Hawaii. Figure 8 is a photograph showing a packaged front-illuminated device. The package is made of aluminum nitride (AlN), which was chosen for its good CTE match to silicon, thermal conductivity, ease of grinding to precise dimensions and flatness, and the fact that thin-film gold interconnects could be patterned on it. The components mounted on the AlN next to the chip are a load resistor for each on-chip MOSFET and low-noise JFETs, which are to be used as second-stage source followers. For thermal reasons the JFET load resistors (or current sources) will be located off the package. The JFET provides the capability of driving the off-chip video line capacitance at a wide bandwidth, and this coupled with the low-noise performance at clock rates considerably faster than 50 kHz means faster readout times for these very large chips. A Nanonics miniature multi-pin connector is used for I/O access to the device.

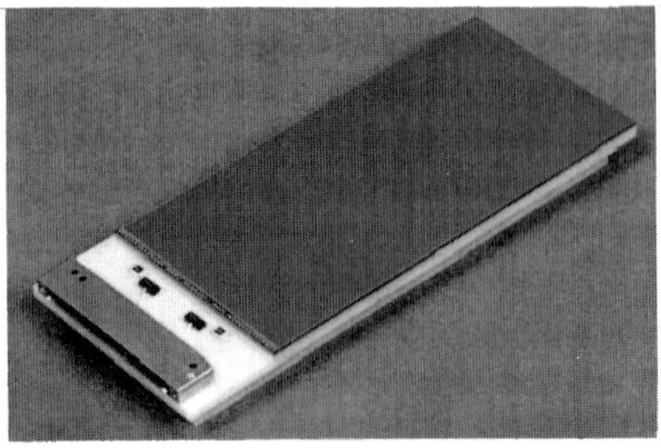

Figure 8. Photograph of a packaged 2k × 4k device.

The first wafer lots have been completed, and the first set of wafers has gone through back-illumination processing. We have tested wafers before and after thinning in our low-temperature wafer probe station at -50°C, and have obtained the following data. The CTI is less than 10^{-5}, but we caution that in our measurement there was a background dark current or slim zero of a few tens of electrons which may contribute to a decrease in CTI. In addition, the CTI may be quite different at the lower temperatures typical of astronomical applications. The well capacity is at least 125 ke$^-$, but we have not yet optimized the clock voltages for maximum capacity. The output-circuit responsivity lies in the range of 10-15 μV/e$^-$ depending on the output MOSFET biasing, and the output-circuit linearity is within ±1% up to the well capacity. The device noise cannot be measured in our probe station because of extraneous pickup. A more complete characterization will be made after the devices are packaged.

4. Blooming Control

Blooming is the response of a CCD to a bright image that produces charge in excess of the well capacity, and manifests itself as a spreading of the charge to adjacent pixels within the same column. Clocked antiblooming, which is basically charge pumping or recombination of excess electrons via interface states, is often accompanied by spurious charge and even pocket pumping. Conventional antiblooming approaches found in consumer CCDs use a reverse-biased pn junction beneath the pixel and are therefore incompatible with back illumination. Using a 512 × 512 test imager we are pursuing a blooming control, which has its origins at RCA Laboratories [5] and is illustrated in Figures 9 and 10. In this approach, the conventional LOCOS channel stop is replaced with a reverse-biased n$^+$ blooming drain straddled on both sides by narrow barrier regions. The blooming control threshold is set by the n$^-$ blooming-barrier doping and the gate potential.

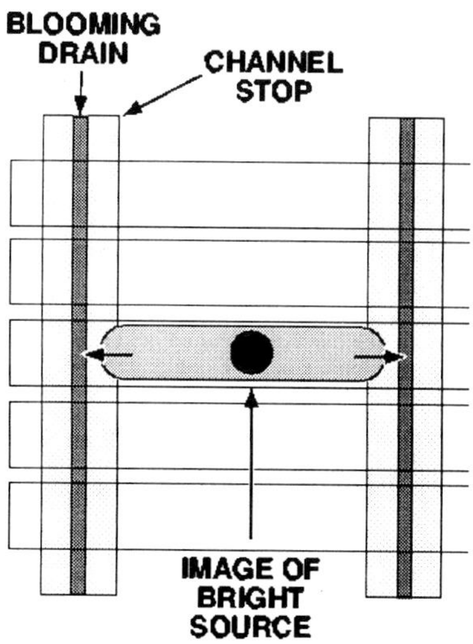

Figure 9. Depiction of a blooming control in which the blooming drains are embedded within the channel stops

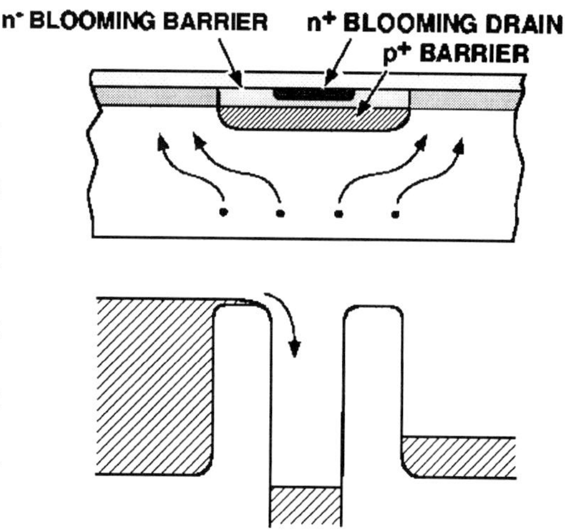

Figure 10. Illustration showing the structural details of the blooming control

An important question regarding the blooming control is to what extent it may compromise device performance. To preserve quantum efficiency a buried p$^+$ layer is placed beneath the drain and barrier regions. This layer forms a high-low junction with the lightly doped p-substrate and presents an electrostatic barrier to photoelectrons, thus preventing loss of carriers to the blooming drain. Another issue concerns the possible reduction of well capacity because of the greater width of the blooming control compared to a simple LOCOS channel stop. In our experimental device, we have used barriers of 1.5 and 2.0 μm and blooming drains of 2.0 μm width. The total width then is 5 to 6 μm, compared with our standard LOCOS channel stop width of 2.0 μm. The latter, however, suffers from enlargement due to lateral oxidation, so that the final LOCOS width may be closer to 3 to 4 μm. Comparative measurements of well capacity and quantum efficiency are planned for these devices.

Measurements of blooming suppression have been made, and an early result on a front-illuminated device with 2-μm blooming barriers is shown in Figure 11. An LED is pulsed on during the image integration period, and then kept off while the image is transferred to the frame store and read out. The optical overload is 1000×, and the pulse duration was about 10 ms. The image is acquired with the phase-2 electrodes held high and phases 1 and 3 low. When the phase-2 voltage is not sufficiently high the blooming barrier depicted in Figure 10 is raised to a higher level than the potential barriers of phases 1 and 3, and blooming occurs. This is shown by Figure 11a where the phase-2 (high) voltage was 3.0 V. By raising this voltage to 4.0 V the blooming barrier is lowered enough to allow carrier overflow into the blooming drain and suppress blooming (Figure 11b). We have been able to suppress blooming even with optical overloads of more than 10^6× with pulsed illumination of several hundred msec duration.

Figure 11. *Experimental imagery demonstrating blooming control in the presence of a 1000× overload from a bright LED. In left), the imaging clock voltages are set to disable the blooming control, while in right) they are raised to a level that enables blooming control.*

5. ORTHOGONAL-TRANSFER CCD

The second of the test devices on the 2k × 4k mask set is a 512 × 512 frame-transfer CCD which permits charge transfer in all four directions, and builds on a smaller 64 × 64 - pixel version previously described [6]. The pixel structure of this orthogonal-transfer or OTCCD is shown in Figure 12, and it is fabricated with four gate

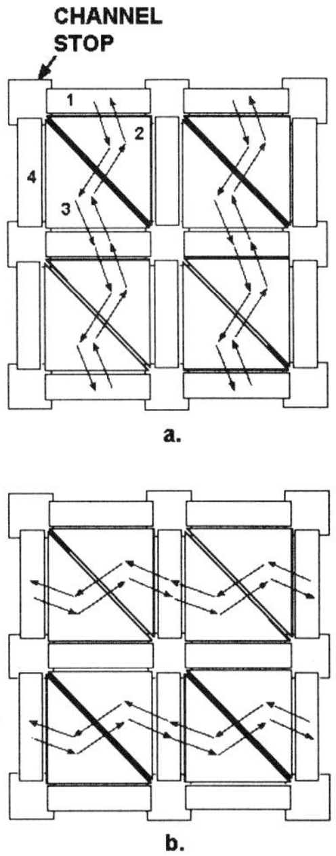

Figure 12: Depiction of the gate structure of the orthogonal-transfer CCD. In a) the phase-4 gates are low and thereby provide vertical channel stops, while in b) the phase-1 gates are low to provide horizontal channel stops.

levels. With phase 4 held at low potential (shaded) to act as a channel stop, gates 1, 2, and 3 can be clocked in a conventional three-phase manner to transfer charge vertically. Likewise, phase 1 can be biased low and 2, 3, and 4 can clock charge horizontally. Figure 13 is a photograph showing the details of the 15 × 15-μm pixels. The gates are fabricated in the usual three poly levels plus aluminum metallization for the fourth level. Other details of the chip include a conventional three-phase frame store with 15 × 15-μm pixels, and the same split bi-directional serial register and dual output ports as the 2k × 4k device.

The primary motivation for this device was to compensate electronically for image motion caused by the atmosphere. Such an application makes sense when the image motion is fast in comparison to the frame time, and this is clearly the case in most astronomical viewing since the atmospheric effects occur on time scales of milliseconds. The lowest order image distortion caused by the atmosphere is a simple random translation, and this (together with telescope vibration) is typically removed by a so-called tip-tilt correction using a moving mirror. With the OTCCD the tip-tilt correction can be performed electronically without the need for the complex and cumbersome mechanical systems currently employed.

In order to maintain registration between the moving image and the charge packets we need to track the motion of a guide star and use this information to set the imaging clocks. For this purpose we omit the usual opaque light shield over the frame store and use the latter as a high-frame-rate tracker.

This requires that a sufficiently bright guide star be located near the object to be studied.

This technique as well as test results on this device are described in more detail in the paper by Tonry and Burke [3] in these proceedings. Other applications can be envisaged for this device, such as imaging from a moving platform. Satellites need attitude control systems which are costly in terms of weight and power in order to maintain a fixed orientation. If an imager can tolerate and compensate for a moderate amount of satellite spin then much smaller attitude control systems can be used. An inertial measurement unit or IMU can provide the information needed to cue the OTCCD clocks. We anticipate an ongoing development of this device, in

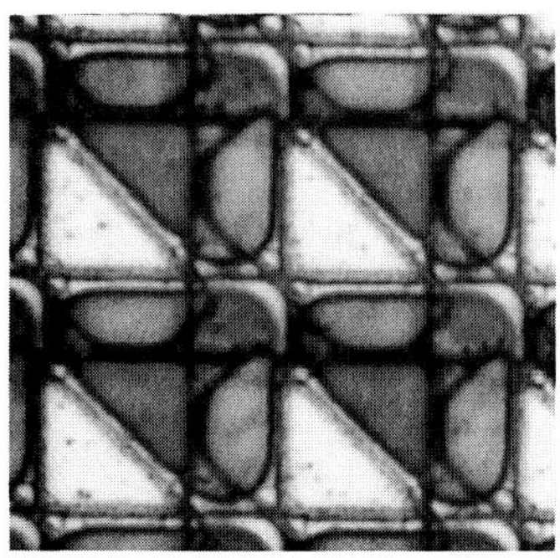

***Figure 13**: Photograph showing details of the imaging pixels. The four gate levels comprise three levels of polysilicon and one metal level.*

particular to larger devices with back illumination.

6. SUMMARY

Recent work in large, back-illuminated CCDs at Lincoln Laboratory has included a 5-Mpixel imager for broad-area space surveillance that filled a 100-mm wafer. With the recent conversion to 150-mm wafers in our fabrication facility, we offer a much higher device throughput as well as the possibility of single devices with active areas up to 80 cm^2. A $2k \times 4k$ imager has been fabricated as part of a collaboration between Lincoln Laboratory and the U. of Hawaii, and the first devices are just now through back-illumination processing. One of the test devices on these are wafers is a so-called OTCCD that can transfer charge in all four directions, and which has been used in a novel way to reduce atmospheric blurring. In another test imager, we have demonstrated that blooming control can be incorporated into our imager technology.

Acknowledgments: This work was sponsored by the Department of the Air Force and the U. of Hawaii Consortium.

7. REFERENCES

[1] B. E. Burke, R. W. Mountain, P. J. Daniels, M. C. Cooper, and V. S. Dolat, "CCD soft X-ray imaging spectrometer for the ASCA satellite," IEEE Trans. Nucl Sci, vol. 41, pp. 375-385, February 1994

[2] J. A. Gregory, B. E. Burke, M. J. Cooper, R. W. Mountain, and B. B. Kosicki, "Fabrication of large-area CCD detectors on high-purity float-zone silicon," Nuclear Instruments and Methods in Physics Research A 377 (1996) 325-333.

[3] J. Tonry and B.E. Burke, "The orthogonal-transfer CCD", these proceedings, p.281.

[4] S. Miyazaki, G. A. Luppino, and M. R. Metzger, "Recent CCD development activities at the University of Hawaii," Proc. SPIE Vol. 2654, pp. 239-247, February 1996.

[5] P. A. Levine, D. J. Sauer, F. L. Hsueh, F. V. Shallcross, G. M. Meray, G. C. Taylor, G. W. Hughes, J. Pellegrino, D. Simon, L. Harrison, and W. Lawler, "Performance of a high frame-rate CCD imager," Proc. SPIE, Vol. 1693, pp.65-74, 1992.

[6] B. E. Burke, R. K. Reich, E. D. Savoye, and J. Tonry, "An orthogonal-transfer CCD imager," IEEE Trans. Electron Devices, vol.41, pp. 2482-2484, December 1994.

CCDs FOR ASTRONOMY

P.J.Pool[1], *R.Holtom*, *D.G.Morris*

Abstract

In astronomy there is a growing need for very high performance CCDs. These CCDs require large area, high quantum efficiency and low read noise. The necessary technologies have been developed at EEV.

1. Large area

The use of a wafer stepper for photolithography gives the ability to define small features with excellent resolution and alignment and with very little optical distortion, but with limited image size. To overcome this limitation, the alignment capability of the stepper is used to "stitch" component parts of CCDs together to form a large device. This system allows the construction of a series of devices of different size from a single mask set. The CCD42 and CCD44 series are designed as large area CCDs for astronomy, differing only in their pixel size.

The CCD42 and CCD44 Series

Each series photocomposed from a single mask set.

Pixel Size		13.5 μm	15μm
2048 × 512		CCD42-10	
2048 × 2048		CCD42-40	CCD44-42
2048 × 4096		CCD42-80	CCD44-82
Butting loss:	sides	210μm	500μm
	top edge	100μm	150μm

Split register: read through either or both outputs
Register dump gate for fast clearing

CCD42 is of conventional *stitched* [1] format with a *middle* section comprising 2048(H)× 512(V) square pixels on 13.5μm pitch and an *ends* section comprising the top and bottom terminations for the array, including 4 extra lines of pixels, the read-out components and the wire bonding pads. The primary variant has eight middle sections (hence the type number CCD42-80) giving a total of 2048(H) × 4100(V) pixels and an image area of approximately 27.6mm × 55.3mm. Other variants with different numbers of middle sections can of course be produced (Figs.1&2). To facilitate thinning for back face imaging, gate protection circuits are designed close to the device and the bond pads are spaced off. If a guide chip is required for an array, the CCD42-10 (2k × 512) is a good candidate, having the same drive characteristics as the main chips in the array.

The common dimension (2048 pixels) of CCD44 is larger than the optical field of the stepper, making it necessary to stitch in two dimensions (Fig.3). CCD44-82 has an image area of approx. 30.7mm × 61.4mm.

[1] EEV Ltd., Waterhouse Lane, Chelmsford, Essex, U.K.. Email: **pp@eev.com**

J. W. Beletic and P. Amico (eds.), Optical Detectors for Astronomy, 29–36.

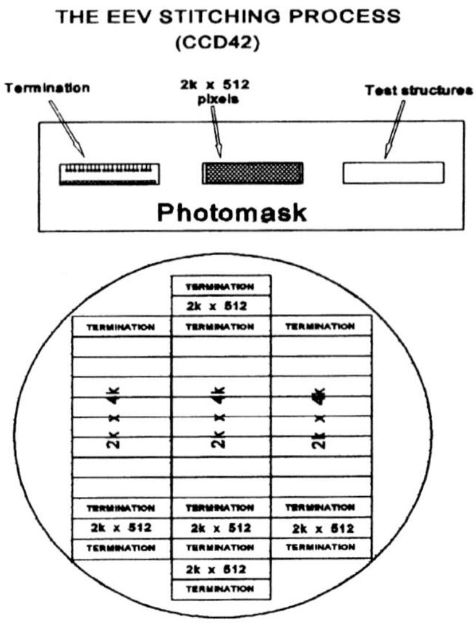

Figure 1

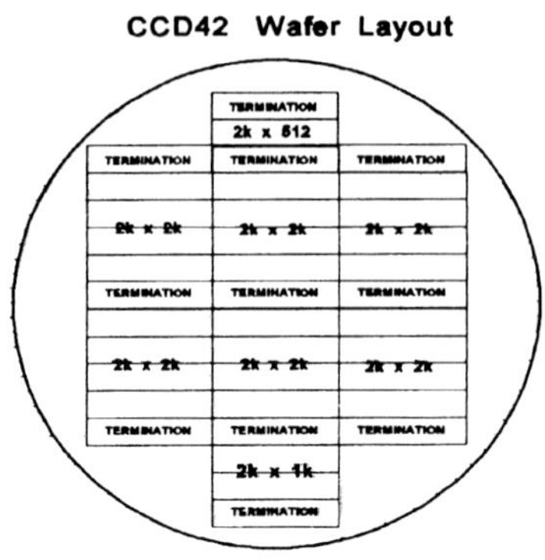

Figure 2

2. READ-OUT REGISTER

In both series of devices separate charge detection circuits are incorporated at each end of the read-out register, which is split such that a line of charges can be transferred to either output, or split between the two. The register is provided with a drain and control gate along the outer edge of the channel for charge dump purposes. The application of about +10V (not critical) to this control gate causes charge to flow from the register directly to the drain, allowing fast clearing of charge from all or part of the image section. Measurements of a similar structure on CCD25-20 show that the charge level in the register falls from saturation to 0.1% in 4 μs [2]. The charge storage capacity of the register is approximately four times that of the pixel. To reduce butting loss at the sides of the device, the register channel is bent through 180° such that the detection circuitry can be placed well away from the edge of the chip. A total of 50 blank "run off" elements are included before each detection circuit.

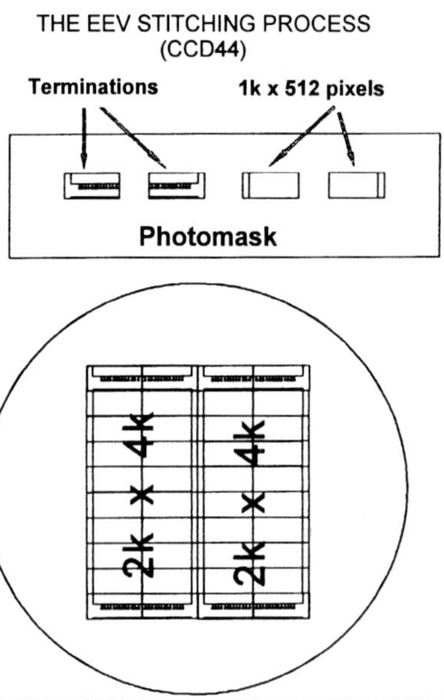

Figure 3

3. Detection Circuitry

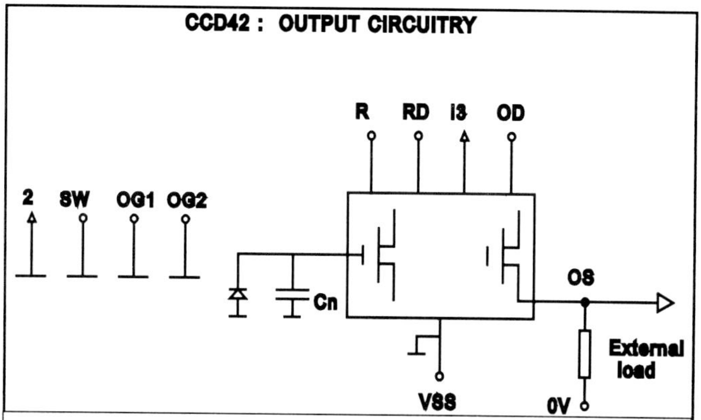

Figure 4

Figure 4 shows a schematic diagram of the output circuitry. This comprises three separately connected electrodes (to be described later) and the charge detection amplifier.

The detection node and reset switch are conventional, but the amplifier is an advanced two-stage type now successfully employed on other EEV devices, such as the CCD30 and CCD35. A two-stage circuit enables the first stage transistor to be very small, thereby maximizing the responsivity and minimizing the noise, with only minimal loading from a much larger second stage transistor which provides a high level of drive capability. Although a.c. coupled, the circuit has d.c. restoration activated by the transfer of charge into the read-out register via an internal connection to the last image section clock phase. It may be noted that the connections to the circuit are identical to those of a single stage type, the only difference being a standing current (<1 mA) now flowing in the substrate connection. There is no light emission to cause generation of spurious charge. The external load is non-critical, and can be a resistor of about 5-10kΩ or a constant current source of about 2-5mA.

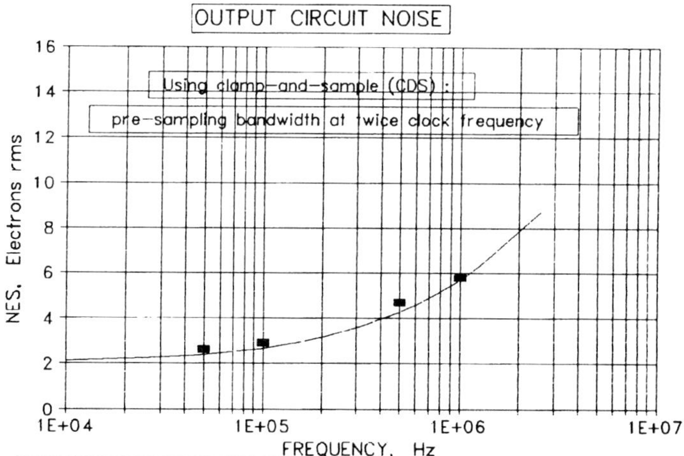

Figure 5

Figure 5 shows the predicted and measured noise performance for CCD42. These characteristic curves show the usual low frequency "floor", a consequence of 1/f noise in the first stage transistor, with the level increasing at higher frequencies through the white noise component becoming dominant.

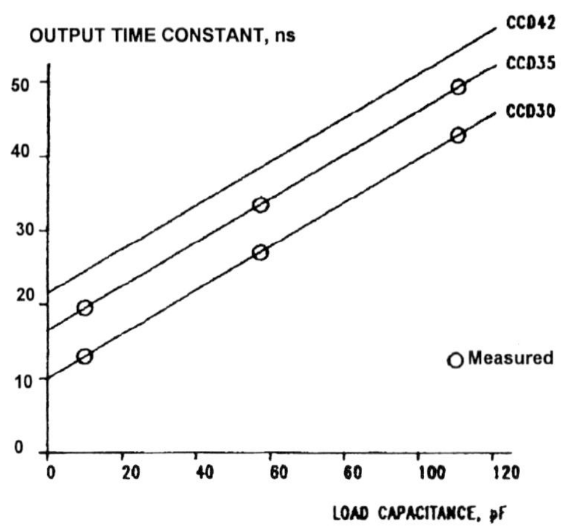

Figure 6

Back-illuminated devices may be slightly inferior in performance to their front-illuminated equivalents on account of extra node capacitance arising from the dielectric constant of the face-down chip-mounting materials being higher than that of air, thereby increasing the magnitude of the fringing fields between components.

It should be noted that the theoretical curves and the EEV measured performance figures shown in Figure 5 are for a relatively conservative pre-sampling bandwidth (twice the clock frequency) and that superior performance should be possible with reducing the bandwidth to nearer the theoretical limit.

Figure 6 shows the predicted output settling time constant in comparison with CCD30 and CCD35 results. Provided the load capacitance is not too large, the settling time should be sufficiently short to permit operation at frequencies up to at least 1mhz.

The peak charge detection capacity is approximately 300k electrons, as set by the voltage swing at the node being within the 4 volt nominal maximum. In cases where this is insufficient, the output node capacitance can be increased to approximately 36fF by taking the bias on the last register electrode (designated OG2) to about +20V (not critical), thereby increasing the charge capacity by a factor 3 and decreasing the responsivity by the same amount. The noise will also increase by a factor 3, not √3 as the size of the first stage transistor remains unchanged, i.e. not optimized for the larger node.

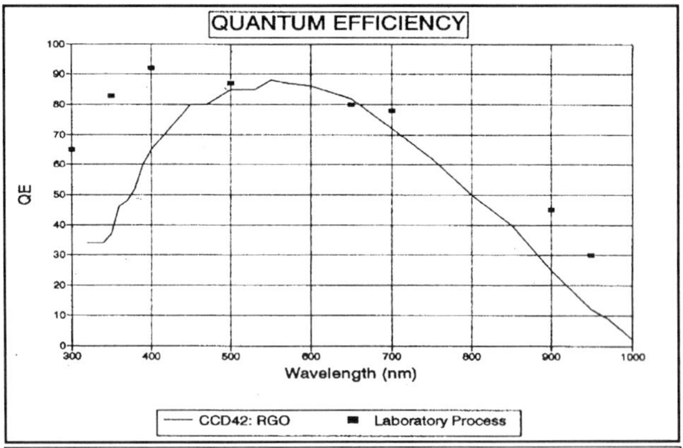

Figure 7

4. Quantum Efficiency

The quantum efficiency of a back illuminated CCD is determined by the quality of the back surface passivation and the anti-reflection coating. As a commercial supplier, EEV have always believed that quantum efficiency should be a stable parameter and consequently back surface passivation by boron implantation and laser anneal has been

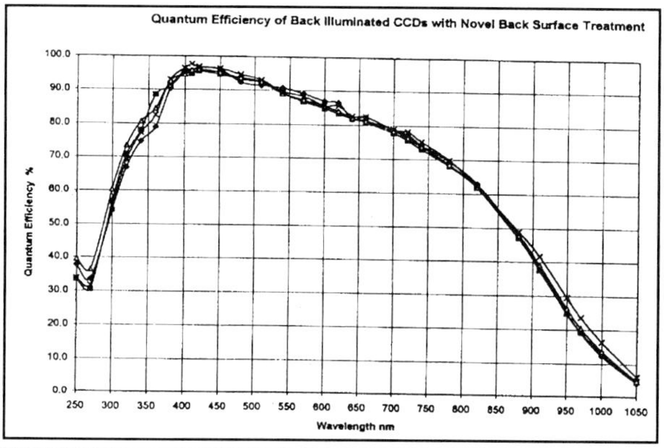

Figure 8

favored. By this process it is more difficult to achieve high quantum efficiency at ultra-violet wavelengths but during the last year, the laboratory process giving the QE curve shown in Fig.7 (AR coating peaked at 350nm) has been developed to become a production process with the performance shown as the solid line in Fig.7. This device has an AR coating peaked at 500nm and analysis shows that the internal QE of the two devices is very similar.

Passivation processes are still being developed and Fig.8 shows the results of the current best laboratory process. Attempts will now be made to transfer this process to production.

Much of the emphasis in QE development concentrates on U.V. response. As a spin-off from work on CCDs for X-ray imaging spectroscopy for the Jet-X, CUBIC and XMM space telescopes, experience has been gained in the processing of CCDs on thick, high resistivity (2000 Ω-cm) epitaxial silicon. Back illuminated CCDs with a thickness of 40 μm and the AR coating peaked at 800 nm have the QE shown in Fig.9. Note that the measurements shown were made at 20 °C so the figures at longer wavelengths will fall at lower temperatures due to band broadening.

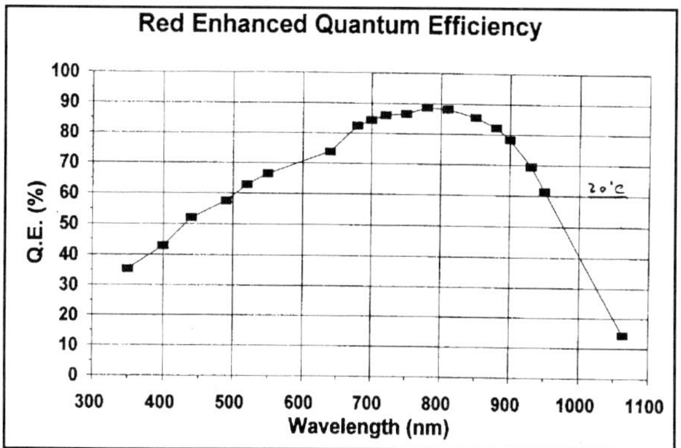

Figure 9

4.1 CCD Flatness/Packaging

To best utilize the optical performance of a telescope it is necessary that the focal plane be assembled flat to a few microns and that flatness maintained at the operating temperature. Fig.10 shows an interferogram of a thin CCD42-80 with a peak to valley flatness of 4 μm. It also shows that, after bonding to the package, this flatness is essentially unchanged.

The sketch in Fig.11 is a longitudinal section of a package designed for 3-side butting, with the silicon bonded to 2.5mm thick Invar. This package is for use on an aluminium

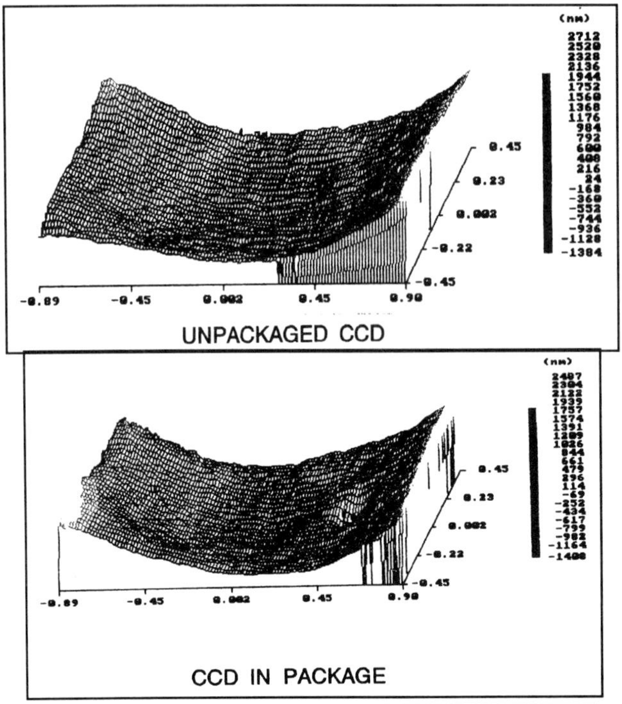

Figure 10

cold table, the thermal mismatch between aluminium and Invar being accommodated within the package. The shim is machined to standardize the thickness of each package and to make the base of the package parallel with the surface of the CCD.

Invar has the best available thermal expansion match with silicon, but the variation of silicon expansion coefficient with temperature is such that the match is only perfect at -70 °C with the result that the assembly bends in one direction as the temperature is reduced to -70 °C and begins to relax back below -70 °C.

A good fit has been obtained between measurements and a simple model and the model has allowed us to predict the effect of altering the Invar thickness (Fig 12).

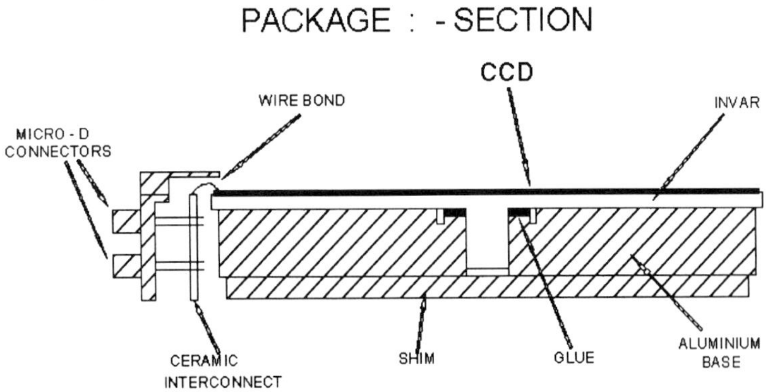

Figure 11

This prediction has been used in the design of a new package (Fig 13). The design addresses a desire among astronomers for a minimum footprint package, enabling the assembly of large arrays and also utilizes a willingness to construct cold tables from Invar. To minimize pick-up the custom PGA separates clock tracks from signal and DC tracks by a metal layer held at substrate potential. Provision is also made within the PGA to accommodate load resistors and FETs to further buffer the outputs.

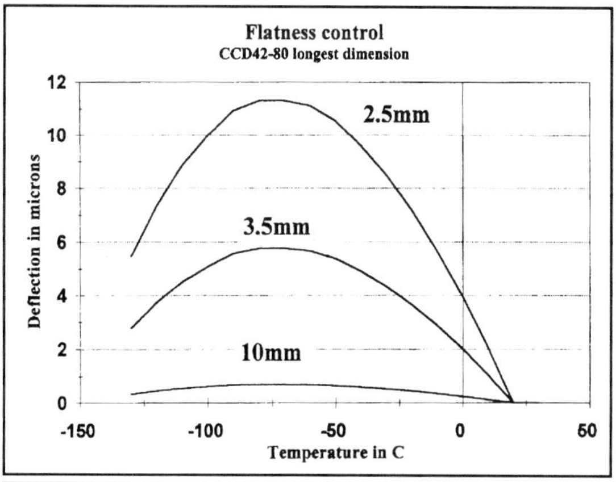

Figure 12

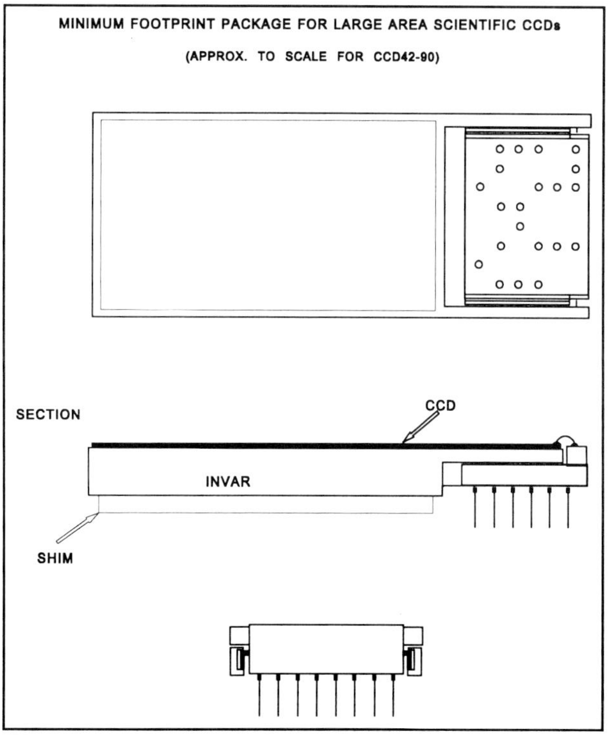

Figure 13

4.2 CCDs for Adaptive Optics

The development of output circuits with good noise performance at 1MHz allows the design of small CCDs with the capability of imaging at 1000 frames/sec with 6 electrons read noise and with only a moderate level of complexity.

The first of these designs is CCD39 (Fig. 14). In order to read 1000 frames/sec with 6 electrons noise it will be necessary to run at 2MHz pixel rate and reset only once in each line, allowing the same sampling time as is available at 1MHz.

CCD50 (Fig.15) has an image area of 128 × 128 pixels, will operate as a split frame transfer device into 16 outputs, using tapered store sections to generate the space for the first stage of the output circuits on simple, straight serial registers. With this number of outputs, 1000 frames/sec is available at 1MHz pixel rate/output and with reset on each pixel.

4.3 General Purpose

CCD47 and CCD48 are similar devices, differing mostly in their output circuits: the CCD47 output is as that on CCD42 whereas that on CCD48 is a

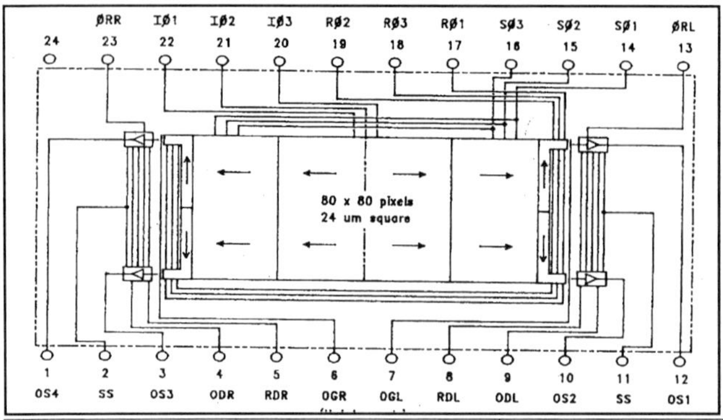

Figure 14

30MHz design for high speed imaging. Both devices will be available front or back illuminated.

5. SUMMARY

To meet the needs of astronomy, EEV have developed designs and technologies, which have advanced the performance of CCDs in all the relevant areas.

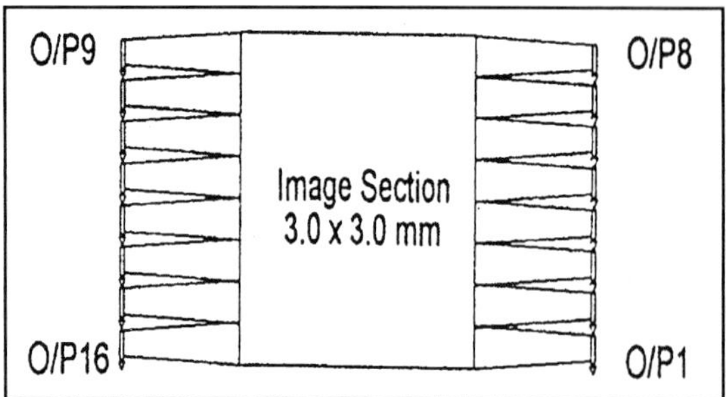

Figure 15

6. REFERENCES

[1] P.J.Pool, S.R.Bowring, W.A Suske, J.E.U.Ashton.: *Proc. SPIE* **Vol 1242** (1990)

[2] T. Eaton: private comm. ref. device assessment for the MERIS instrument on the ESA satellite ENVISAT 1.

[3] D J Burt, *CCD Performance Limitations: Theory and Practice Nuclear Instruments and Methods in Physics Research.* **Vol A305** (1991) pp 564 -573.

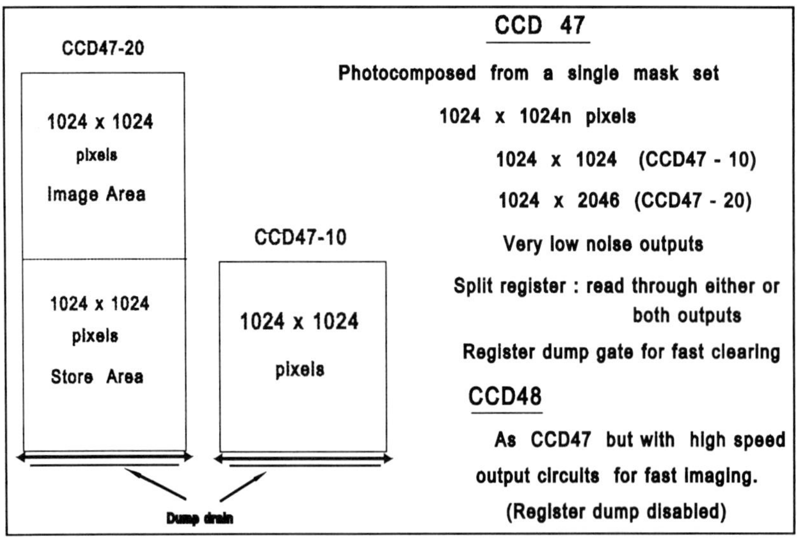

Figure 16

PHILIPS IMAGING TECHNOLOGY AND THE MK×NK IMAGE SENSOR

ALBERT J.P THEUWISSEN[1]

GHOSTWRITTEN BY J.W. BELETIC & P. AMICO

ABSTRACT

Through timely research, development and production of application-specific image sensors, Philips Imaging Technology wants to supply quality products, including adequate support, to its customers according to their needs and satisfaction.

1. INTRODUCTION

Philips Imaging Technology is located at the Philips Research plant in Eindhoven, The Netherlands. The production clean room is shared with IC technology R&D for Philips Semiconductors and image sensor R&D is carried out in the same clean room as production.

In the last few years CCD image sensors have become more specialized and their production more complicated. To reduce development costs, Philips Imaging Technology has created a process to build the sensors out of basic modules or building blocks. In this way it is possible to adapt the CCD to almost every imaginable application in a very short time.

Due to this design philosophy, the modular CCD can outperform any competitor in price as well as performance, as far as custom-designed or application-specific imagers are concerned.

2. CLEAN ROOM

The Philips Imaging technology fabrication facilities in Eindhoven are among the best in the world, including:

- 1200 m^2 total area, class I
- Diffusion, test, assembly, color filter capabilities
- Processing on 6" wafers, with migration to 8" in 1998
- Two shift operation
- Processing capability beyond 0.5 μm
- All activities are computer controlled, including analysis
- ISO 9001 certified

3. FT AND IL IMAGERS

The Frame Transfer Charge Coupled Device (FT-CCD) consists of four functional parts:

1. The image area (CCD array)

[1] Philips Imaging Technology, Nederlandse Philips Bedrijven B.V., Prof. Holstlaan 4 (WAG14), 5656 AA Eindhoven, The Netherlands. Email: **theuwiss@natlab.research.philips.com**

J. W. Beletic and P. Amico (eds.), Optical Detectors for Astronomy, 37–44.

2. The storage area (Memory array)
3. The horizontal register
4. The on chip amplifiers

The location of each of these parts is illustrated in Fig. 1a. The FT device is able to take an image and to transfer it to the storage part of the sensor, where it is shielded from further exposure to light. Afterwards the storage section can be read out without significantly corrupting the data by continued exposure. In comparison to frame transfer, the read out is a slow process.

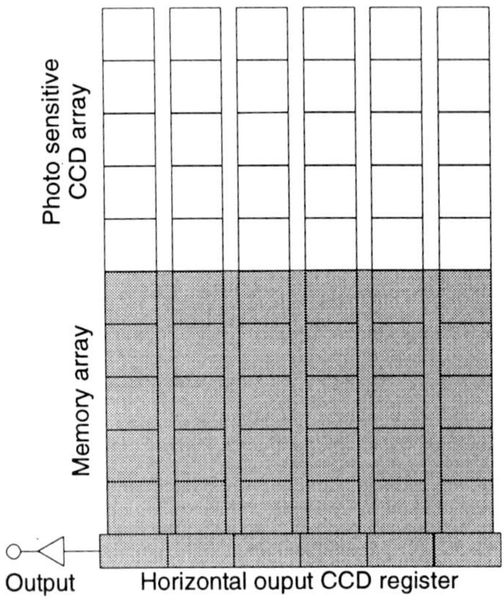

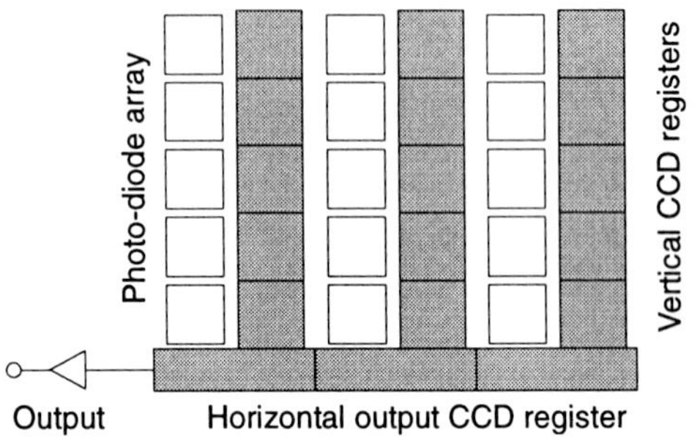

Figure 1: a) left: schematic of the Frame Transfer (FT) device. b) top: schematic of the Interline Transfer device (IL)

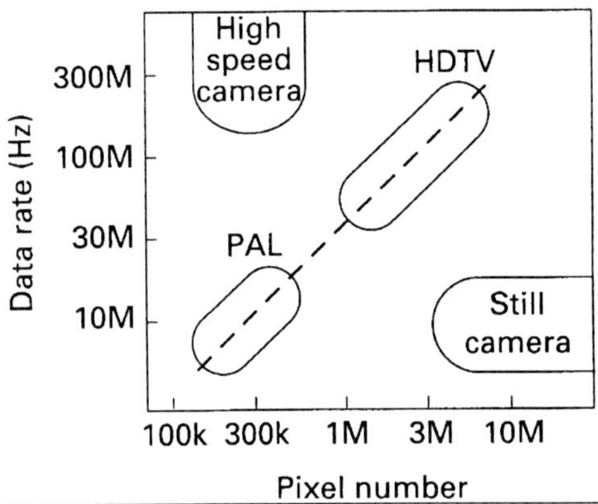

Figure 2: Comparison between the required performances of different technologies using CCD sensors: PAL (standard TV), HDTV, High Definition TV, Still Cameras and High Speed Cameras.

A similar device, used for TV applications, is the Interline Transfer (IL) device, shown in Fig. 1b. The requirement of minimal image blurring is fulfilled by building an imager composed of an array of photo-diodes, which collect the light, coupled in parallel by storage columns (vertical CCD registers), where the image is quickly transferred for successive read-out from the horizontal output register.

The diagram in Fig. 2 compares the performances required for different CCD applications, in terms of total sensor area (Number of pixels) versus required read rate. As can be inferred from the diagram, there is a growing demand for large area sensors, but, on the other hand, the technical difficulties for building large area sensors make the process slow and expensive. Philips' answer to customers requirements is the modular CCD, a mK×nK image sensor, built up from smaller blocks.

4. BUILDING BLOCKS CONCEPT

The idea is to create a large sensor built up from smaller blocks of 12 μm 1024×1024 pixels. The fact that the sensor's block size is fixed allows quick development from idea to product. These blocks can be placed next to each other during fabrication, by stitching, without significant distortions to the uniformity of the pixel array, allowing arbitrarily large sensors. With this approach, the size of a CCD sensor is limited to be smaller than the size of the wafer and is multiple of the 1K pixel module. Other than these, limits the amount of blocks the CCD is built from is based only on the kind of application, i.e. what resolution is needed.

Two different types of sensors can be made with this concept, the already mentioned frame transfer (FT) device and the full frame (FF) device. The decision between them is also application dependent. The maximum size for a mK×nK sensor with 6 inch wafers pixel size of 12 μm is 7K×9K for FF devices. The limit for a FT device is lower since the storage section is not included in the mK×nK specification. That means a maximum width of 7168 (7K) and a maximum height of 9216 (9K) pixels (total of 66,060,288 pixels) for a FF device. The maximum size of a FT device is a 7k×4k image area (29,360,128 pixels).

Fig. 3 shows an example of large sensor built up of smaller blocks.

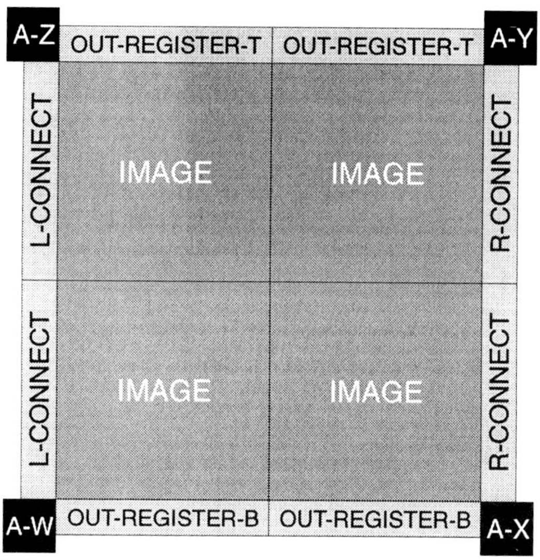

***Figure 3**: gates A-Z, A-Y, A-W and A-X positioned respectively in the upper left, upper right, lower left and lower right parts of the image area*

READ OUT OPTIONS

By clocking the gates of the FT or FF imager it is possible to read out an image in many different ways. The possibilities for shifting an image are illustrated in Figure 4.

- *Full Frame device*: the entire image can be shifted to the upper or lower horizontal register or the upper half of an image can go to the upper register and the lower half to the lower.

- *Frame Transfer device*: the entire image can be shifted to the storage section (Frame Shift, quick process) or to the upper horizontal register (slower process, due to the fact that each shifted line has to be read out).

On both type of sensors there is an output amplifier on each end of the horizontal register so the image can also be read entirely through one amplifier (left or right) or split between the two of each horizontal register. All four amplifiers can read a quarter of the image (FF only), or two

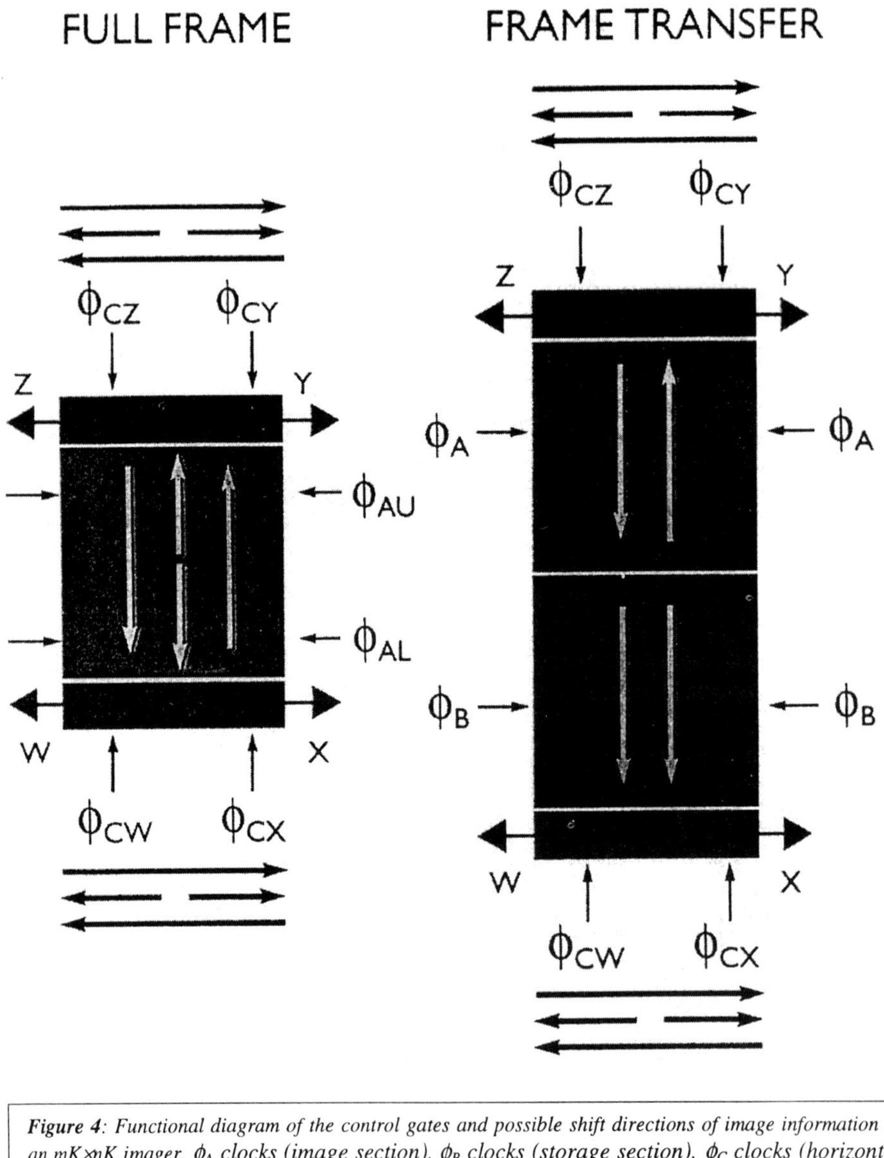

Figure 4: Functional diagram of the control gates and possible shift directions of image information in an mK×nK imager. ϕ_A clocks (image section), ϕ_B clocks (storage section), ϕ_C clocks (horizontal register).

can read half, or one can read an entire picture. In other words, the image of a FF sensor can be split along two axes (one horizontal and one vertical) and the image from a FT sensor can be split along one axis (vertical). However, both can also be treated as a continuous array over the mid-lines.

As the illustration shows, the possibilities for the frame transfer device are even greater. For example, two pictures can be taken with only a small time between them by snapping one, shifting it to the storage section, snapping a second image and reading both out at the same time after protecting the image area from light.

Almost every structure of the mK×nK CCD is possible. The choice between the different types and sizes of sensors depends on the application and on the topological limits. Note that a frame transfer CCD needs a storage area at least as big as the entire image area and that in general a mK×nK FT device can be treated as an mK×nK FF device.

5. TECHNOLOGY

5.1 Pixel

A section of the pixel structure is shown in Figure 5: the left image shows a y-distorted view of a single pixel. The four polysilicon layers stripes are connected to the corresponding gates, which perform the four-phase charge transport. The polysilicon layer has been etched to obtain a T-shaped profile. Its purpose is to enhance the quantum efficiency of the layer-free part of the pixel. The right side of Figure 5 shows the two horizontal cross sections (AA' and BB'): as the polysilicon layer absorbs part of the incident light, the Blue quantum efficiency through the BB' cross-section (lower right panel) is lower than the one through the AA' cross-section. The shape of the p^+ doped silicon helps diffusion of the electrons generated by blue photons (which have a small penetration depth) towards the depleted region.

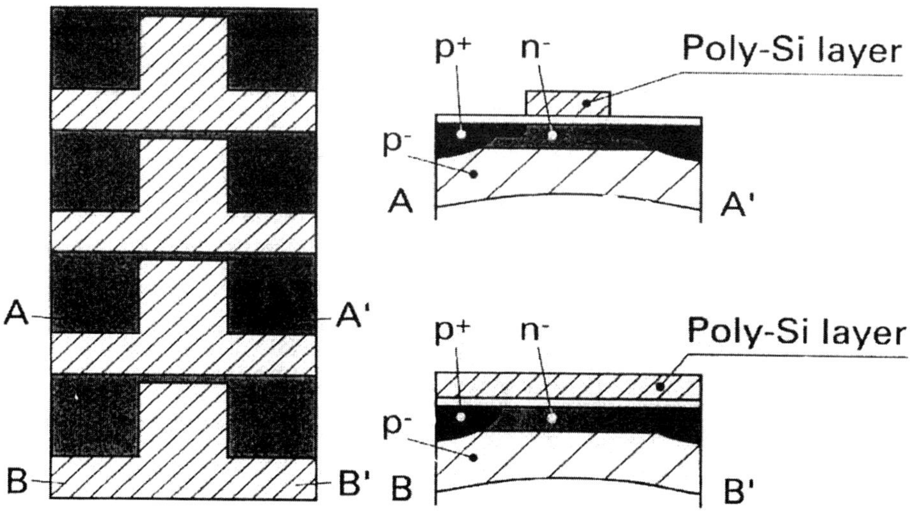

Figure 5.: Layout of polysilicon wires of 4-phase pixel (Note: the vertical axis of the left hand side is stretched, as this is a square.)

Figure 6 shows three potential curves as a function of pixel depth: the vertical n-p-n structure enables anti-blooming. The lower curve (10V clock bias) shows that only the photons penetrating the first 2.5 microns are transferred to the potential well. The

0V clock bias works as an *electronic shutter*, allowing complete wiping of charges towards the n-substrate when applied.

The shifting of charges (e.g. a vertical frame shift in the image storage sections for a FT device) is achieved with a four-phase charge transport. Vertical binning can be performed easily by shifting two rows instead of one into a horizontal register.

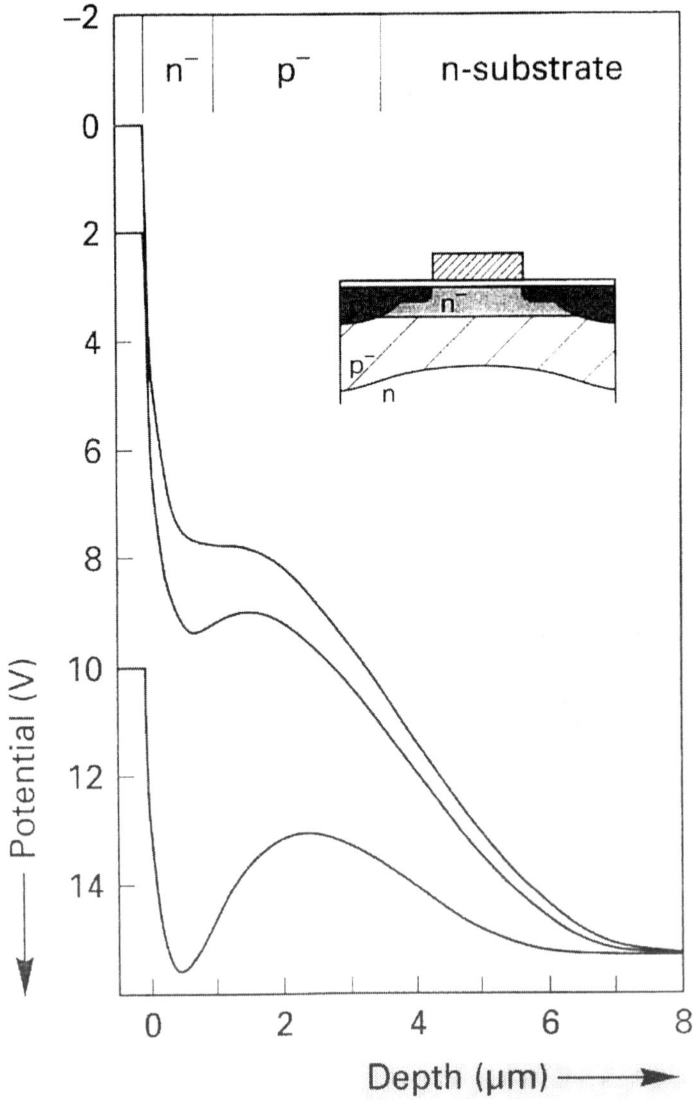

Figure 6: *Potential curves of the Philips CCD as function of clock phase voltage and penetration depth. This structure has vertical anti-blooming and a very fast image clear capability.*

Afterwards the line resulting from two vertical adjacent lines is read out as a single line. Thus each pixel charge in the register results from the addition of two vertically neighboring pixels.

5.2 Output

The shift in the horizontal registers is achieved with a three phase (ϕ_1, ϕ_2, ϕ_3) charge transport gate control.

Figure 7 shows the functional diagram of the output amplifier with floating diffusion:

- OG (Output Gate), it is the only gate supplied with a DC voltage. It separates the floating diffusion in front of the output amplifier from the horizontal register.
- SG (Summing Gate), it makes horizontal pixel binning possible. In normal read out mode (no binning) it is controlled with the same clock signal waveform as the ϕ_3 gate; in case of binning the summing gate must have a higher clock swing than the other ϕ_C gates, because up to four pixel charge packets have to be accumulated without spilling.
- RG (Reset Gate). A pulse on this gate clears the floating diffusion to be able to accept the next pixel. When it is high the charge of the last pixel is drained off to reset drain (RD). The timing of the RG depends on the read out mode.
- RD (Reset Drain) has two functions: first it drains pixel charges, second, it adjusts the potential of the amplifier's input.
- SF(D/S). The three stage source follower (drain/source), first two stages have on chip loads, the third one has its load off chip.

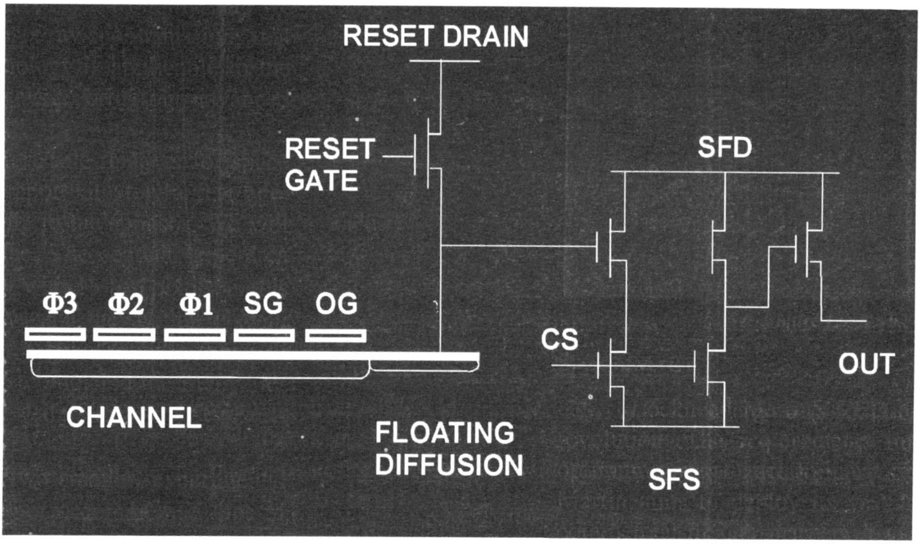

Figure 7: functional diagram of the output amplifier. Key to symbols:
SG Summing gate
OG Output Gate
CS Current Source
SF(D/S) Source Follower (Drain/Source)

A summary of the sensor's performances is given in the following table.

Performances	
Q_{max}	> 200,000 electrons
I_{dark}	< 0.5 nA/cm^2 (60°C, non –MPP)
Quantum Efficiency	13% @420 nm
	27% @530 nm
	7% @ 800 nm
Vertical Anti-Blooming	200 x
Conversion Factor	6μV/electron
Bandwidth	110 MHz
Amplifier Noise	23 electrons (5 MHz)
Summing Gate Capacity	> 600,000 electrons
Floating Diffusion Capacity	> 600,000 electrons
Random Non-Uniformity	< 1.0%
Fixed Pattern Noise	< 50 electrons (60°C, 20 ms $T_{integrtion}$)

***Table 1**: Typical performances for the mK×nK image sensor.*

CONCLUSIONS

The advantages of the mK×nK image sensor can be summarized as follows:

- High Performances
- Large area sensors design
- Modular solution
- Reduced mask costs and development time
- Easier design changes.

SECTION II: OBSERVATORY PLANS FOR OPTICAL DETECTORS AND INSTRUMENTS

CFHT - WIDE FIELD IMAGING AND "CCD-WORLD" HOST

T. M. C. ABBOTT[1]

1. WIDE FIELD IMAGING

CFHT is committed to a future of wide field imaging in the optical, in order to take full advantage of its unique high quality site and to support the coming age of 8-10 m class telescopes. The UH8k mosaic [5] is currently in frequent use at the prime focus - it consists of an array of 2k × 4k thick Loral CCDs, or 8192 by 8192, 15 micron pixels, covering 0.23 degree square. Another mosaic of 2 × 6 thinned CCDs (or 8192 by 12288, 15 micron pixels) to be selected from the Lincoln Labs/MIT consortium [3] is currently under construction at the Institute for Astronomy, The University of Hawaii at Manoa [4]. This camera will cover 0.34 degrees square and is expected to see first light before the end of 1997. On the longer term, CFHT is entering into a collaboration with the Commissariat à l'énergie atomique, Saclay, France to build a 16k by 16k thinned CCD array, named MEGACAM, which will cover up to 0.96 square degrees and will represent the flagship of CFHT's imaging programme [2]. First light is anticipated before the end of the millennium. For further information on these and other CFHT projects, please visit the CFHT Web homepage at http://www.cfht.hawaii.edu/.

2. THE CCD-WORLD MAILING LIST

At the previous CCD workshop hosted by ESO in 1993, it was resolved that an electronic forum should be set up for the discussion and dissemination of information relevant to the discipline of CCD development for astronomy. This mailing list was originally hosted by ESO, but now has a home at CFHT. At the time of writing the list has nearly 200 members and its traffic, while sporadic, is always of high quality. The list has proven itself a valuable resource for the transmission of useful information to the community. Requests for subscription should be sent to ccd-world-request@cfht.hawaii.edu. See also [1].

3. REFERENCES

[1] Abbott, T. M. C., 1997, http://www.cfht.hawaii.edu/~tmca/CCD-world/.

[2] Boulade, O., et al., 1997, this volume.

[3] Burke, B., 1997, this volume.

[4] Luppino et al., 1997, this volume.

[5] Metzger, M. R., Luppino, G. A., Miyazaki, S., 1995, BAAS, **187**, 7305.

[1] Canada-France-Hawaii Telescope and Institute for Astronomy, University of Hawaii at Manoa, Box 1597, Kamuela, HI 96743 U.S.A. Email: **tmca@cfht.hawaii.edu**.

J. W. Beletic and P. Amico (eds.), Optical Detectors for Astronomy, 47.

THE ASTRONOMICAL CCD DEVELOPMENT PROGRAM AT COPENHAGEN UNIVERSITY

JOHANNES ANDERSEN[1]

ABSTRACT

A CCD camera/controller system has been developed for use at the 2.5m NOT and the Danish 1.5m telescope on La Silla. Based on a standard industrial dewar, the system is designed to accommodate a variety of commonly used CCDs up to the Loral 2k×2k devices in up to 2 ×2 mosaics, read through up to 8 parallel channels. The PC/MC68020 based controller performs in all essential respects on a par with other systems currently in use at major observatories worldwide. The deployment and planned upgrades of the system are briefly described.

1. WHY START ANOTHER IN-HOUSE DEVELOPMENT OF CCD CAMERAS?

Danish geography and climate being what they are, the bread and butter of Danish ground-based observational astronomy is our membership of ESO, the scientific returns of which are much enhanced by our 50% share of the observing time on the Danish 1.54m telescope on La Silla. An essential recent northern-hemisphere addition to our facilities is our 20% share of the 2.5m Nordic Optical Telescope (NOT) on La Palma, Spain.

Both of these are Ritchey-Chrétien telescopes with direct imaging as their main asset. Both are being equipped with EFOSC-type focal reducers offering wide-field direct imaging and multi-object grism spectroscopy in an efficient, versatile configuration. A standby CCD camera offering direct imaging at higher spatial resolution is already implemented (NOT) or at the planning stage. At the NOT, additional imaging instruments and standby (fiber-fed) or dedicated spectrographs exist or are under way. The scientific competitiveness of both telescopes is thus critically dependent on their being equipped with the best large-field CCD detectors that the state of the art can supply.

At the time (1990) when strategic decisions on future detector policy at our two telescopes had to be taken, the obvious - indeed only - access to large-format CCDs was to invest in a foundry run of the Loral 2048×2048 CCDs designed by John Geary, allowing us to build mosaics of at least 2×2 such devices. In addition, the thinning process being developed by Michael Lesser at Steward Observatory promised to result in detectors of outstanding performance. With several partners, we were fortunate to obtain funding to invest in a total of three runs of these Loral CCDs, yielding a large number of potential science-grade devices. Commercial camera systems capable of operating (mosaics of) CCDs in these sizes were not on the market, and after considerable soul-searching we had to conclude that none of our host organizations was likely to provide cameras of competitive performance on a scientifically interesting

[1] Astronomical Observatory, Niels Bohr Institut für Astronomie, Physic & Geophysic, University of Copenhagen, Denmark. Email: **ja@astro.ku.dk**

J. W. Beletic and P. Amico (eds.), Optical Detectors for Astronomy, 49–54.

schedule. While the effort to develop our own system in only one or two copies would not have been justified, the projected total needs of the two telescopes was in fact 5-10 cameras. The prospect that these would achieve state-of-the-art performance, appear to users and staff as uniform, well-known systems, and offer considerable economy in hardware and (especially) software development and maintenance effort, spare part investment, etc., led us to the decision to undertake a full-scale development programme. As an important spin-off, we would also acquire the know-how to maintain these cameras at optimum performance and upgrade them as detector technology progressed.

Our first large-format camera was ready for field tests after about a year, and several copies have now been built and brought into operation as suitable CCDs have become available. Since one of the actual cameras is on display at this meeting and a detailed technical description of the system was published recently ([2]), the following comments will be limited to a brief discussion of our overall strategy and experience with the cameras so far.

2. INITIAL SYSTEM PERFORMANCE SPECIFICATIONS

While realizing that neither our telescopes nor our resources are in the 8-16m class, our goals were ambitious:

- Large detector format, 2048 $\times$ 2048 pixels (15 microns) per CCD, and possibility to accommodate mosaics of 2 $\times$ 2 of these devices. Easy implementation of existing and future CCDs in other formats.
- State-of-the-art performance as regards quantum efficiency, readout noise, linearity, readout time, immunity from electrical interference, parallel readout through up to 8 amplifier channels, X-Y binning and windowed readout, communication with telescope control system, etc.
- Economical, modular design allowing permanent installation of multiple parallel systems and minimum effort in changeover and maintenance, rational planning of spare part inventory, etc.
- Built-in maintainability and upgradability.

Given the very limited manpower that we would be able to deploy on the project, it was decided to concentrate the development effort strictly on the camera system as such, making no attempt to duplicate any of the tens of man-years of effort invested in image processing systems such as IRAF, MIDAS, and others. The goal would be for the system to set up and execute the sequence of exposures best suited for the user's scientific programme and deliver the result, fully documented and as rapidly and transparently as possible, into the image processing system of the user's choice for further analysis with the most sophisticated tools of the trade. Thus, the end product of the system would be a series of FITS files of raw CCD exposures, with all relevant information documented in the file headers.

3. AS-BUILT CAMERA DESIGN AND PERFORMANCE

As seen from the camera system on display at the meeting and described in more detail by [2], our CCD cameras are based on standard IR Labs LN2 dewars with a mechanical module containing the CCD proper and its front-end electronics and connections, designed and built in our workshops. In the laboratory, the front flange with the dewar entrance window can be replaced by one containing an Fe55 source for CTE and gain measurements using monochromatic X-ray photons.

Next to the dewar is a small electronic rack containing the complete CCD controller, with on-board CPU, phase drivers, and amplifiers and ADCs. One controller can handle up to 4 CCDs, read out through 8 channels. During readout, the image is transferred via a dual optical fiber link to a Frame Store card in a Pentium PC, containing enough on-board memory to store a complete image. This solution allows us to interface the camera to a (cheap) PC for high-level control with satisfactory performance, instead of using much more expensive workstations.

The original system used a linear power supply for minimum noise generation, housed in a second small rack close to the dewar. Recently, it has been possible to implement switched-mode power supplies with satisfactory noise performance while saving one rack and reducing heat dissipation appreciably.

The PC, running under LINUX (UNIX for PCs), handles the user interface, including all exposure control and status display information, controls the CCD temperature, communicates with the Telescope Control System, and monitors a variety of environmental parameters. In parallel with the readout itself, the incoming image is displayed on the PC monitor, stored on disk, and transferred to a workstation disk and/or DAT tape in FITS format with all header information. At our two main telescopes, all further image processing is done on a dedicated workstation connected to the PC, but the higher-end LINUX Pentiums can run suitable subsets of the full IRAF or MIDAS systems in smaller-scale installations.

Currently, we have two cameras with Loral 2k $\times$ 2k CCDs in operation at the NOT on La Palma, one on the Danish 1.5m at La Silla, and one in our lab for chip test and optimization. A further camera is operated at the NOT with a SITe 1k CCD and another at Ondrejov Observatory (Czech Republic) with a spectroscopic-format CCD (256 $\times$ 2688 pixels). A prototype Peltier cooled camera is also running with a CRAF/CASSINI 1k CCD (courtesy of Jim Janesick) at the Schmidt telescope at Brorfelde Observatory in Denmark. Finally, two small Peltier cooled cameras with quad-readout 64 $\times$ 64-pixel Loral CCDs running at up to 100 Hz frame rate (binned) are used with the same controller for wavefront sensing in a prototype tip/tilt correcting system at the NOT, foreseen for later upgrade to low-order adaptive optics. User feedback is generally favorable and is being recycled into regular hardware and software updates for all cameras.

Briefly stated, all key design goals have been met or exceeded. Readout time for a 2k×2k CCD is about one minute through one amplifier; system noise is below 1 e^{-}rms, and actual readout noise thus dominated by CCD amplifier noise; and linearity and other key parameters are generally competitive with other systems of the "pre-FIERA" generation. An adequate test facility has been developed for system integration and calibration and performance optimization and verification.

4. CCD PURCHASE AND THINNING

As was to be expected, we have re-learnt a number of lessons in vacuum and cryogenic technology and ultra-low noise electronics the hard way, and contributed a few original mistakes of our own along the way. However, the single factor that has caused us the most frustration and delay is the supply of the key component of the cameras, the CCDs. While we are also not unique in this experience, setting down the basic facts may be useful to others who have gone down similar paths, or consider doing so. Below, a "production run" refers to a batch of 22 4-inch wafers containing a theoretical maximum of 88 2k × 2k CCDs plus a variety of smaller chips, all nominally with 15-micron pixels.

Our first procurement, in 1990, was half a run of the "SAO" two-side buttable Loral 2k × 2k CCDs, designed by John Geary and delivered in 1992. The run, shared with Princeton Scientific, Inc., yielded some 25 working 2k CCDs. A few of these were mounted thick, two being coated for UV sensitivity by ESO, and one was in successful operation at the NOT for several months in mid-1993. However, due to a problem in manufacture, these chips gradually developed various malfunctions, and by 1994 the whole lot had essentially self-destructed on the storage shelf.

In 1994 we were able to purchase another two production runs of the "Edgar Smith" three-side buttable 2k × 2k CCDs (an improved Geary design), one with Instituto de Astrofisica de Andalucia (IAA), Granada, another with the EROS consortium. The two lots contained over 60 and 40 functional 2k devices. Euphoria ran high; numerous silicon-hungry friends put their names (and money) on our ball-card, and half-promises were made with a good conscience.

However, due to a variety of CTE, bonding, and other problems, filling the two EROS II 2×4 mosaic cameras with respectable (thick!) 2k CCDs consumed the entire run intended for that purpose plus a couple, with nothing left for later thinning as we had originally hoped. Of the IAA run, we have so far committed over 20 of the large CCDs - of good quality, judging by the warm Loral test images - to bonding and thinning at Mike Lesser's laboratory. Of these, four (4!) have reached the telescope thinned and AR coated. One of these was damaged in a mechanical accident and another contaminated with oil from a poor vacuum pump. One died this year, along with six others that never reached the telescope, of a delamination problem in the substrate that causes loss of contact with the CCD after a few to several thermal cycles. The rest were lost to an impressive variety of other causes.

Nonetheless, the CCDs we did get into operation on the telescopes had tremendously sensitivity (see QE curve in [1]), and we are pushing on boldly with our plans for the further camera systems sorely needed by our impatient users, but keeping a nervous eye on the silicon supply and consumption. While waiting for the large CCDs, one camera is operated at the NOT with a SITe 1k CCD, which has performed flawlessly, but neither size nor QE is quite what we were hoping for.

And the larger-size CCDs now offered from several sources are still somewhat off our financial scale.

A final unpleasant discovery was the "blue MTF problem", extensively discussed at this meeting, i.e. the charge diffusion taking place in the field-free region of the epitaxial layer, and apparently particularly severe in the relatively thick, low-resistivity epi layer of the Loral CCDs which was designed to give maximum QE in the red - which it does admirably. However, having CCDs with effective pixel sizes around 25 microns in focal reducer and spectrograph optics designed for 15-micron detector resolution degrades both the effective seeing and the spectral resolution, at considerable loss in scientific performance. And given the general lack of quantitative information, this is likely another case where the lack of evidence for the existence of the effect in CCDs from other manufacturers cannot be taken as evidence for its absence. An easy fix for this vexing problem would be much appreciated!

5. FUTURE DEVELOPMENT PLANS

As hinted above, our top priority in the near future is to supply good-quality 2k × 2k CCDs to the several cameras that we and our partners have built or under way, with chip upgrades at the 1.5m and the NOT as the most urgent. On a medium-term time-scale (1-2 years), we are committed to another 2k camera for a fiber-fed spectrograph and a 2 × 2 mosaic of 2k chips for the NOT, and possibly a direct standby camera for the 1.5m telescope. And as part of the Heidelberg-led consortium building the fiber-fed Echelle spectrograph FEROS for the ESO 1.5m telescope on La Silla, we are committed to operate one of ESO's new-generation 2k × 4k EEV CCDs in their continuous-flow dewar with our controller system, an interesting challenge that will also give us valuable experience with the modern design of CCD with fast-readout capability.

Finally, as part of the comprehensive instrumentation upgrade program at the NOT (see C. Aspin's poster at this meeting), we intend to modify our controller system to operate a Rockwell 1k × 1k "Hawaii" array in the new near-IR camera for high-resolution imaging at the NOT, hoping in this way to both save development effort and ease maintenance at the telescope through maximum commonality between optical and near-IR detector systems.

In view of these perspectives, we are planning a moderate upgrade program for our controller system to keep it competitive within its class, if not with FIERA-type machines. Apart from "standard" desiderata like even better stability, noise immunity, flexibility, and user-friendliness, and the capability to handle near-IR arrays, we would like to improve the readout speed of the camera by perhaps a factor 5 or so. This would imply a limited amount of hardware and software reorganization at the controller level, and possibly replacement of the "remote" control PC and fiber cable connections with a second on-board CPU and memory for high-level control, interfacing directly via a high-speed Ethernet link to the data acquisition workstation. These improvement are currently in the drawing-board stage.

Acknowledgements: Credit for the successful developments described here belongs to the technical staff of Copenhagen University Observatory who did the real job. Special but, sadly, posthumous thanks must go to Ralph Florentin Nielsen who, as the guiding light of instrument development at CUO for three decades inspired and directed the work described here, until his untimely death in 1995 robbed us of his wisdom, encouragement, and friendship. Key roles in the development are also played by Preben Nørregaard and Jens Klougart, with major contributions from Hans H. Larsen, Niels Michaelsen, Anton N. Sørensen, and Michael I. Andersen. Along the way, we have been much helped by the friendly collaboration of John Geary, Dick Bredthauer, Mike Lesser, and not least the ESO Optical Detector Team. And finally, without the very substantial financial support of the Danish Natural Science Research Council and the Carlsberg Foundation, the programme outlined here would not have existed at all.

6. References

[1] Andersen J., Andersen M.I., Klougart J., Knudsen P., Larsen H.H., Michaelsen N., Nielsen R.F., Nørregaard P., Olsen E., Rasmussen P.K., Seifert K.E.: 1995, *The ESO Messenger,* **No. 79**, 12

[2] Klougart, J.: 1995, *SPIE Conf. Proc.* **Vol. 2416**, p. 65

FUTURE INSTRUMENTATION FOR THE NORDIC OPTICAL TELESCOPE

COLIN ASPIN[1]

ABSTRACT

The Nordic Optical Telescope is being equipped with new proprietary instruments, specifically a modern low-noise high-quantum efficiency CCD camera, Brocam-I and a faint object spectrograph, ALFOSC, In addition, new instruments for observations in the optical and near IR, have been planned, like the camera, NOTCam for wide-field and high spatial resolution direct imaging. The characteristics of its modes and detector are briefly outlined.

1. INTRODUCTION

The Nordic Optical Telescope (NOT) on the island of La Palma in the Canary Islands is a 2.56m (Super) Ritchey-Chretien alt-az telescope. First light at the NOT came in late 1988, and regular observing started during 1989.

Until recently, instrumentation at the NOT consisted of mostly visiting instruments from both Nordic and other institutes and Universities. However, this has begun to change.

A modern low-noise high-quantum efficiency CCD camera, Brocam-II, built by the IJAF Copenhagen, has been commissioned. Brocam-II is based on a Loral-Lesser thinned 2048 × 2048 CCD and works at 0.11"/pixel.

Soon to be commissioned (Oct 96) is a new faint object spectrograph, ALFOSC, owned by the Instituto de Astrofisica de Andalucia and built by the (Instrument center for Jordbaseret Astronomisk Forskning (IJAF) Copenhagen. ALFOSC is a copy of the ESO DFOSC spectrograph and will be based on a Loral-Lesser thinned 2048 × 2048 CCD working at ~0.2"/pixel and hence having a field of view (fov) of ~6.5' × 6.5'.

In addition, a proposal has gone into the four Nordic Research Councils for funds to establish a "Core Instrumentation Program" (CIP) whereby a new set of instruments will be built for NOT. These instruments will be locally supported, maintained and, in some cases, improved.

2. CURRENT PLANS

The current plan for the CIP is to provide NOT with FOUR state-of-the-art instruments for both optical and near-IR observations. Specifically, these instruments are:

1. A 2 × 2 Loral-Lesser ultra-thin PPtF coated CCD mosaic camera (MOSAIC) 4096 × 4096 pixels in total imaging area (to be built by IJAF Copenhagen).
2. A focal reducer (FOCRED) optical unit to allow direct imaging over ~17' × 17' fov (to be built by Opteon OY, Finland). The FR will operate with the 2

[1] Nordic Optical Telescope, Apartado 474, E38700 S/C de la Palma, Canarias, Spain. Email:**caa@not.ica.es**

J. W. Beletic and P. Amico (eds.), Optical Detectors for Astronomy, 55–56.

× 2 CCD mosaic camera for wide-field direct imaging (at 0.25"/pixel, fov~17' × 17').

3. A near-IR 1-2.5μm array camera, NOTCam (to be designed and built by several institutes under contract to NOT) for wide-field (~0.25"/pixel, fov ~ 4' × 4' LORES mode) and high spatial resolution (~0.08"/pixel, fov ~1.4' × 1.4' HIRES mode) direct imaging. Additionally, NOTCam will be able to perform near-IR spectroscopy (grism point source, long-slit and multi-object), imaging polarimetry and spectropolarimetry observations.
4. A 5-mode low-order adaptive optics unit, HiRAC-II (to be built by IJAF Copenhagen). Hirac-II will operate with the 2 × 2 CCD mosaic camera for high-resolution direct imaging (at 0.11"/pixel, fov ~7.5' × 7.5') and NOTCam for diffraction-limited imaging in HIRES mode.

Since the optical CCD program for NOT has been presented elsewhere (in a talk by Johannes Andersen, IJAF), here I relate details of the proposed near-IR camera.

The 1-2.5μm near-IR camera for NOT, NOTCam, will have the following characteristics:

- **IR detector:** Rockwell International 1024x1024 HgCdTe/Al2O3 SWIR focal plane ("HAWAII") array with readout noise ~6-10e$^-$ double correlated sampling, dark current < 0.1e$^-$/pixel and quantum efficiencies ~50% at 1.2μm, rising to > 80% at 2.2μm.
- **Pixel scales**: ~0.25"/pixel (fov ~4' × 4') and ~0.08"/pixel (fov ~1.4' × 1.4').
- **Imaging mode:** J, H, and K filters, short K filter, numerous narrow band (1%) interference filters.
- **Imaging Polarimetry:** performed using a focal plane mask and a Wollaston prism for spatial separation of orthogonal polarization components on the array. An external half-wave plate will be used to select the polarization component to be analyzed.
- **Spectroscopy mode:** three near-IR grisms optimized for acquisition of the complete J, H and K windows over the full 1024 pixel wavelength axis of the array with an effective resolution of R~4000. Point source, long-slit and multi-object spectroscopy will be selected via the aperture wheel slits/slit combinations.
- **Spectropolarimetry:** the combination of aperture wheel slits, blocking filters, grisms and polarimetry optics will give accurate spectropolarimetry of compact sources over the 1-2.5μm spectral regime at R~4000.
- **Readout electronics:** a modified IJAF Brocam-II/ALFOSC type CCD controller for Rockwell array readout. This work will be undertaken at IJAF Copenhagen.
- **Motor Control:** a modified IJAF Brocam-II/ALFOSC type motor controller unit for control of all motor internal to NOTCam. This work will be undertaken at IJAF Copenhagen.

The project to build NOTCam is underway and we anticipate commissioning NOTCam in early 1998.

ESO's PLAN FOR OPTICAL DETECTORS

JAMES W. BELETIC[1]

ABSTRACT

ESO presently maintains 12 scientific CCD systems at the La Silla Observatory. During the next few years, ESO's Very Large Telescope facility, with four 8.2-meter telescopes, will come on-line. The VLT instruments and the upgrades for La Silla require a large number of new scientific CCD systems - a total of 15 new systems with 35 CCDs are to be implemented through 1999. This article presents ESO's plans for fulfilling these needs, with focus on the three main technologies of CCD systems: (i) scientific quality CCDs with new generation amplifiers, (ii) high speed CCD controllers, and (iii) long hold time cryostats.

1. INTRODUCTION

The European Southern Observatory (ESO) combines the talents and resources of eight European countries* for astronomical research. Central to the function of ESO is the provision of state-of-the-art telescopes and instrumentation for use by the astronomers of ESO's member states. For nearly 30 years, the telescopes of the La Silla Observatory in Chile, with optical apertures up to 3.6-meter diameter, have filled this role. On these telescopes, ESO now operates 12 scientific CCD systems, 11 of which incorporate thinned, high efficiency detectors. See Table 1 for a list of these systems.

For more detailed data please access the ESO Web pages at:

http://www.hq.eso.org/odt/generalinfo.html

Recognizing the need for more powerful telescopes, ESO is constructing a facility with four 8.2-meter telescopes, which can operate independently or together with three auxiliary 1.8m telescopes as a coherent interferometric facility. Located in northern Chile on Cerro Paranal, this facility is called the Very Large Telescope or VLT. The first of the VLT telescopes will see "first light" in early 1998.

2. NEW CCD SYSTEMS TO BE DELIVERED DURING 1997-1999

Each of the VLT telescopes has three main foci, Cassegrain and two Nasmyth platforms, with a possibility to send light to a Coudé focus. With these many foci, there is a broad plan for instrumentation development, half of which are optical instruments and half of which concentrate on the infrared portion of the spectrum. (See *http://www.hq.eso.org/vlt* for information on VLT instrumentation.)

In addition to new developments for the VLT, ESO will continue to upgrade the instrumentation capabilities of the La Silla Observatory. The new CCD systems that are now scheduled for delivery during 1997-1999 are listed in Table 2 and the delivery schedule is shown graphically in Fig. 1. This amounts to 15 systems and 35 CCDs to be delivered during these 3 years.

---

[1] European Southern Observatory, Karl-Schwarzschildstr. 2, D-85748 Garching, Germany.
Email: **jbeletic@eso.org**

* Belgium, Denmark, France, Germany, Italy, the Netherlands, Sweden and Switzerland.

J. W. Beletic and P. Amico (eds.), Optical Detectors for Astronomy, 57–64.

Telescope	Instrument	Use	CCD
3.6-m	EFOSC1	spectroscopy, imaging	Tektronix 512 × 512, 27 μm
	CASPEC	spectroscopy	Tektronix 1k × 1k, 24 μm
	Imager	imaging	Tektronix 512 × 512, 27 μm
3.5-m NTT	SUSI	imaging	Tektronix 1k × 1k, 24 μm
	EMMI Red	spectroscopy, imaging	Tektronix 2k × 2k, 24 μm
	EMMI Blue	spectroscopy, imaging	Tektronix 1k × 1k, 24 μm
2.2-m	EFOSC2	spectroscopy, imaging	Loral/Arizona 2k × 2k, 15 μm
1.52-m ESO	B&C	spectroscopy	Loral/Arizona 2k × 2k, 15 μm
1.4-m CAT	CES	spectroscopy	Loral/Arizona 2688 × 512, 15 μm or Loral 2k×2k, 15 μm (frontside)
1.54 Danish	DFOSC	Spectroscopy, imaging	Loral/Arizona 2k × 2k, 15 μm
90 cm Dutch	Cass. Cam.	Imaging	Tektronix 512 × 512, 27 μm

***Table 1**: ESO's La Silla Scientific CCD Systems*

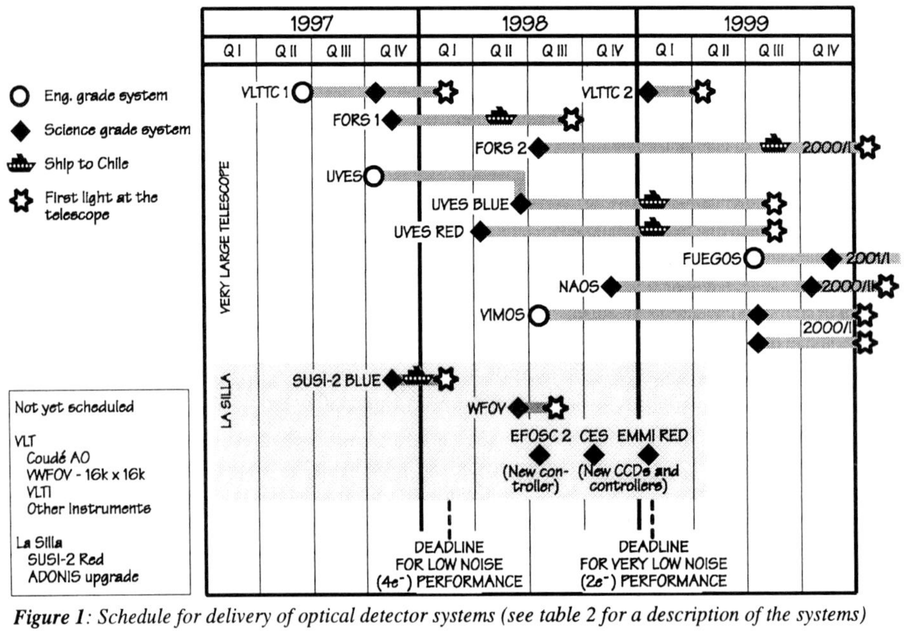

***Figure 1**: Schedule for delivery of optical detector systems (see table 2 for a description of the systems)*

Instrument	CCDs	CCD Manufacturer	Telescope
VLTTC1 & VLTTC2 VLT Test Cameras (imagers)	2k × 2k, 24 μm and 4 tracker chips	Loral / Arizona	VLT Unit Tel. #1,2
FORS1 & FORS2 Focal Reducer/Low Dispersion Spectrograph (imager, low resolution spectrograph)	2k × 2k, 24 μm	Tektronix / SITe	VLT Unit Tel. #1,2
UVES Ultraviolet Visible Echelle Spectrograph (high res. spectrograph) Red arm Blue arm	4k × 4k, 15 μm (two 2k ×4k) 2k × 4k, 15 μm	MIT/LL and EEV EEV	VLT Unit Tel. #2
VIMOS Visible Multi-Object Spectrograph (imager, low resolution spectrograph)	four 2k × 4k, 15 μm (four cryostats and two controllers)	EEV or MIT/LL	VLT Unit Tel. #3
NAOS Nasmyth Adaptive Optics System (Hartmann-Shack wavefront sensor)	128 × 128, 24 μm	EEV	VLT Unit Tel. #1
FUEGOS Fiber Unit for European General Optical Spectroscopy (fiber fed low to medium resolution spectrograph)	4k × 4k, 15 μm (two 2k × 4k)	EEV or MIT/LL	VLT Unit Tel. #2
SUSI2 Superb Seeing Imager - generation 2 (imager)	4k × 4k, 15 μm (two 2k × 4k)	EEV	La Silla NTT 3.5m
EMMI Red ESO Multi-Mode Instrument (imager, low to med. res. spectrograph)	2k × 2k, 24 μm	Loral / Arizona	La Silla NTT 3.5m
WFOV Wide Field of View (imager)	8k × 8k, 15 μm (eight 2k × 4k plus 2k×4k tracker chip)	EEV	La Silla 2.2m
CES Coudé Echelle Spectrometer (medium to high res. spectrograph)	2k × 4k, 15 μm	EEV	La Silla 1.4m

***Table 2**: ESO's Optical Detector System Deliveries during 1997-99*

3. PHILOSOPHY FOR ESO'S NEW CCD SYSTEMS

In planning the fabrication of these new CCD systems, we have established the following goals:

- system performance limited only by the CCD (and the imagination of the user) - the hardware and software should be transparent to the user
- design lifetime of at least 10 years
- ease of integration and implementation
- reliability and ease of maintenance

To achieve these goals and to have a system that serves ESO well for many years, we have designed our systems to be ready for any CCD advancement that is foreseen in the next decade.

The biggest development that is occurring now and will continue to improve is the ability of CCD amplifiers to read fast with relatively low noise. Presently, very low noise (<2 e^- rms) is achievable at slow readout speeds and moderate noise levels (4-6 e^-) are being obtained at MHz pixel rates. We expect this trend will continue and it may be feasible in a few years time to obtain 1 e^- noise at MHz pixel rates.

The amplifiers that can now achieve moderate noise at high speed are the same ones that achieve the lowest noise at slow readout speed. At ESO, we refer to these as "new generation" amplifiers and we are now attempting to only procure CCDs that have this capability. Of course, we also wish to get as high quantum efficiency as possible, and ESO places particular emphasis on high quantum efficiency in the ultraviolet.

In order to produce, implement and maintain a large number of systems, we must develop a standard system that can be easily duplicated and installed into a variety of instruments. It is mandatory for ESO to move away from individualized system design that requires lengthy design and implementation cycles, and is difficult to maintain.

4. ESO'S NEW CCD CONTROLLER - FIERA

As we procure CCDs that can read out fast, we require electronics that can run many ports at high speed. Our goal at ESO is to have a CCD controller that can operate all of our CCDs at the best performance possible, whether the CCD has fast or slow amplifiers. Thus, ESO is putting major effort into development of a new generation CCD controller, which we call FIERA. See papers [1,2] in these proceedings for an overview of the FIERA hardware and software.

In making FIERA, we recognize that initially the system will not be optimal at all readout speeds. Since the first instruments using FIERA are direct imaging cameras that require high speed and moderate noise, we are concentrating on this goal during 1997. We anticipate reading out the 4k × 4k SUSI2 system in 5 seconds and will be able to read out the 8k × 8k WFOV system in 10 seconds.

During 1998, we plan to optimize FIERA for slow readout speeds and the lowest noise possible. Thus, in Fig. 1, we show that the deadline for "very low" noise is set for the beginning of 1999, when the first spectrograph (FORS 1) becomes available to the user community.

5. ESO's CCD Procurements

ESO's CCD procurements have concentrated "new generation CCD detectors" that match the new standard format: 2k × 4k, 15 μm three-side buttable with 2 readout ports along a short (2k) side.

EEV

ESO has a contract with EEV for the procurement of twenty scientific grade 2k × 4k, 15 μm devices. EEV is also making a custom designed 128 × 128, 24 μm pixel split frame transfer CCD with 16 readout ports; ten of these devices are being procured for adaptive optics applications. The contract for the EEV CCDs specifies a readout noise of <2 e^- at 50 kps and <6 e^- at 1000 kps, where kps means "thousands of pixels per sec per readout port". The guaranteed minimum quantum efficiency for these devices is shown in Fig. 2. As can be seen in Table 2, ESO is counting on the EEV procurement for many of its new instrument deliveries. The delivery of the twenty 2k × 4k devices will commence in the second half of 1997 and is scheduled for completion in the second quarter of 1998.

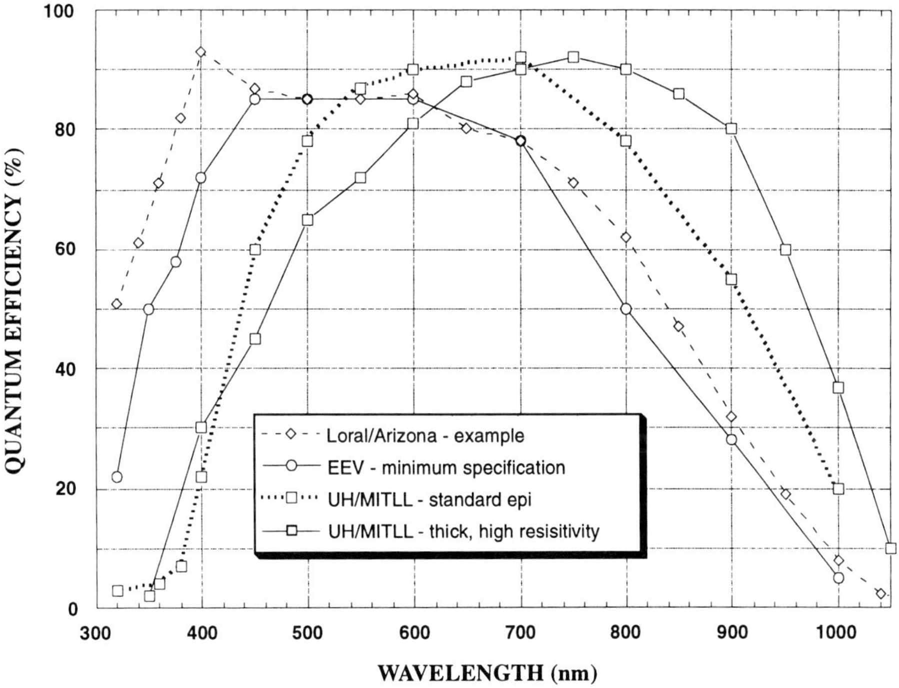

Figure 2: Quantum efficiency of EEV, MIT/LL and Loral/Arizona CCDs

MIT Lincoln Laboratory (MIT/LL)

ESO is part of the U. Hawaii consortium for the MIT/LL 2k × 4k, 15 μm CCD detectors. ESO is planning to use these devices for instruments that concentrate on the near infrared portion of the spectrum. ESO's main interest in this effort is the (50 μm) thick, high resistivity CCD that has quantum efficiency of 90% at 800 nm and 80% at 900 nm (see Fig. 2). Combined with readout noise of ~1 e^- at 20 kps, the MIT/LL CCDs are nearly perfect detectors in the red portion of the spectrum. ESO hopes to receive about 10 of these CCDs during 1997 and 1998.

Loral foundry / U. Arizona thinning

ESO has a good supply of quality silicon from the Loral 2k × 2k, 15 μm foundry runs and we have 5 additional three-side buttable devices being thinned by U. Arizona. These devices are not yet allocated to an instrument, but will serve as a backup to the EEV and MIT/LL procurements. In addition, the U. Arizona is thinning the VLT Test Camera CCD. This device, which is a 2k × 2k, 24 μm CCD with four "tracker" CCDs placed at the four corners, shown in Fig. 3, was fabricated at Loral. These CCDs will be used in the two VLT Test Cameras, the systems that will see "first light" at the VLT and be used for commissioning and science verification.

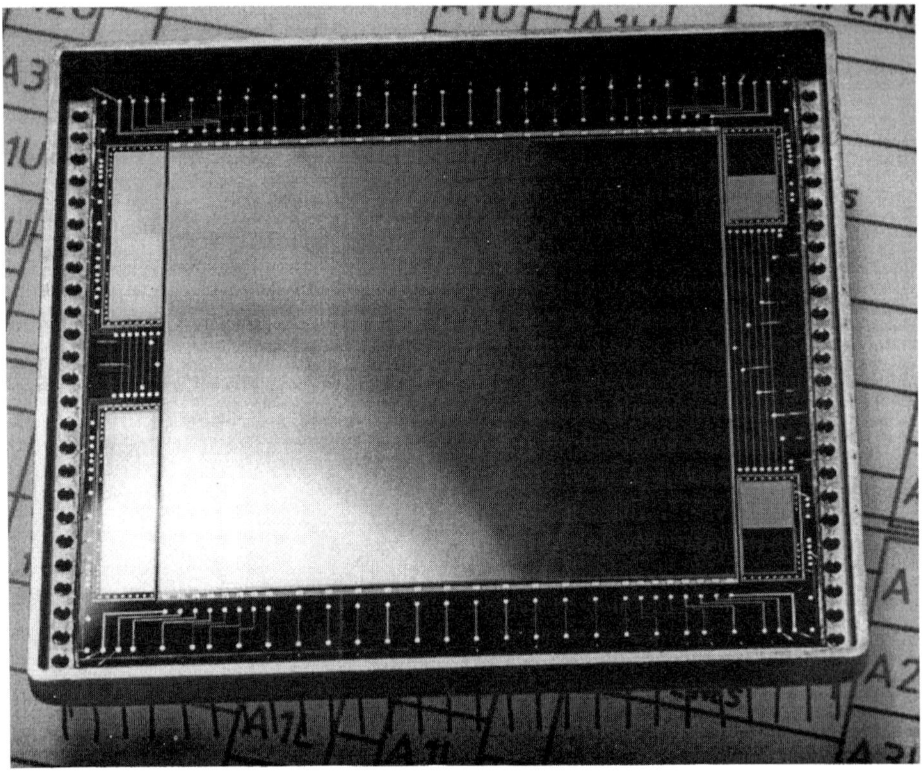

Figure 3: *VLT Test Camera CCD - 2k × 2k, 24 μm with four tracker chips which are of two sizes: 180 × 200 and 180 × 400 pixels in the image area (frame transfer devices)*

Large CCD development

As ESO has concentrated on the procurement of 2k x 4k, 15 μm three-side buttable devices, we maintain a strong interest in the development of larger detectors. A natural next step is a 4k x 4k, 15 μm device since many instrument designs use a 4k x 4k area, which we now construct from two 2k x 4k devices (see Table 2). Beyond 4k x 4k, ESO has great interest in the largest detectors possible.

With this motivation, ESO joined American Digital Imaging and the Max Planck Institute for Astronomy in the procurement of the Philips 7k x 9k, 12 μm pixel CCD, the largest area CCD ever made. That project demonstrated the ability of Philips to manufacture full wafer devices, but unfortunately the low quantum efficiency (peak of 28%) is not useful in ESO's new instruments. Presently, in order to get the highest quantum efficiency, we must invest the labor and expense to "pave" a focal plane with an array of 2k x 4k devices.

New CCD testbench

In order to be able to characterize the new CCDs, ESO is developing a new CCD testbench [3]. This new testbench is required to interface to the new FIERA controller and be able to accommodate very large CCDs or mosaics. With automated equipment and routines, we hope to be able to undertake a broad range of thorough tests with minimal labor involved.

6. ESO'S NEW CRYOSTAT AND DETECTOR HEAD DESIGN

In parallel with CCD procurement and CCD controller development, ESO has developed a new cryostat and detector head for the "VLT era" [3]. The new detector head is for CCDs that are 4k × 4k pixels or smaller, and the head can be used with both of ESO new standard cryostat designs: (1) a "bath" design that holds 6.5 liters of LN_2 or (2) a "continuous flow" design, which is attached to a tank of LN_2. Figure 4 shows the detector head with the Nasmyth version of the bath cryostat.

***Figure 4**: Bath cryostat and detector head, Nasmyth version with a frontside version of the VLT Test Camera CCD. The 6.5 liter LN_2 capacity of the cryostat has a hold time of 40 hours or longer.*

The two basic functions of a cryostat and detector head are: (i) cooling of the CCD for long periods of time, and (ii) holding the CCD in a very stable position as the telescope moves across the sky. The new ESO design achieves both of these goals very well. The 6.5 liter bath cryostat has a hold time of 40 hours and the continuous flow cryostat has a hold time of nearly 1 month, when attached to a 120 liter tank. As for holding the CCD steady, the new detector head has been demonstrated to have CCD motion of less than 1 μm as the telescope moves across the sky.

For large mosaics, such as the 8k×8k Wide Field of View instrument, we are developing a large detector head that can mate directly to the existing cryostat. We expect hold time of 24 hours with the large detector head and 6.5 liter bath cryostat.

A cryostat system requires a number of "housekeeping" functions for its operation, such as: temperature sensing, heater operation for temperature control, vacuum sensing, and liquid nitrogen empty alarm. ESO has developed a small electronics module, which communicates via RS-232, that provides all of this functionality and more, including a shutter driver and timer. This module, named PULPO, is described in paper [5], which can be found in these proceedings.

7. Possible future CCD systems

Even though we are full of commitments for the next few years, there are other projects being planned, including:

- A 4k × 4k mosaic of two 2k × 4k thick, high resistivity MIT/LL CCDs for the red arm of the SUSI2 imager.
- A 16k × 16k Very Wide Field of View Imager for a new telescope that may be constructed at Paranal.
- Adaptive optics at the Coudé foci of the VLT. Needed for the VLT Interferometer, these AO systems will use CCDs for curvature wavefront sensing if CCDs can achieve less than 1 e^- noise when reading out a small number of pixels at kHz frame rate.

8. Summary

As you can see, ESO's Optical Detector Team has a very ambitious schedule, especially through the end of 1998. Due to the heavy workload, we regret that we will not have time to host a major meeting in 1998, but instead are planning the next ESO CCD Workshop to take place in August or September of 1999. In the meantime, we hope to survive our schedule and be able to provide a positive report on our activities in 1999.

9. References

[1] Beletic, J.W., R. Gerdes and R. DuVarney (1997), "FIERA: ESO's new generation CCD controller", these proceedings, pp. 103 - 114.

[2] Cumani, C. and R. Donaldson (1997), "The architecture for two generations of ESO VLT CCD controllers", these proceedings, pp. 115 - 122.

[3] Amico, P. and T. Böhm, "ESO's new CCD testbench", these proceedings, pp. 95 - 102.

[4] Lizon, J.L. (1997), "New cryostats for scientific CCD systems in the VLT era", *The ESO Messenger*, n° 88, pp. 6 - 7.

[5] Haddad, N. and P. Sinclaire (1997), "PULPO: Temperature, vacuum, shutter, LN2 level, all in one box", these proceedings, pp. 131 - 134.

AAO DETECTORS - PRESENT STATUS AND FUTURE PLANS

RUSSELL CANNON[1] *AND JOHN BARTON*[1]

ABSTRACT

The AAO operates the 3.9m Anglo-Australian Telescope (AAT), which is equipped with a wide range of instruments for different types of astronomy, and the 1.2m UK Schmidt Telescope (UKST). We summarize the current suite of detectors and controllers at the AAO and go on to describe plans for the acquisition of new CCDs, IR arrays and controllers.

1. PRESENT CCD USAGE

Most applications of CCDs at the AAO are for spectroscopy rather than direct imaging; the AAT is used in spectroscopic modes for as much as 90% of the time. This is partly a consequence of the site in Australia, which is truly photometric for only a small proportion of the time. It also reflects the scientific importance of spectroscopy, which in general gives much more astrophysical information than can be obtained through imaging alone.

1.1. Tek 1024 CCD

The principal current detector on the AAT is a thinned Tektronix 1024 CCD, which is used for the following purposes:

Instrument	Application	Requirements
f/3.3 prime	imaging	large CCD, fast readout, dynamic range
RGO spectrograph	$R \leq 20{,}000$	long narrow strips, lower noise
Coude UCLES	$R \sim 40{,}000$	largest CCD, lowest read noise
Coude UHRF	$R \sim 10^6$	long strip, lowest read noise
Taurus	scanning FP	lowest read noise
2dF	multi-object	low noise, dynamic range
MAPPIT	interferometry	high speed, low read noise

UCLES is a multi-order Echelle spectrograph; Taurus is a scanning Fabry-Perot interferometer which generates data cubes with two spatial directions and one spectroscopic dimension; 2dF is the new multi-object fiber spectroscopy system on the AAT, with 400 fibers in a 2° field at the prime focus; MAPPIT is a unique multi-aperture interferometer. There are other ultra high-speed applications such as seeing tests, adaptive optics experiments and speckle interferometry. For some of these, a small CCD window is collapsed to 1-D to give maximum speed.

[1] Anglo-Australian Observatory, PO Box 296, Epping, Australia. Email: **rdc@aaoepp.aao.gov.au**

J. W. Beletic and P. Amico (eds.), Optical Detectors for Astronomy, 65–70.

1.2. Other arrays in use at the AAO

The second most-used array is the Rockwell NICMOS HgCdTe array in the infrared imager and spectrograph, IRIS. This is a now outdated 128 × 128 array with 60μm pixels, sensitive out to 2.5μm (J, H and K bands).

The fixed-format Faint Object Red Spectrograph (FORS) has a dedicated small front-illuminated GEC CCD (type P8603/A with 385 × 576, 22μm pixels); suggestions to replace this with a more sensitive thinned CCD have been rejected because that would introduce too much fringing.

The Schmidt FLAIR fiber spectroscopy system uses a small thinned, coated EEV CCD (type CCD02-06 with a nominal thinned area of 320 × 500, 22μm pixels) for maximum blue sensitivity and low read-out noise.

The AAO also has two Thomson-CSF 1024 × 1024 CCDs, one of them dye-coated for blue response. The red Thomson is used with the f/1 focal reducer to give a 15-arcmin field for imaging at prime focus. Although their sensitivity is not much more than half that of the Tek 1024, both Thomson CCDs are still sometimes preferred because they have much lower levels of fringing in the far red and near infrared, and because they have somewhat smaller pixels (19μm compared with 24μm).

2. FUTURE REQUIREMENTS

Some existing AAO instruments, such as the RGO spectrograph, could immediately make good use of larger format CCDs, especially if they had smaller pixels. However, none can fully exploit the much larger arrays soon to become available. Our future detector requirements and our instrumentation plans are therefore intimately related.

2.1. Instrumentation plans

New instruments currently under consideration for the AAT include the following:

- IRIS 2, a new infrared imager and spectrograph, based on a Rockwell "Hawaii" 1024 × 1024 HgCdTe array. This requires fast multi-channel low-noise readout.
- A new camera for UCLES, using a pair of 2k × 4k CCDs, one optimized for blue/UV response, which would cover the entire optical wavelength range in a single exposure at high spectral resolution. This requires CCDs that can be butted together, with a small footprint to minimize vignetting, and with the lowest possible read-out noise and dark current. Fast readout is also needed, for twilight flat fields.
- A new f/3.3 prime focus-imaging camera, using eight butted 2k × 4k CCDs to give a field 8k pixels square covering half a degree of sky. These should be thinned, multi-output, fast readout CCDs with maximum dynamic range.
- SPIRAL, a novel spectrograph concept which uses a close-packed array of micro-lenses feeding optical fibers for area spectroscopy; this, in common with other spectroscopic applications, requires a single 2k × 4k CCD with ~ 15μm pixels, low noise and fast readout.

2.2. Current plans for new detectors

The AAO is a member of two consortia for the purchase of large format thinned CCDs: the University of Hawaii contract with MIT Lincoln Laboratories for 2048 × 4096 CCDs with 15μm pixels and one led by the Royal Greenwich Observatory with EEV for 2048 × 4096 CCDs with 13.5μm pixels. Both types will be 3-edge buttable, have two readouts and should be capable of read-out noise below two electrons. Both contracts are expected to yield some useful devices by early 1997 and at least one large CCD should be in use on the AAT by the middle of the year.

The AAT Board has agreed that the highest priority new instrument for the AAT is IRIS 2 and has approved the purchase of a Rockwell 1k × 1k square infrared array.

2.3. Possibilities for the future

Three current developments are potentially of great interest to the AAO:

- The use of very large area devices for imaging in the UK Schmidt Telescope, such as the Philips 7k × 9k CCD with 12μm pixels which would cover a field 2° across with 0.8 arcsecond pixels, or the Loral (now Lockheed Martin) 4k × 4k CCD with 7.5μm pixels covering half a degree with half arcsecond pixels (or of course the much larger 9k × 9k devices announced at this meeting). Having large CCD arrays with small pixels in the UKST would open new fields of research and be used for objective prism spectroscopy as well as for imaging.
- Deep depletion devices, optimized for maximum QE at about 1μm, would open up a spectroscopic region which has been little explored up to now, between the classical optical and near-infrared regions. Such devices should have optimized anti-reflection coatings and be as free as possible from fringing.
- The promised development of 2k × 2k Rockwell infrared arrays would be a big advance and is already influencing the design of the AAO's IRIS 2 instrument.

3. The Current AAO CCD Controllers

The AAO currently has four identical controllers, henceforth referred to as type AAO-1, which were designed in 1983 and built in-house. Three of them are used at the AAT and one in the Laboratories in Epping. FLAIR on the Schmidt uses a different controller, provided by Paddy Oates while at the University of Durham. The AAO-1 controllers can drive either CCDs or the NICMOS array in IRIS.

3.1. Strengths of the AAO-1 controllers

The current controllers have the following advantages:

- the in-house design permitted extensive adaptation and development over the years as new features were required;
- they can achieve 2.3 electron RON at telescope (1.7 electrons in the laboratory);
- they provide flexible windowing and binning facilities;

- conversion to run the NICMOS IR array for IRIS was feasible;
- they provide an ideal test bed for laboratory optimization and characterization of CCDs;
- there is a wide variety of read options, trading speed *versus* RON;
- they give an easy interface with the telescope and instrument control systems, such as DRAMA, which were also developed in-house.

3.2. Applications of the AAO-1 controllers

A number of special modes of operation have been developed for specific astronomical applications. These include:

- time series readout for taking sequences of spectra of variable stars;
- very fast modes for the MAPPIT aperture mask interferometer;
- shuffling readout (CSR), enabling images to be repeatedly moved backwards and forwards on the CCD in phase with instrument changes, allowing images to be built up in various beam-switching modes;
- in phase with frequency switching, creating the Taurus Tunable Filter (see below);
- possibility of drift scanning, although this has not yet been implemented.

3.3. Limitations of the AAO-1 CCD controller

The AAO-1 controllers do however have some serious disadvantages, being:

- poorly matched to the large, low-noise fast readout CCDs now becoming available;
- capable of processing only a single video output;
- incompatible with the Rockwell 1k × 1k square IR arrays, which cannot be read out through one corner;
- built to an old design with components which are obsolete, bulky and relatively power-hungry;
- based on old technology, including a slow microprocessor with limited memory and instrument interfacing capability.

4. Plans For New AAO CCD Controllers

The development of a new controller was proposed in 1989 and work commenced in 1992 but soon stalled under pressure from the very large 2dF project. Work started again earlier this year but cannot now be completed in time for the expected arrival of the new 2k × 4k CCDs. Two interim possibilities are to modify the AAO-1 controllers for faster read-out, or to acquire a stand-alone system based on an SDSU-2 controller, probably through Mount Stromlo Observatory.

Meanwhile, work will continue on the prototype AAO-2 controller, which should be complete towards the end of 1997. The first engineering model of the AAO-2 controller should be available for use on the AAT by the end of 1998.

4.1. Specification of the AAO-2 CCD controller

- four output ports for standardization

- synchronized controllers for larger mosaics
- 0.5 Mpix/sec/port is the immediate goal
- 2MHz ADC so that 2 Mpix/sec/port is not precluded
- dual video processor scheme:
    - ◊ dual slope integration for lowest read noise
    - ◊ range of clamp and sample speeds provided
- fully programmable readout, timing and voltages
- 50 MHz 32-bit TMS32C40 DSP
- 32-bit pixel data allowing:
    - ◊ more than 16 data bits
    - ◊ encoding of pixel source
    - ◊ encoding window source
    - ◊ encoding the readout mode
- 20nsec maximum sequencer resolution
- compact dewar mounting unit
- 250MHz fiber optic link
- VMEbus interface to the AAT external memory (XMEM)
- like AAO-1, the ability to:
    - ◊ operate the existing suite of detectors
    - ◊ perform the current specialized readouts
    - ◊ operate any future devices
    - ◊ perform new innovative readouts
    - ◊ interface with telescope and control systems

5. A Recent Innovation - The Taurus Tunable Filter (TTF)

In closing, we illustrate the power of a modern CCD detector with a versatile controller through one application developed recently at the AAO. Emission nebulae are usually studied by taking a series of images through different narrow band filters, but then each image is taken in different "seeing" conditions and detailed matching is difficult. By combining Taurus with the "charge shuffling" CCD mode this problem can be overcome. The CCD detector is masked so that the central third of its area is exposed to the sky while the top and bottom thirds are masked off. The principle is to take a short exposure at one wavelength, then move *(shuffle)* that image up on to one of the masked storage areas on the CCD while simultaneously changing the Taurus FP spacings to the second desired wavelength. A second short exposure is made, then the two images are moved down on the CCD, Taurus is set back to the first wavelength, the initial exposure repeated, and so on. Thus, alternate exposures through the two filter bands are accumulated in a pair of CCD images. When sufficient total exposure time has elapsed, the full CCD is read out.

Effectively this amounts to a beam-switching technique, so that "seeing" and transparency variations average out. Furthermore, the two images are obtained through identical optical paths. The widths of the bandpasses can be chosen almost at will, as can the relative exposure times. The process can be extended to include many separate narrow bands in a single sequence of observations. The key ingredients of the TTF are a

CCD controller able to do repeated charge shuffles, say up to 100 times, without any smearing or other degradation of the image, and an interface to the Taurus control system so that the bandpass changes can be done synchronously with the image shuffling. More details of the TTF can be found in an article by Joss Bland-Hawthorn, who was the driving force behind the whole project, in the January 1996 issue of the *AAO Newsletter.*

A DRIFT SCAN CAMERA FOR ASTROMETRY

M. GAI[1], M.G. LATTANZI[1], M.D. GUARNIERI[2], G. MASSONE[1]

ABSTRACT

The Astronomical Observatory of Torino plans an upgrade of the CCD camera at its 1.05 m telescope, implementing the drift scan observing mode. The targets of the project are both scientific (galactic populations statistics, parallaxes) and technological (acquisition of expertise on drift scan technique for high accuracy astrometric ground based and spaceborne surveys). In particular, the issues of field distortions and timing accuracy are discussed in view of the astrometric performance.

1. SYSTEM AND DETECTOR SPECIFICATIONS

The Reosc astrometric telescope of OATo (1.05 m diameter) provides a focal plane optical scale of 20.89"/mm. The camera used in the last three years (a commercial system from AstroCam, UK, implementing only pointed observations) has a detector from EEV, UK, model CCD05-30, format 1242 (H) × 1152 (V), with pixel size 22.5 μm.

This results in an on-chip scale of 0.47"/pixel; the field of view in the North-South direction is Φ_{NS} = 9'44" and in the East-West Φ_{EW} = 9'2" (pointed mode), corresponding to an exposure time T_0 = 36 s at $\delta = 0°$ in drift scanning mode. The expected limiting magnitude, from present performance figures, is $V_{lim} \cong 19$ at SNR ≥ 3.

The sidereal drift speed decreases for increasing declination, thus extending the exposure time, which depends on the vertical size of the CCD; at $\delta = 0°$, $V_0 \cong 15''/s$, and $T_0 \cong 36$ s , whereas at declination δ (fig. 1 and 2)

$$V = V_0 \cos\delta \qquad T = \frac{\Phi_{EW}}{V} = \frac{T_0}{\cos\delta} \qquad T_0 = \frac{\Phi_{EW}}{V_0} = 36s \qquad (1)$$

[1] Osservatorio Astronomico di Torino, Italy. Email: **mario@otoxd3.to.astro.it**
[2] European Southern Observatory, Garching

J. W. Beletic and P. Amico (eds.), Optical Detectors for Astronomy, 71–74.

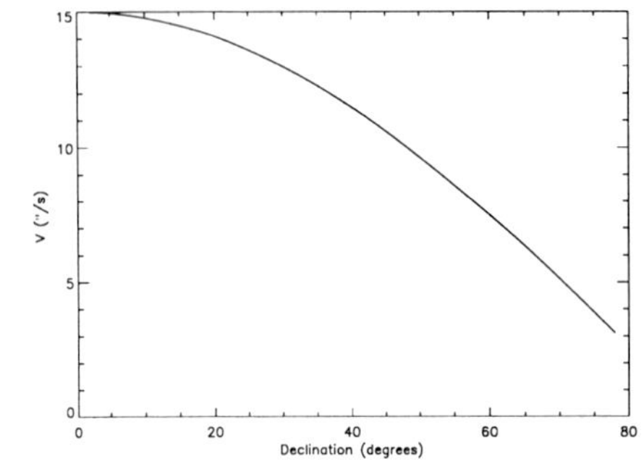

Figure 1

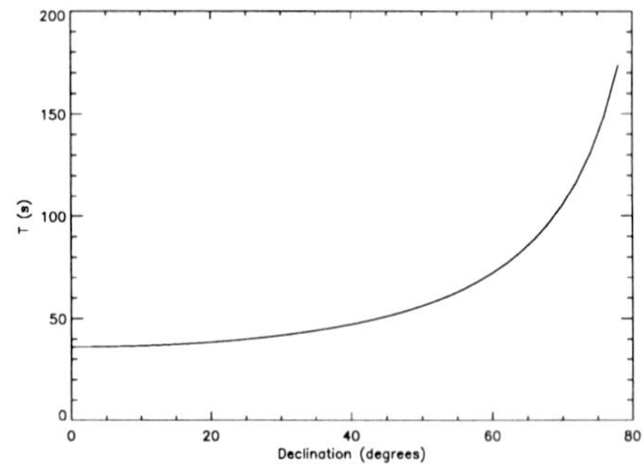

Figure 2

2. FIELD DISTORTIONS FROM SPEED MISMATCH

The declination dependence of the sidereal drift speed results in field-dependent image distortions, because the on-chip charge pattern is clocked at the same electronic drift speed, common to the whole CCD. A perfect matching between them is possible only for a single declination value δ_m.

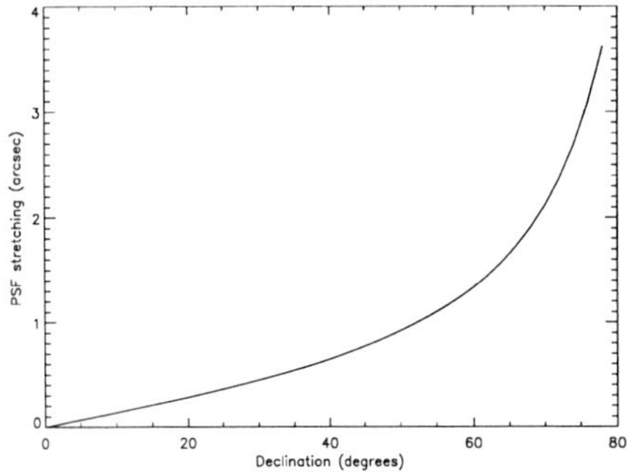

Figure 3.

The speed mismatch generates an artificial PSF stretching in the East-West direction, since the stars travel faster, on one side, and slower on the opposite, than the CCD potential wells. Thus, the charge distribution, resulting from the integration of the photocurrent all along the transit, has a minimum E-W width, corresponding to the "natural" PSF extension, only at δ_m, whereas *above* and *below*' the images are enlarged by an amount depending from the position in the field.

At a displacement ρ from δ_m, the speed error ΔV provides a PSF width increase ΔW; the N-S edges feature the largest distortion:

$$\Delta V = V(\delta_m) - V(\delta_m - \rho) = 2V_0 \sin(\delta_m + \rho/2)\sin\rho/2 \cong \rho V_0 \sin(\delta_m) \quad (2)$$

$$\Delta W = T\Delta V = \rho T_0 V_0 \tan(\delta_m) = \rho\Phi_{EW} \tan(\delta_m) \tag{3}$$

in our case: $\Delta W_{MAX} = 0''.77 \times \tan\ \delta_m$ (fig. 3). The effect is deterministic, and correctable for positions, but it introduces errors in both astrometry (the centroid shift is known, but the effective PSF spread is larger) and photometry (reduced SNR because of noise contribution from more pixels).

3. PROCESSING TIME AND THROUGHPUT

The drift scan mechanism is based on the transfer of the CCD charge pattern in step with the sidereal speed. During an interval corresponding to one pixel displacement (0".47), the whole row of N_R = 1242 pixels must be acquired, resulting in a pixel processing time $T_{PP} = T_{PP0} / \cos\ \delta$, $T_{PP0} = T_{PT0} / N_R = 25\ \mu s$.

The readout provides a 16-bits data for every pixel, on the same time scale. This results in a declination dependent data throughput

$$DT = DT_0 \cos(\delta);\ DT_0 = 2\,N_R / T_{PT0} = 79.36 KB/s = 285.70 MB/hr \tag{4}$$

A full-night observation (8-10 hours) yields 2.3-2.9 GB.

4. TIMING REQUIREMENTS FOR ASTROMETRY

Small timing random errors (jitter) result in a further enlargement of the PSF in the along scan direction, degrading local astrometry.

Customer electronics usually resorts to the stability of a low-cost crystal oscillator, in the range ~10^{-5}. Thermal stabilization of the electronics can improve the stability to S=10^{-6}. Assuming a stability S = 5×10^{-5}, the cumulative timing error for the whole transit is

$$\Delta T = S \times T = \Delta T_0 / \cos(\delta)$$
$$\Delta T_0 = S \times T_0 = 1.8 ms \tag{5}$$

corresponding to a PSF width enlargement *independent* from declination:

$$\Delta W = \Delta T \times V = \Delta T_0 \times V_0 = 27 marc\sec \tag{6}$$

comparable with atmospheric aberrations.

Besides, the long-term effect (relevant for determination of wide angles) may become significant: over a 30° sky strip (δ = 0°, 2 hours scan), the timing error (worst case) is $\Delta T = 0.36$ s, i.e. $\Delta W = 5.4\ arcsec$.

Such error can be corrected *post-hoc* (e.g. by verification of bright star positions onto a suitable catalog), but it degrades the intrinsic measurement simplicity and efficacy. The timing error can be reduced both by improvement of the internal time reference (thermal stabilization and optimal circuit design), and by synchronization to a high accuracy external time base, e.g. GPS (Global Positioning System).

This improves the timing stability to $S \sim 10^{-8}$, i.e. ΔW $6/1500$ $\mu arcsec$, as errors on local and global astrometry, well below other error sources for standard ground based instruments.

5. Future Developments

A single, large area CCD is a convenient trade-off for a simple first implementation of the project. The image distortion can be reduced by using many smaller format CCDs, as it depends on both device dimensions. Besides, shortening the exposure time (Φ_{EW}) also decreases the SNR: the best solution is a mosaic of CCDs featuring a small number of very long columns, i.e. large vertical and small horizontal format.

Suitable devices are commercially available, e.g. from EEV (CCD05-10, 298×1152, 22.5 μm pixels and CCD15-11, 256×1024, 27 μm). Conversely, for higher performance programs, optimal devices might be custom designed, with proper geometry.

The electronics complexity increases linearly with the number of devices, but field distortion and consequent astrometric and photometric errors can be strongly reduced, up to very high declination, with a very simple and stable optical and mechanical assembly.

Moreover, although the number of analog signal lines increases with the number of CCDs, it is not mandatory to use a separate sequencer for each device: accepting minor additional image distortion, the drift speed can be matched to each device by proper tailoring of the clocking scheme.

RGO (AND ING) CCDs - CURRENT DEVELOPMENTS

P.R. JORDEN [1] *& A. P. OATES* [1]

ABSTRACT

All three ING telescopes (at the Roque de los Muchachos Observatory in La Palma) have a suite of CCD detectors; these are reviewed with a presentation of their focal plane requirements. The current status of the RGO programmes of CCD-upgrades is also discussed. We present details of 'recent' RGO detector work (since the previous 1993 ESO workshop), together with plans for the future.

1. INTRODUCTION

The Isaac Newton Group of telescopes, at the international Roque de los Muchachos Observatory in La Palma, use CCD arrays for over 95% of their observing time. The 4.2m William Herschel telescope, the 2.5m Isaac Newton telescope, and the 1m Jacobus Kapteyn telescope offer a variety of imaging and spectroscopic instruments. The RGO installed the original systems, dating from 1984, and has been involved in supporting and upgrading them since then. Here we shall present details of these systems, together with a description of current CCD activities at RGO, and planned developments.

2. ISAAC NEWTON GROUP TELESCOPES AND CCD SYSTEMS

The 'flagship' WHT, as well as the two smaller telescopes, offer a set of observing options - including interchangeable CCD detectors, or optimized sensors in some cases. Table 1 shows the main instruments and detector types that are used on ING telescopes.

Currently, we use the TK1024 (1024×1024, thinned) CCD on most of the instruments; we operate six of these sensors, which were installed as replacements for earlier EEV (smaller or unthinned) CCDs. As indicated later, many instruments are being upgraded, and hence column three in the above table indicates the type of array that the instrument requires. We identify several categories of CCD:

- 4k × 2k. This is the spectroscopic CCD, of approximate size 50 × 25 mm.
- 2k × 2k. This is the square format (imaging or spectroscopic) CCD of size ~30×30 mm.
- *embedded.* This is typically a small 8 × 14 mm sensor in the fixed-format FOS instruments.
- *large.* This is generally a 4k × 4k or larger sensor (or mosaic), to cover image planes >60 mm diameter.

[1] Royal Greenwich Observatory, Cambridge, England. Email: **prj@ast.cam.ac.uk** .

J. W. Beletic and P. Amico (eds.), Optical Detectors for Astronomy, 75–81.

Focus	Instrument	Detector
WHT		
Cassegrain	ISIS double spectrograph	4K×2K CCD
Cassegrain	Faint Object spectrograph	embedded CCD
Cassegrain	TAURUS Fabry-Perot	2K×2K CCD
Cassegrain	TAURUS direct imaging	2K×2K CCD
Cassegrain	Auxiliary port imaging	2K×2K CCD
Cassegrain	Low Dispersion Survey spectrograph	2K×2K CCD
Nasmyth	Utrecht Echelle spectrograph	4K×2K CCD
Nasmyth	GHRIL	various CCDs
Nasmyth	WHIRCAM infra-red imaging	(IR array)
Nasmyth	Martini-3 adaptive optics	IR or CCD
Prime	CCD direct imaging	*large* CCD
Prime	WYFFOS/ Autofib multi-object spectrograph	embedded CCD
INT (2.5m)		
Cassegrain	Intermediate dispersion spectrograph	4K×2K CCD
Cassegrain	FOS (Faint Object Spectrograph)	embedded CCD
Prime	CCD direct imaging	*large* CCD
JKT (1.0m)		
Cassegrain	CCD direct imaging	*large* CCD
Cassegrain	Richardson-Brealey spectrograph	4K×2K CCD

Table 1. *Isaac Newton Group telescopes and CCD systems- List of currently scheduled instruments.*

3. Current CCD procurements/ upgrades

Most of our current detectors were installed in the period 1992-1994. Since then we have reviewed the requirements (see [5]), and initiated a programme of upgrades. The aim has been to acquire larger arrays, with smaller pixels as well as low noise and high quantum efficiency (of course!). It is constructive to examine this process, and so we summarize below the two main procurement programmes that are nearing completion (A, B) as well as a few other CCD purchases (C) that we have engaged upon.

3.1 'Recent' RGO CCD procurements - a historical summary.

A. Loral/Lesser 2048×2048 2K3EB CCDs.

- September 1993 Order placed for Loral batch of 2K3EB devices.
- February 1994 Loral completed batch, wafers shipped to Lesser.
- February 1994 Order placed with Mike Lesser for thinning (6 devices).
- February 1995 1st science-grade device delivered to RGO.

- May 1995 - Oct 1996 Further devices delivered. One installed on telescope. Many are not useful as science-devices and require replacement.

B. EEV Contract for 2048×4096 CCD42 devices.

- April 1993 Started contract discussions, established consortium, tested samples.
- Oct 1994 Concluded specifications, after demonstration samples had been evaluated.
- Nov 1994 Contract placed with EEV (jointly by RGO, AAO, TNG).
- March 1995 Stage-1 complete (design).
- Nov 1995 Stage-2 complete (samples).
- Nov 1996 Stage-3 complete (3 thin engineering chips).
- Mar 1997? Stage-4, conclusion of contract (12 science chips) anticipated.

C. Additional purchases.

- 1995 Additional TK1024 device for WYFFOS instrument.
- 1994 to date INT PF Mosaic. 'Lick3' (Lesser-thinned) chips.
- Jul 1995 - Oct 1996 Three Loral/Lesser 512FT (512×1024) devices for FOS upgrades.
- 1996 *Sample* EEV CCD39 WFS chips (80×80).

3.2 Discussion

Our experience of procurement-A is that it has been extremely lengthy and time consuming. One contributing factor, has been the lack of good quality test data of the originally fabricated chips (on wafer). This meant that the yield of useful thinned chips has been low (and unpredictable). Although the apparent cost is quite low, the effort involved has been high, and the yield of useful devices has proved to be low so far. We anticipate that this may improve with the recent introduction of cold-probing by Lesser prior to thinning [4].

Procurement-B started later because we spent some time establishing specifications, and verifying expected performance before initiating a contract with EEV. As usual with the production of a new custom-made device, time-to-completion would seem to be about double the original estimate! Details of the EEV CCD42 devices may be found in the references ([8], [6])

The main thrust of our programme has been to obtain suitable sensors for our spectrographs (2k × 4k format), and to obtain suitable imaging arrays (2k × 2k). There was an element of 'redundancy' in our procurement programme - but it can be seen that both main programmes (A & B) have taken some time in completion. In addition to the primary interchangeable detector heads, we also have requirements for embedded sensors, and these have been met by using a TK1024 and Loral 512FT devices. The INT PF Mosaic is discussed in the next section.

4. Recent RGO CCD work

At the 1993 ESO workshop, we presented details of RGO CCD work at that time, so here we will briefly present an update since that time.

4.1 ING detector installations

1994	TEK3 (INT)	TK1024	1024×1024 thin
	TEK4 (JKT)	TK1024	1024×1024 thin
	TEK5 (WHT aux.)	TK1024	1024×1024 thin
Feb 1995	EEV8 (WHT UES)	EEV CCD05-50	2186×1152 frontside
March 1995	WYFFOS (WHT)	TK1024	1024×1024 (embedded)
May 1996	LORAL1 (WHT)	Loral/Lesser 2K3EB	2048×2048 thin
early 1997	EEVT1 (WHT)	EEV CCD42-80	4096×2048 thin (planned)

It can be seen that most of our instruments are currently served by the TK1024 devices, although these await upgrades as discussed above. The introduction of the LORAL1 head proved very popular, and we certainly await delivery of more of these chips so that we can complete more such heads. Similarly, we have cryogenic heads constructed, and awaiting installation of suitable EEV devices.

4.2 RGO detector work

Here we will highlight the main areas of activity at RGO within the last few years, with pointers to more information.

- INT PF Mosaic. A 2x2 mosaic of 'Lick3' CCDs - giving 4096×4096 FOV, with integral autoguider. This major instrument is nearing completion, and it is outlined below.
- WYFFOS. This fibre-fed multi-object spectrograph was commissioned in 1995, and has been in regular use since then on the WHT. It incorporates an embedded TK1024 CCD. See [11] for details.
- Lots of testing of spectral response etc. of Loral/Lesser & EEV CCDs. As part of our programme of Loral/Lesser and EEV thinned chips, we have engaged upon a significant amount of device testing and characterization. (Some of this is presented in the WWW-page, [6]).
- Development of a flatness measurement tool for mosaics & single chips. ([10] this workshop).
- Development of LED calibration sources - for flat-fields & spectral response. ([10] this workshop).
- Measurements of spatial resolution of thin CCDs. Several years ago we analyzed the modulations of sensitivity within the pixels of a CCD [3]; more recently we have studied the spatial resolution of thinned CCDs at various wavelengths (N. Falleur, RGO internal report 1996).
- Introduction of 'Low-smear-drift mode', fast readout CCD on WHT spectrograph. ([2] this workshop).

4.3 Isaac Newton Telescope Prime Focus Mosaic

The INT Prime Focus Camera (INT PFC) is designed to provide a large field survey and supernovae search capability for the prime focus of the 2.5m Isaac Newton Telescope. It is a joint collaboration between the Royal Greenwich Observatory (United Kingdom), Kapteyn Sterrenwacht Werkgroep (Holland) and the Lawrence Berkeley

Figure 1. The 5-chip INT Mosaic assembly

Laboratories (California). The unit will be used as a 'common-user' instrument on the telescope.

The INT PFC consists of a 4 chip mosaic utilizing thinned and anti-reflection coated CCDs. These are LORAL devices of the LICK3 design. They are operated cryogenically in a purpose-built camera assembly. A fifth CCD, of the same type, is co-mounted with the science array in the cryostat to provide autoguider functions. This cryostat then mounts to the main camera assembly at the prime focus. This assembly includes standard filters and a novel shutter wheel, which has been specifically designed for this application. The camera has an unvignetted field of 40 arcminutes and a focal ratio of f/3.3. This results in a very tight mechanical specification for coplanarity and flatness of the array of CCDs and quite stringent flexure tolerance of the camera assembly. (See [9] and [10] for flatness measurement technique). One of the main requirements is to read the whole array out within 100s, with less than 10e rms readout noise and very low CCD cross talk.

A fuller description of the instrument may be found in the paper by [1]. Here, we shall merely highlight the main features, and present up-to-date status information. The early ('Lick3') Loral CCDs were not designed for butting, but have been assembled into the most compact layout that their custom-packages allow (after thinning by Lesser). Figure 1 illustrates the focal plane layout- in the form of a preliminary (frontside) mosaic of the same construction.

The instrument uses two similar CCD controllers- one for the autoguider chip, and one for the four science detectors. The science mosaic can be read out in < 100 sec., with a readnoise < 10 e^-. The system has been fully assembled, and is scheduled for first commissioning on the INT in January 1997. At the time of writing, we do not have a full set of science-grade thin CCDs; however the system has been functionally tested and will go to the telescope with only two science-CCDs and the autoguider-CCD if necessary.

5. Future developments- Ongoing/ future RGO detector work

- *Higher-speed, multi-channel implementation- particularly for larger arrays.* The existing RGO CCD controller in use on the ING telescopes allows multi-channel operation, at normal 'slow-scan' speeds (of ~10-100kHz). We are undergoing a programme of work to extend it's capability to operate at ~1MHz pixel rates, allowing faster readout times. This will be an operational improvement (less dead-time) for large scientific CCDs, and will be beneficial for 'quick-looks' of the field of view. Furthermore, we require an enhanced speed for operation of infrared arrays (see below).
- *Testing (at RGO) and commissioning (on ING) of EEV & Loral large-format CCDs.* This programme of work is expected to continue through 1997, as devices undergo final testing at RGO, followed by installation on all three ING telescopes. Particularly in the case of the guaranteed-performance EEV devices, we shall be characterizing their performance and comparing with specifications.
- *WHT 4K×4K mosaic for PF imaging (2×EEV CCD42-80).* We plan to build a modest-sized mosaic, as a 'fast-track' upgrade to imaging on the WHT; however, most wide-field imaging is expected to be performed with the INT PF wide-field mosaic.
- *INT 8K×8K mosaic (proposed); could be 12K×12K, or 16K×16K.* A proposal has been made for a very wide-field imager to utilize the maximum possible FOV on the INT PF (awaiting funding approval).
- *CCD39 WFS work (Gemini WFS).* The new EEV CCD39 (80×80) wavefront sensor chip has been evaluated, and adopted for use on Gemini telescopes. We anticipate more work as this device is integrated into use as part of the 'A&G' systems.
- *ING 1024×1024 IR array imaging camera.* We have ordered a Rockwell Cadmium-mercury-telluride 'Hawaii' array for use on the WHT. The current programme of work entails an adaptation of the existing CCD controller to drive this IR array - thus enabling it to be integrated into the telescope instrumentation systems rather readily. Design of a 'fast-track' imaging camera is underway, with completion expected in 1998.

Acknowledgements: many colleagues at RGO are acknowledged for their contributions to the work described; these include - Tony Bennett, Derek Ives, Ian Richardson, Percy Terry, and Simon Tulloch.

6. References

[1] Ives, D.J., Tulloch, S.M., Churchill, J.E., 1996, Proc SPIE **2654,** 266. *'INT Prime Focus Mosaic Camera'.*

[2] Ives, D.J. & Bennett, A.V., 1996 ESO CCD workshop. *'The low-smear-drift mode of rapid CCD readout'*.

[3] Jorden, P.R., Deltorn, J-M., Oates, A.P., 1994 Proc. SPIE, **2198**, 836. *'The nonuniformity of CCDs and the effects of spatial undersampling'*.

[4] Lesser, M., WWW 1996. http://bilbo.as.arizona.edu/ccdlab.html

[5] Oates, A.P. & Jorden, P.R., Nov 1993, ING/RGO Technical Note 92. *'New Large Format CCDs for the ING'*.
[Also visible at: http://www.ast.cam.ac.uk/~apo/docs/ingccds/ingccds.html]

[6] Oates, A.P. & Jorden, P.R., 1996 ESO CCD workshop. *'The EEV thin CCD42-latest performance data'*.

[7] Oates A.P., WWW 1996. http://www.ast.cam.ac.uk/~apo/docs/ccds.html

[8] Pool, P.J., Burt, D.J., Oates, A.P., Jorden, P.R., 1996 in press, Royal Society of Chemistry Special Publication No. 194, Proc of 1995 ICSOI-3 Grand Cayman meeting. *'A new large-format, very low noise scientific CCD'*.

[9] Tulloch, S.M., 1996 ESO CCD workshop. *'A technique for the determination of co-planarity in mosaic CCD cameras'*.

[10] Tulloch, S.M., 1996 ESO CCD workshop. *'CCD calibration & testing using compact LED light sources'*.

[11] Bingham, R.G., Gellatly, D.W., Jenkins, C.R., Worswick, S.P., 1994, Proc. SPIE, **2198**, 56. *'Fibre-fed spectrograph for the 4.2m William Hershel Telescope'*.

CCD DEVELOPMENT AT THE UNIVERSITY OF HAWAII

GERARD A. LUPPINO[1]

ABSTRACT

The Institute for Astronomy at the University of Hawaii has an active program in detector development, both in optical CCDs and Infrared Arrays. Below, I briefly describe the optical detector program and the existing scientific CCD cameras in place on UH facilities on Mauna Kea. I also list present and future development efforts.

1. OPTICAL CAPABILITIES OF IFA DETECTOR LAB FACILITY IN MANOA

The IFA Detector Lab contains several complete CCD systems (dewars, controllers, and computer workstations) along with multiple custom camera heads for testing a variety of detectors from different manufacturers. Over the course of the past 3 years, we have evaluated numerous CCDs from Loral, SITe, MITLL, and Hamamatsu. In addition to the test cameras, we also have the following capabilities:

- Grating monochromator for QE measurement
- Packaging and wire bonding equipment
- X-ray test measurement facility

The primary responsibility of the Detector Lab is to provide the focal planes for instruments for our Mauna Kea (and Haleakaka) Observatory facilities. Although UH operates its own 2.2m telescope on Mauna Kea and must provide the instrumentation for that telescope (including detectors), we also are intimately involved in detector development with the other large optical telescopes on Mauna Kea (CFHT, Keck, Subaru and Gemini). In the next section I list the CCD cameras that are currently in use on Mauna Kea.

2. EXISTING CCD CAMERAS ON MAUNA KEA

1. Thinned SITe 2048^2 CCD - commissioned Spring 1992. Presently the primary, general-purpose CCD for the UH 88-inch telescope.
2. Thinned (by Lesser), blue-AR-coated Loral 2K (only half the imager is operational). Commissioned 1994. Used on the 88-inch for projects that require high blue/UV sensitivity.
3. UH 8K×8K CCD mosaic camera. Used on the 88-inch at f/10 (20′× 20′ field of view), and on the CFHT prime focus (30′×30′ field of view).

[1] Institute for Astronomy, University of Hawaii, Honululu, 96822 USA **Email:ger@hokupa.ifa.hawaii.edu**

J. W. Beletic and P. Amico (eds.), Optical Detectors for Astronomy, 83–84.

3. FUTURE DEVELOPMENT

- A consortium of observatories led by UH is working with MITLL to develop a 3-edge-buttable, thinned 2K×4K CCDs with ultra low read noise ($< 2e^-$) and low read noise at relatively high speed ($< 5e^-$ at 500 kps), on high resistivity (deep depleted) silicon for increased NIR response and reduced interference fringing (see paper by B. Burke in these Proceedings).
- We plan to upgrade the UH 8K×8K CCD camera with thinned CCDs.
- We are presently building an "augmented" clone of the UH8K CCD camera for CFHT. This camera will have a 50% larger focal plane that will measure 8K×12K pixels. In addition, the CCDs will be thinned and the controllers and CCDs will allow much faster readout speed. This camera is scheduled for first light in Fall 1997.
- We are also building an 8K×8K CCD focal plane for a large coudè spectrograph being built by UH for US Air Force AEOS 3.6m telescope on Haleakala (Maui). This instrument will be used primarily in the far red and near infrared regions and will require CCDs with ultra low noise (1.5 e^-) that as well as low tendency for interference fringing.
- Now that John Tonry has joined the faculty at the IFA, we will be pursue new technology for image motion compensation "on chip" using multi-direction charge transfer OTCCD architecture (see paper by Tonry and Burke in these Proceedings, p. 281).

HIGH PRECISION SHUTTER FOR THE T.N.G. CCD CAMERA

F. PEDICHINI[1], *M. COMARI*[2], *R. COSENTINO*[3] *G. FARISATO*[4], *W. DE BACO*[5], *G. SANDRE*[5]

ABSTRACT

The requirement to have detectors able to cover large fields of view on the focal plane of telescopes, demands the use of new generation CCDs with large dimensions or the combination of CCDs to form mosaics. The TNG optical imager, for example, will use a 6 cm squared CCD detector (2 X 2 Loral chips). Furthermore, modern fast optics telescopes, such as the NTT or the TNG, use, for standard star calibration, exposure times of the order of 1 s.

These requirements make the design and realization of the shutter for the CCD camera a very delicate and important task. Large linear dimensions, to allow a proper illumination of the whole detector, and high accuracy in exposition time, to avoid systematic photometric errors during short exposures, are the fundamental characteristics that the shutter has to comply with.

We present the main characteristics of the shutter built for the TNG CCD camera. Its working area is 8 × 8 cm. Thanks to a peculiar double blade mechanics and a two step-motors driving system, the shutter assures a uniformity of exposure time of ± 0.8 ms over the whole field when exposed for 1 s. Finally, the control electronics provides a simple interface to any type of CCD controller and a telemetry read-out feedback of the 'real exposed time' to avoid problems when using non real-time control software.

1. TECHNICAL SPECIFICATIONS

Shutter working area: 80 x 80 mm
Dimensions: 220 x 150 x 35 mm
Weight: 800 gm - (control electronic 300 gm)
Control cable: 2 meters of flat cable between shutter and control electronics
Power supply requirements: 18V, standby 0.25 A, working 1.2 A
Minimum exposure time: 10 ms
Blades driving: two step motors
Position and state feedback: two on-axis custom optical encoders.
Host interface:

opto-coupled parallel TTL input control signals: open/close, reset, arm
opto-coupled parallel TTL output control signal: readout of real exposed time

[1] Osservatorio Astronomico di Monte Porzio, Roma (Italy); Email: **pedik@coma.mporzio.astro.it**
[2] Osservatorio Astronomico di Trieste, Trieste (Italy)
[3] Osservatorio Astrofisico di Catania,Catania (Italy)
[4] Osservatorio Astronomico di Padova, Padova (Italy)
[5] C.F.R. Padova (Italy)

J. W. Beletic and P. Amico (eds.), Optical Detectors for Astronomy, 85–88.

Reset and loading: automatic on power-up (it is possible to configure the control electronics for auto loading after the last exposure).

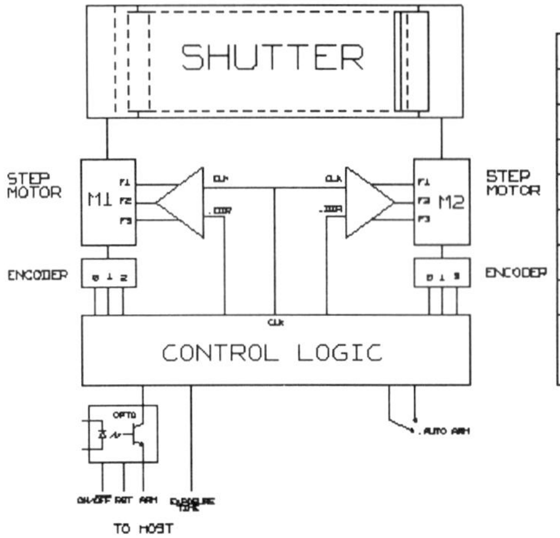

CLK	CLOCK FOR THE STEP MOTORS
F1, F2, F3	PHASES FOR THE STEP MOTORS
DIR	STEP MOTOR DIRECTION
ON/OFF	H=SHUTTER OPEN, L=SHUTTER CLOSED
RST	RESET THE ELECTRONICS
ARM	LOAD THE SHUTTER
AUTO ARM	IF CLOSED THE SHUTTER IS LOADED AUTOMATICALLY AFTER EXPOSURE
EXPOSURE	IF OPEN THE CONTROL LOGIC WAIT THE ARM SIGNAL TO LOAD THE SHUTTER
TIME	THE EXPOSURE TIME IS THE ELAPSED TIME BETWEEN ''UPPER BLADE HALF OPEN'' TO ''LOWER BLADE HALF OPEN''

Figure 1: TNG shutter block diagram

2. TEST RESULTS

2.1. Exposure uniformity

Exposure uniformity measurements on the 8x8 cm field has given the following result: ± 0.5 ms during 1 second exposure mean time.

2.2. Stress tests

The stress tests have been performed at a temperature of 22°C and at an altitude of 200 m above sea level.

The tests consist in:

- 2 series of 2000 exposures of 2 sec with a frequency of 0.1 Hz: the test has shown an increase in the device temperature less than 1°C.
- 1 series of 2000 exposures of 1 sec with a frequency of 0.2 Hz: the test has shown an increase in the device temperature of 1°C.
- 1 series of 1000 exposures of 0.4 sec with a frequency of 0.5 Hz: the test has shown an increase in the device temperature of 2°C.
- 2 series of 500 exposures of 0.2 sec with a frequency of 1 Hz: the test has shown an increase in the device temperature of 4°C.
- The control electronic working temperature was in the 30 - 35°C range. All the tests have not shown any significant stress.

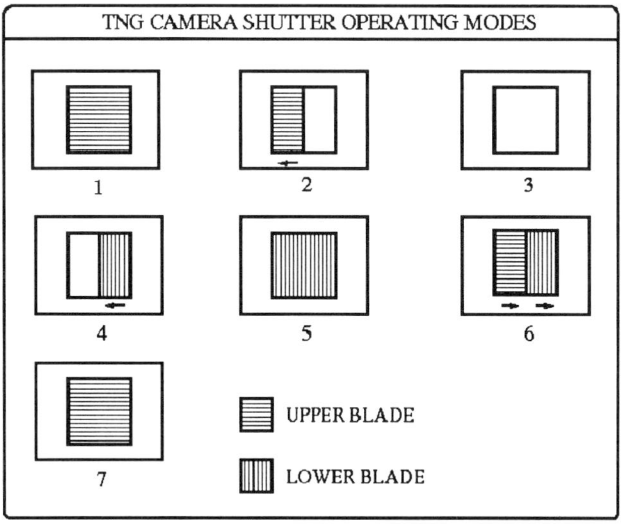

Figure 2: Operating modes

1. THE SHUTTER IS READY TO OPEN [both blades are loaded]
2. THE SHUTTER IS OPENING [upper blade is moving while lower blade is loaded]
3. THE SHUTTER IS OPEN [upper blade is open while lower blade is loaded]
4. THE SHUTTER IS CLOSING [upper blade is open and lower blade is moving]
5. THE SHUTTER IS CLOSED [both blades are open]
6. THE SHUTTER IS CLOSED [both blades are loaded]
7. THE SHUTTER IS READY TO OPEN [both blades are loaded]

3. CONCLUSIONS

The shutter has shown a very good reproducibility and uniformity of exposure with opening times of the order of ten milliseconds and is useful as a shutter for CCD cameras that mount large area detectors or CCD mosaics. No failure or malfunctions have been detected during test session of 8000 cicles. A final inspection revealed no degradation in the mechanic assembly and the moving parts. Future tests at 2200 m above sea level and at a temperature of -20° C are foreseen.

At the time of publishing three shutters have been produced by the C.F.R. in collaboration with ESO and the Osservatorio Astronomico di Roma for the SUSI 2 imager of ESO. Several improvements in the mechanics and the control electronics enhance the performances and the reliability of the system.

A black anti reflection painting (reflectivity < 0.3% from 300 nm to 1100 nm)has been applied o both blades and the body of the shutter is black anodized.

The 5 volt linear regulator has been substituted by means of a switching device to reduce the dissipated power.

All the mechanical parts have been machined with CNC tools and laser cutting to achieve a high production standard.

Simple self regulating springs have been added on the link between the blades and the pulling cables to reduce backlash.

The three shutters are now under intensive testing at ESO, C.F.R. (Padova) and in Rome to acquire detailed statistical performance.

Fabrication of Thinned CCDs at UCO/Lick Observatory

R. J. Stover[1], *W. E. Brown*[1], *D. Booth*[2], *M. Wei*[1], *D. K. Gilmore*[1]

Abstract

The University of California Observatories/Lick Observatory (UCO/Lick) has undertaken to produce science-grade CCDs for current and future instruments to be used on UCO/Lick and UCO/Keck telescopes. In this paper we describe the techniques developed and the progress achieved to date.

1. Introduction

Both UCO/Lick and UCO/Keck have immediate need for a large number of high-quality CCDs. At Keck, we are building two new spectrographs. One spectrograph, DEIMOS, will use two 8k × 8k CCD mosaics produced from 2k × 4k CCDs. The second spectrograph, the Echelle Spectroscopic Imager or ESI, will use a single 2k × 4k CCD. At Lick we would like to upgrade a variety of direct imaging and spectroscopic instruments with either 2k × 2k CCDs or 2k × 4k CCDs. Lick and Keck are funding a number of different CCD development efforts in order to help insure that we can obtain the devices we need. One of those efforts is the CCD thinning facility being developed at UCO/Lick.

In developing the thinning facility, we have several long-range goals: first, we want to develop the technology and facilities to produce the science-grade thinned CCDs required by Lick and Keck Observatories.

Second, we want to use the thinning facility along with the knowledge and experience gained as a step toward new detector fabrication technologies. We have three immediate objectives. The first is to produce 2k × 4k three-side buttable, thinned CCDs for the 8k × 8k mosaics of the Keck II DEIMOS spectrograph. The second is to produce 2k × 2k and 2k × 4k thinned CCDs for Lick Observatory instruments. The third objective is to produce 2k × 4k buttable, thinned CCDs for a 4k × 4k mosaic upgrade of the Keck I HIRES spectrograph.

We have been able to undertake this effort now because we have a large project, the DEIMOS spectrograph (the need), important institutional support outside the instrument budget (the funding), and we have been able to draw significantly on the capabilities of the University of California Microfabrication Facility in Davis, California and the many nearby companies of "Silicon Valley" (the resources).

2. CCD Fabrication

We have produced our CCDs at a silicon foundry called Orbit Semiconductor, which is located in Sunnyvale, California. We have two wafer designs, with each design producing several different CCDs on the wafer. Design #1 produces two 2k × 2k CCDs

[1] University of California Santa Cruz, CA 95064 USA. Email: **richard@ucolick.org**
[2] Department of Electrical and Computer Engineering, University of California, Davis CA 95616 USA

J. W. Beletic and P. Amico (eds.), Optical Detectors for Astronomy, 89–91.

and one 2k × 4k CCD. Design #2 produces two 2k × 4k CCDs. The 2k × 4k CCDs are three-side buttable, with the serial register along the 2k edge. In all cases, the pixels are 15 micrometers square. To date Orbit Semiconductor has fabricated at least eight lots of design #1 and four lots of design #2. Recently we have seen yields as high as 69% science-grade 2k × 4k CCD in cooled probe testing. Design #2 was developed specifically for the Keck Observatory DEIMOS spectrograph and our CCD thinning project.

3. THINNING STRATEGY OUTLINE

Our CCD thinning strategy consists of six steps:

1. Select the CCDs for thinning based on imaging tests in Lick's cooled wafer-probe test facility.
2. Bond the CCD wafer face-down onto another silicon wafer called the handle wafer.
3. Lap and polish the CCD wafer to a thickness of about 50 micrometers.
4. Chemically thin the CCD wafer to the final device thickness, typically about 20 micrometers.
5. Mask the CCDs with a protective layer and etch away the silicon above the bond pads.
6. Cut the wafer to obtain the individual CCDs, mount them on an aluminum nitride (AlN) substrate, and make wire-bond connections to the CCD.

These steps are described in more detail below.

4. WAFER BONDING

Wafer bonding has three objectives: To protect the CCD during thinning, to support the thinned CCD, and to insure a flat CCD (<10 micrometers peak-to-valley variations). We bond the CCD wafer face-down onto the handle wafer using Epotek 301-2 epoxy. To the epoxy we add 0.5% by volume glass microspheres which have a 20 micrometer diameter (+/- 3%). The microspheres insure that we have a very uniform bond line and the flatness of the CCD and handle wafers are maintained. After bonding, the wafers are checked with an infrared camera to look for voids in the epoxy layer. The handle wafer has a silicon nitride layer on it to protect it during the chemical etch step.

5. WAFER LAPPING AND POLISHING

We have experimented with both commercial grinding (with a grinding wheel) and lapping (with powdered diamond slurries). Lapping produces far less stress on the wafer and is compatible with the strength of the epoxy bond. Lapping followed by polishing with 1-micrometer diamond produces a mirror-like finish with very low surface damage. Wafer lapping removes the bulk of the silicon, leaving only a thin layer to be removed in the next step.

6. CHEMICAL WAFER THINNING

Chemical thinning reduces the thickness of the CCD wafer until the epitaxial silicon layer is reached. The epitaxial layer is the photosensitive layer of the CCD and the surface of this layer represents the final backside of the CCD. We are experimenting with two types of silicon. In the first type, we use a standard epitaxial layer for the CCD. The second type has an additional, very thin epitaxial layer, which is very highly doped with boron. The boron layer acts as a very good barrier to the chemical etchant. In addition, we hope the boron layer will provide a backside surface enhancement for improved UV response. All chemical thinning procedures are done in the University of California, Davis Microfabrication Facility.

7. Bond Pad Opening

The last chemical step is to etch away the silicon above the bond pads, to expose the backside of the pads where we make electrical contact to the CCD. This is the most complex part of the thinning procedure.

First, a mask is produced which covers the entire surface of the CCDs, except for the area of the bond pads, and a thin *scribe line* around the perimeter of the devices. The mask consists of a layer of titanium followed by a layer of gold. The silicon in the unprotected areas is etched away until the silicon dioxide layer just above the bond pads is reached. Another chemical is used to strip the first layer of the mask. A third chemical removes the rest of the mask and the silicon dioxide, finally exposing the bond pads.

8. CCD Mounting

The thinned CCD and its handle wafer are glued to an aluminum nitride (AlN) substrate. Development of the substrate mechanical design was done by Gerry Luppino at the University of Hawaii, and is the same as that used for the Lincoln National Laboratory 2k × 4k CCD project. This will make the Lincoln CCD interchangeable with our CCD - an important requirement since either or both may be used in the Keck Observatory DEIMOS spectrograph. AlN is chosen as the substrate material because its thermal expansion coefficient is similar to silicon, and this will help to reduce mechanical stresses and maintain a flat CCD when the device is cooled to operating temperature.

9. Current Status

This is very much a work in progress. We have achieved good success in the wafer bonding. Our objective was to maintain wafer flatness to a few micrometers, and we have achieved that. Mechanical thinning of the bonded wafers has produced good results, and we are working on a lapping jig that will give us even better control to reduce wedge in the lapped wafer. We have had good success in the chemical wafer-thinning step, although we still need to fabricate some better wafer handling fixtures.

Our only remaining major problems have been in the last chemical step, Bond Pad Opening. Here we have had problems with gold-mask failure and silicon etching inconsistency. However, useful discussions at the conference, notably with Barry Burke of Lincoln National Laboratory, have given us new direction and possible solutions.

SECTION III: CCD DEVICES, CONTROLLERS, TESTBENCHES & INSTRUMENTS

ESO's NEW CCD TESTBENCH

PAOLA AMICO[1] & TORSTEN BÖHM[1]

ABSTRACT

During the forthcoming years, ESO will implement a large number of CCDs in optical instruments for the Paranal and La Silla observatories. An important step in this process is the CCD characterization and optimization that is done after the device is received from the manufacturer. In order to accommodate the new large CCDs and to interface to the new FIERA controller, ESO is building a new CCD testing facility in Garching. Among the measurements that will be made with this new facility are: quantum efficiency, conversion factor, charge transfer efficiency, dark current, read out noise and cosmetic quality.

1. INTRODUCTION

ESO's plan for optical detectors [1] foresees a large number of systems for deployment to the optical instruments at the Paranal and La Silla observatories. During 1998 alone, we expect to implement 21 new CCDs in 8 instruments. During the next 5 years, we may produce 30 scientific CCD systems that use nearly 100 CCDs. All these new systems will use the new FIERA CCD controller [2,3].

While routine diagnostic testing is done at the telescope site, complete characterization of each device is initially carried out at ESO's headquarters in Garching. For example, all the CCDs operating at the La Silla Observatory (12 mounted at the telescopes plus backup devices) have been tested with the existing ESO testing facility, which is equipped to operate with the VME controller.

The same procedure is foreseen for all the detectors needed by the new VLT and La Silla instruments. The introduction of the new FIERA controller as well as the growing size of CCDs, which cannot be handled by the existing testing facility, have made the construction of a new testbench essential for the characterization of the future generations of scientific detectors.

The new testbench prototype for FIERA controlled CCDs, named the TAMER (Test And Measurement Evaluation Rig), is presently being developed at ESO/Garching and will begin operation in 1997. The new design provides the possibility to test very large detector sizes (8k $\times$ 8k 15 μm pixels devices or 200 mm diameter) in a one-step process (e.g.: a uniform flat field is obtained with a single exposure). Fundamental CCD parameters are tested in fully automated mode, while specific additional tests can be performed interactively. The testbench software is integrated with the VLT Standard Central Control Software (CCS, [4]) and is fully compliant with VLT standards. Some of the software modules and testing procedures will be identical for both the Garching testbench and the diagnostic testing tools at the telescope.

The major requirements we fixed during analysis phase of the project are listed in table 1.

---

[1] Optical Detector Team, ESO, Karl-Schwarzschildstr. 2, 85748 Garching, Germany. Email: **pamico@eso.org**

J. W. Beletic and P. Amico (eds.), Optical Detectors for Astronomy, 95–102.

System Requirements
Uniform illumination over a 20 cm diameter.
Wavelength range 320-1100 nm.
Use of monochromator instead of filters to provide narrow band measurements capability and to reject stray light
Use of double monochromator to minimize red leak when testing blue wavelength.
Light source with smooth continuum and no spectral lines.
Availability of a work area between the integrating sphere and the detector for experiments (e.g. image projector, point spread function measurement).
Fe_{55} measurement setup.
Full compatibility with FIERA's hardware and software.

***Table 1**: CCD Testbench Requirements*

2. TESTBENCH HARDWARE AND DESIGN

The selection of the hardware components was based on ESO's present testbench design combined with the ideas of other observatories. A useful source of information for preliminary design has been provided by Lesser and McCarthy's work for the Steward Observatory [5]. A subsequent step has been to create a testbench simulator (by means of Microsoft Excel) to predict the performance of the system. One of the questions we had to solve was whether a double monochromator can provide enough UV flux at the detector's position. Although it provides major improvement in signal/stray-light ratio with respect to a single monochromator, the double monochromator introduces twice as many reflections, with consequent loss of signal. The entire system's efficiency is lowest at the blue end of the spectral range of interest (320 – 1100 nm). Therefore, the selection of many crucial hardware components (precisely the monochromator and its parts), as well as the choice of smaller items, such as the lamp type or the coating for the integrating sphere, have been taken into consideration and simulated.

A schematic of the testbench design is given in Fig. 1.

The system is mounted on a Newport 300 cm × 90 cm optical table top, which provides static rigidity and flatness, in addition to the standard configuration of sealed mounting holes. All hardware components are GPIB driven and controlled through a GPIB controller, from National Instruments, that can operate up to 14 devices. The GPIB controller is connected to the SPARC board of the FIERA controller via SCSI connection. A PULPO monitoring unit [6] for environmental variable control (temperature, humidity, etc.) will also be part of the system.

The major system components are:

- **Lamp housing and lamp**: a Quartz Tungsten Halogen (QTH) 100 W lamp is held in a standard convection cooled housing, equipped with a f/1 condenser, producing a ~3 cm diameter collimated beam. This beam is then focused on the monochromator input slit by means of a secondary f/4 plano-convex lens, which matches the acceptance pyramid of the monochromator. The light system maximizes the total power into the monochromator and, being a thermal source, it provides a smooth continuum without strong emission or absorption features, within the desired wavelength range. The

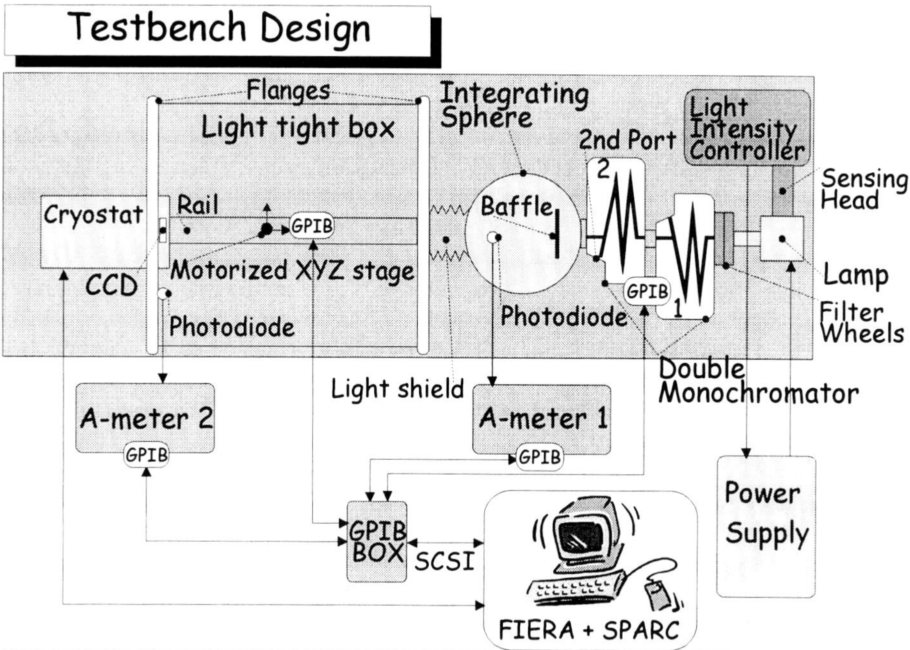

Figure 1: *Schematic of the testbench design. The elements shown on the table top (300 X 90 cm) are in scale (except for the cryostat). From right to left: lamp and light intensity controller connected via the power supply, the double monochromator system with two output ports. The secondary port will input light into an optical fiber, used by PSF experiment; the primary port feeds the integrating sphere. The flanges are screwed onto the table top and the space in between is enclosed in a light tight box. The CCD head is mounted on a metallic plate fixed to the left flange and the cryostat hangs outside the box. All GPIB driven instruments are connected to the GPIB controller, which in turn interfaces via SCSI to the FIERA system. Arrows indicate direction of data flow.*

wattage has been chosen in order to have enough flux in the UV part of the spectrum. This condition is especially critical to achieve a good S/N for the calibrated diode used for QE measurements.

- **Lamp power supply and light intensity controller:** The light intensity controller (Oriel) is directly connected to the lamp housing through a light sensing head, which monitors light variations, and interfaced to the power supply. It maintains constant light levels, for the duration of an exposure regardless of lamp aging, line voltage variations or changes in the ambient temperature (exposure times vary from few seconds to about ten minutes).
- **Monochromator**: Oriel Multispec 257 Double Monochromator in subtractive dispersion configuration. In the present setup, the output from the first unit is dispersed in the reverse direction by the second unit, thus homogenizing the light across the output slit. The net dispersion remains as that produced by the first monochromator, but the amount of stray light is greatly reduced, quoted by Oriel to be of the order of 10^{-7} of the unblocked signal. That is, almost three orders of magnitude smaller that the measured

stray light for a single monochromator of the same kind. This ensures that the light produced by the system in the UV-blue part of the spectrum does not suffer from the presence of red leak. The latter can cause significant errors in the measurements of quantum efficiency because CCD devices are usually more efficient at red wavelengths. The two monochromators can be controlled via GPIB in either an independent way or together (using the first one as master). Both are equipped with microstepping motor driven slits and 600 lines/mm ruled gratings, whose peak efficiency is at 400 nm. The usable wavelength region (where the grating efficiency is > 20%), is 250 to 1300 nm - well beyond our requirements. With this configuration, a minimum bandpass of ~0.1 nm can be reached. Two motorized filter wheels, which hold up to five 1-inch diameter filters each, are attached at the input of the first monochromator. The filter wheels contain order sorting filters (2 filters, with cut-off wavelength respectively at 450 and 665 nm) and neutral density filters (4 filters with optical densities in the range 1-4). The first monochromator is also equipped with an integrated shutter, which can be controlled both via external TTL signals and through GPIB commands. The minimum exposure time setting is 20 msec, the transition time ~2 ms. Positioning the shutter before the integrating sphere, instead that putting it at the sphere exit, has the advantage of eliminating the shutter pattern problem.

- **Integrating Sphere**: a 50 cm diameter "custom made" Labsphere. Its 8 inch (200 mm) exit port provides uniform illumination over an area bigger than the size of a 8k × 8k CCD or mosaic (a typical 8k × 8k with 15 μm pixels has a diagonal of about 17 cm; we will refer to this example throughout the paper). The primary output port is aligned with the input port. A secondary output port (∅ 1.3cm), which hosts a photodiode, is positioned close to the primary output port. A baffle situated inside the sphere prevents the output port from directly "seeing" the light source. The internal coating of the sphere is made in Spectraflect, a material that ensures a reflectance better than 98% in the range 400 – 1100 nm and better than 96% in the UV range (320-400 nm). 100% uniformity across the illuminated field is achieved when mounting the CCD in close contact with the exit port. Otherwise, the degree of uniformity, defined as the ratio of the illuminance at the edge of the field to the illuminance on the axis through the center, is a function of the distance of the target from the source. Uniformity goes asymptotically to 100% as the distance target - source increases [7]. The second option has been chosen in order to have enough space between the sphere exit port and the detector to perform experiments; for instance, to put lenses and a target image to be projected onto the CCDs. The detector will be put at a distance of 50 to 75 cm from the sphere output port, so that, for a 8 inch light beam and a 8k × 8k 15 μm pixel CCD, the degree of uniformity of illumination is always in the range 95% - 98%. A better than 1% uniformity is obtained for smaller detectors. Any variation in uniformity will be calibrated and the measured CCD data scaled appropriately.
- **Picoammeters and diodes:** the testbench is equipped with two photodiodes, one permanently mounted at the secondary output port of the integrating

sphere and the other, which is needed for absolute flux calibration of the system, is put at the detector's position. A permanent solution, with the latter diode fixed as close as possible to the detector and sharing with it the same focal plane, is also planned for the future. Separate ammeters are attached to diodes through low-noise triax cables. A Keithley 486 is connected to the sphere's diode: a 5½-digit autoranging picoammeter designed for low current applications where fast-reading rates must be performed. The measurement range is between 2nA and 2mA, with a resolution of 10fA (@2nA range). The diode we use is a Hamamatsu 1cm^2 Silicon Photodiode for precision photometry (NEP ~10^{-15}) with good quantum efficiency in the UV. The second diode is also a 1 cm^2 silicon Hamamatsu photodiode, which has been absolute calibrated by reference to NPL (National Physical Laboratory, England) and to PTB (Physikalisch-Technische Bundesanstalt) standards. At present it is interfaced to a Keithley 619 Electrometer/Multimeter, with the same measuring range and resolution as the 486 model. Both ammeters are controlled via the GPIB controller.

- **Flanges system and light tight zone:** the integrating sphere is attached to a flange, fixed onto the table top, through a flexible light shield, which allows a length span of ~25cm. A second flange, which will hold a custom made plate for each detector head (at least three different systems are foreseen for the VLT detectors systems), is positioned at a distance of 50 cm from the first flange. The dewar itself will be hanging from the outer wall of the flange. The space in between the two flanges will be closed by a wooden light-tight box, with lateral access door and with removable baffles to avoid scattered light. The flanges, the box, and some other minor elements are being designed by ESO's mechanical design office.

Specifications for the hardware components are listed in the table 2 below:

Product	Manufacturer	Model
Optical table + Legs	Newport	M-RT-310-8(0.9x3.00) NN4-28 or 23.5
Integrating sphere	Labsphere	CSTM-US-200-SF
Light source System : Lamp Housing + Power Source+1st Condenser Light Intensity Controller 2nd Condenser	Oriel	 6333+60043 60067 68850 3-40570
Monochromator + accessories	Oriel	Multispec 257/77700
Photodiodes	Hamamatsu	S2387-1010R S1337-101
Picoammeter	Keithley	486
GPIB SCSI board	National Instruments	GPIB-SPRC S 240 V

Table 2: List of hardware components

3. TESTBENCH SOFTWARE

The requirement to test new generation CCDs driven by the FIERA controller has constrained the choice for the software development tools and for the development platform. The latter must be a SPARC Board that is part of the FIERA system, which runs under the Sun Solaris Operating System. The development language is standard C to allow full compatibility with FIERA and, more generally, with the VLT software. LabWindows/CVI for Sun Solaris by National Instruments is the software development environment of choice, since it offers a full ANSI C and GNU C compatibility. Moreover, applications written under LabWindows/CVI are independent from separate commercial run-time engines (which have to be licensed). This requirement is essential because the data analysis modules will be made available for testing at the telescopes.

In addition, LabWindows/CVI provides very attractive features for building instrumentation software applications: run-time libraries, simplified GUI programming, handling of GPIB instrument I/O, use of TCP and a set of instrument drivers.

Figure 2 shows a schematic layout for the testbench software. It consists essentially of a main interface window (GUI), which allows the selection of the desired testing procedure, either a complete automated sequence of all main tests or a single user-defined procedure. In all cases, scripts provide interactions with three major and independent modules:

1. The Hardware Control software, which controls all the hardware peripherals driven via GPIB. It consists of graphical interfaces plus drivers to allow access and manipulation of all the instruments' features. Among these, selection of the wavelength range, light intensity, bandpass and collection of the data sent from the photodiodes to the ammeters. Flux readings from the ammeters, setup data from the monochromators and data from PULPO are stored for later analysis. The PULPO device will be used to monitor various parameters, both CCD auxiliary functions and environmental data: detector's temperature, cryostat pressure, ambient temperature and humidity, light levels (for instance, in the light tight parts), etc. The ammeter readings will also serve to double check exposures times.
2. The CCD Camera Control module, which is in charge of sending to FIERA instructions (exposure start, exposure time, I/O operation with data files, CCD initialization, voltages settings, etc.) for taking sets of exposures. FITS images are stored for later analysis.
3. The Data Analysis module, which takes the data stored by the other modules, processes it and extracts test results.

The modules are independent in the sense they exchange data but in general they can run in standalone mode. This requirement will allow the use of the analysis software at the telescope site, thus providing a standard tool for quick testing of the CCD performance when connected to the instrument.

The main interface also accesses a calibration routines module, used periodically to check the system's hardware responses and perform wavelength/flux calibrations.

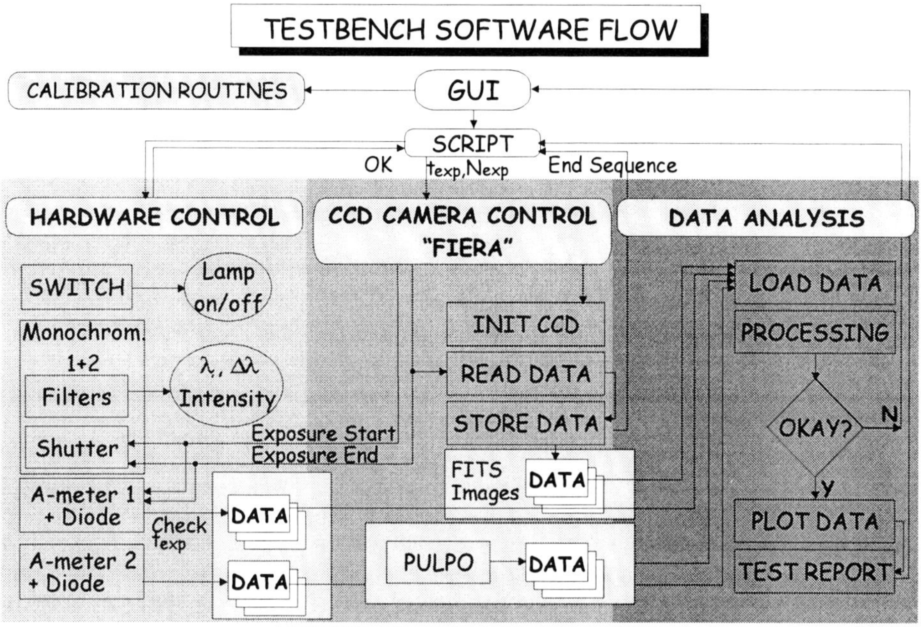

Figure 2: *The testbench software flow. The background shading delineates separate software modules: left) hardware control, which interacts with all GPIB driven hardware; middle) FIERA control, which interacts with FIERA software, i.e. with the CCD, and sets exposures sequences and times; right) data analysis, which collects data and processes them into a final test report. Also shown is PULPO, used here for CCD auxiliary functions and environmental control. All modules are operated by a startup GUI, which allows test sequences, parameters definition and access to the calibration routines.*

4. TEST PROCEDURES

A complete description of all calibration and testing procedures and methods is in preparation. Table 3 lists all test procedures that will be implemented in the first release of the project, together with the expected accuracy in their measurements.

Measured parameter	Expected accuracy
Conversion factor	2 digits
Absolute Quantum Efficiency	2.5%
Linearity	< 1%
Noise	2 digits
Full well capacity	1%
Crosstalk (multiple ports)	1/300,000
Charge Transfer Efficiency	10^{-6}
Homogeneity (Relative Q.E.)	< 0.5%
Other parameters to be measured	
Dark current	Cosmic Ray Sensitivity
Bias level	Remanence (Persistance)
Cosmetic quality	

Table 3:*List of all tests foreseen for first release of the testbench.*

Charge transfer efficiency measurements will be performed by means of both EPER (Extended Pixel Edge Response) and Fe_{55} methods. Additional tests of mechanical flatness, pixel scale, intra-pixel quantum efficiency and CCD PSF will be implemented with the second release.

5. CONCLUSIONS

ESO's new CCD testing facility is currently under development. Design and hardware implementation phases have been completed, with the exception of the flanges and the light tight parts, now under study. Software development and integration with the FIERA system are presently in progress. We expect to have the testbench working with basic functionalities during 4th quarter 1997.

Though in its first release the testbench will perform complete test of all major CCD parameters, further developments are under discussion. We plan to develop automatic CCD optimization procedures that find the best performance as function of voltages and readout speeds. The following procedures have been proposed:

1. Optimize for CCD gain, noise and linearity as function of the CCD amplifier drain voltage.
2. Optimize CCD charge transfer efficiency, dark current, noise and cosmetics as a function of horizontal and vertical clock amplitudes.
3. Optimize gain, noise, and charge transfer efficiency at different readout speeds.

The implementation of such testing procedures in an automated mode will make this testing facility a more valuable tool, for both the characterization and the optimization of scientific CCDs.

Acknowledgments: All members of the Optical Detector Team, both in Garching and Chile, have contributed to design and critique the development stages of the project. We thank Mike Lesser of U. Arizona for sharing with us his experience in building a test facility. We are also grateful to Fernando Pedichini of Osservatorio Astrofisico di Roma, who greatly contributed to the project during his visits to ESO and who is still actively collaborating with us.

6. REFERENCES

[1] J.W Beletic, these Proceedings, pp. 57 - 64.

[2] J.W. Beletic, R. Gerdes and R C. DuVarney, these Proceedings, pp. 103 - 114.

[3] C. Cumani & R. Donaldson, these Proceedings, pp. 115 - 122.

[4] Central Control Software (CCS) User Manual, VLT-MAN-ESO-17210-0619.

[5] Lesser, M.P. and B. L. McCarthy, 1996, "QE Characterization of Scientific CCDs", Proc. SPIE 2654B.

[6] Haddad, N., P. Sinclaire and J. Anguita, A. R., these Proceedings, pp. 131 - 134.

[7] R. Kingslake, *Illumination in Optical Images,* Applied Optics and Optical Engineering, Ed. R. Kingslake, Vol. II, Ch. 5, Academic Press, 1965.

FIERA: ESO'S NEW GENERATION CCD CONTROLLER

JAMES W. BELETIC[1], *ROLF GERDES*[1] *AND RAYMOND C. DUVARNEY*[2]

ABSTRACT

ESO is developing a new generation CCD controller for the Very Large Telescope (VLT) facility on Cerro Paranal. This CCD controller, named FIERA, will also be used at the Cerro La Silla Observatory and in ESO's testing and integration laboratories. A major emphasis in the FIERA design is the ability to run a large number of CCD outputs at fast readout rates, so as to take advantage of the new generation of high speed, low noise amplifiers. The first FIERA prototype was successfully demonstrated in October 1996. The design is undergoing revisions during 1997 and about 20 systems will be produced and deployed through the end of 1998. FIERA will see "first light" at La Silla and Paranal telescopes in early 1998.

1. MOTIVATION FOR NEW CONTROLLER DEVELOPMENT

The quality of CCD detectors continues to improve. CCDs are getting larger (the new standard is 2k x 4k and sizes up to 9k × 9k are being made) and high quantum efficiency devices, with peak q.e. ~90%, are now expected for all advanced instruments. Another significant advance in CCD technology is the development of high speed, low noise readout amplifiers. Two examples are:

a) the MIT/LL amplifier which achieves 4-6 e^- at 1000 kps* (i.e. 1 MHz), ≤ 2 e^- noise at 100 kps, and ≈ 1 e^- at 20 kps,
b) the EEV amplifier which should achieve ≤ 6 e^- at 1000 kps and ≤ 3 e^- at 50 kps.

The combination of large CCDs and the new amplifiers generates a need for CCD controllers that can read many ports in parallel at high speed. The motivation for short readout times is often due to the impatience of a waiting astronomer, but there are more important reasons for high speed, low noise readout:

1. Time is money. It costs about 2 Deutschmarks (~1US $) per second to operate an 8-meter class telescope. Reducing readout time of large mosaics from periods of minutes to 10 seconds will save the equivalent of the cost of the CCD controller every few months, on every instrument.
2. A large fraction of telescope data can be read out at 1000 kps and 6 e^- noise without any degradation to the data quality. This category includes acquisition, focusing, flat-fields, calibration spectra, high light level spectra and nearly all direct sky imaging.

[1] European Southern Observatory, Karl-Schwarzschildstr. 2, D-85748 Garching, Germany. Email: **jbeletic@eso.org**
[2] Emory University, Atlanta, Georgia, USA

* kps - kilopixels per second per readout port

J. W. Beletic and P. Amico (eds.), Optical Detectors for Astronomy, 103–114.

3. Fast frame rate imaging will allow new kinds of science to be done (e.g. fast photometry of rapidly varying objects) and is mandatory for adaptive optics systems.

As ESO is concentrating its CCD procurement on new generation amplifiers, in parallel we are developing a CCD controller that will be able to take full advantage of the speed possible with these CCDs. *(See[1] for further information on ESO's plans for CCD systems.)*

Also, this is an ideal time for ESO to develop and deploy a new generation CCD controller. Starting in 1998, ESO will begin commissioning the Very Large Telescope (VLT) on Cerro Paranal in Chile. This facility will consist of four 8.2 meter telescopes with three 1.8 meter auxiliary telescopes for interferometry. Each of the 8.2-m telescopes has 3 main foci (Cassegrain and 2 Nasmyth) and also a Coudé feed, which means there is a need for a large number of optical detector systems. We project that about 40 scientific CCD systems will be deployed to Paranal, La Silla and ESO laboratories during the next 5 years (through 2001).

Since the VLT control system is a very large software system within which the CCD controllers operate, it is important to develop and maintain a well-defined interface. To provide the highest reliability and to minimize maintenance, it is best to use a single controller for all scientific CCD systems.

For system integration and deployment to instruments, it is also best to have a single hardware design for all systems. By taking this approach, for slow readout CCDs, the "universal" controller will be overbuilt and will not make use of all of its higher priced components. But we are certain that the savings in development, deployment and maintenance will more than offset the total component cost.

In addition, we wish to be prepared for the "next generation" amplifiers. There are efforts to make amplifiers that will be only 1 to 2 e^- noise at 1000 kps, and such amplifiers may be sub-electron noise at 100 kps. This goal may not be that far away, as this type of performance was demonstrated nearly a decade ago ([2]). Since the CCD controller should have a lifetime of at least 10-15 years, it must be prepared to easily integrate the coming generation of ultra-low noise CCD amplifiers.

We have named our new controller FIERA, which is an acronym for Fast Imager Electronic Readout Assembly. As the CCD systems will primarily reside in Chile, it is appropriate that FIERA means "wild animal" in Spanish; an apt description of this controller, for which no holds were barred in the name of speed and flexibility.

2. FIERA Design Philosophy

The overriding design philosophy for FIERA is that it should be transparent to the user. That is, the performance of the entire CCD system should be limited only by the CCD and the imagination of the user.

Also, as stated above, FIERA should be a universal controller that can be used for every CCD or mosaic in ESO's instruments. This includes the following list of CCDs and mosaics:

- Tektronix / SITe 2k × 2k - 4 port
- VLT Test Camera CCD, which is a custom (Loral) designed 2k × 2k, 4 port device with 4 single port tracker chips located at the 4 corners (see Fig. 3 of [1], to see a photo of the CCD),

- 4k × 4k mosaic of two 2k × 4k CCDs - total of 4 ports,
- 8k × 8k mosaic of eight 2k × 4k CCDs - total of 8 or 16 ports used,
- 128 × 128 split frame adaptive optics CCD - 16 ports, all of which must be read at speeds to 1 million pixels per port per second (1000 kps),
- 16k × 16k mosaic of CCDs - design is still under discussion, but could be a mosaic of thirty-two 2k × 4k devices, meaning 32 or 64 readout ports that could be controlled by two FIERAs running synchronously.

FIERA will be run relatively slowly with the Tektronix, SITe and Loral CCDs, but will often need to run at 1000 kps for the mosaics of 2k × 4k devices. For the larger mosaics and the adaptive optics CCD, at least 16 ports need to be read out at speeds to at least 1000 kps. This means that FIERA must be able to handle at least 16 million pixels per second peak data rates.

In addition, FIERA should also have the following qualities:

- synchronous timing of all signals,
- fully deterministic generation of all signals, no random timing of processor operating software should be involved,
- use standard commercial ("off the shelf") components as much as possible,
- no software in the detector head electronics, implementing almost all software in the embedded computer using C and C++,
- design / use lifetime of at least 15 years,
- use of industry standard components, shying away from some components that may not be available on the market in a few years time,
- modular design so that some aspects (e.g. video signal processing) can be upgraded at a future date for better design or to overcome the unavailability of specific components,
- easy to program and adapt to different CCD systems,
- ability to run standalone, within the VLT control system or operate at the non-VLT standard telescopes of La Silla,
- as simple as possible

Of course, we also require FIERA to have system noise that is negligible relative to the CCD noise, to be linear to better than 0.5% over the dynamic range of the A/D, to have no pattern noise, and to have crosstalk between video channels which is less than 1 part in 100,000 (for 16 bit A/D converters).

Note that there are two major items missing from the above list of qualities for FIERA: (a) small size, and (b) low power. For all of the planned ESO systems, size and power usage are not strong constraints and thus we have not imposed them on our design. For removal of heat, the FIERA system will use liquid cooling to satisfy the VLT temperature requirement (outside of the electronics housing within 1.5 ^{o}C of ambient air temperature).

3. Hardware Overview

With the above requirements in hand, the FIERA design process was initiated in 1995 by the three authors of this paper. During 1996, significantly more resources were put to the task, and in October 1996 a prototype was tested demonstrating that the design philosophy and implementation approach were both valid.

An overview of the FIERA hardware design is shown in Fig. 1 and a summary of the system specifications are given in Table 1. There are five main components:

1. Embedded computer with auxiliary boards for sequence timing, data capture and network connection to other computers (monitor and keyboard are optional).
2. Detector head electronics with clock drivers, analog bias generation and video signal processing.
3. CCD detector head with pre-amplifier circuitry.
4. PULPO housekeeping box for temperature control, vacuum sensing and shutter control.
5. DC power supply module to generate the 5, ±15, 24 and 30 volts DC power needed by the detector head electronics and PULPO

Also shown in Fig. 1 is the LN_2 cryostat for cooling the CCDs, but that is not considered part of the FIERA system. Typically, all of these components will be mounted at the instrument and the connection to the telescope control room is via the network connection from the embedded computer. However, in some cases, the embedded computer will be placed either in the control room, or in a rack which is located on the Nasmyth platform or telescope floor.

Fundamental clock frequency	50 MHz
Number of readout channels	Up to 32
Maximum total pixel rate	50 million pixels/sec
Number of bits/pixel	Up to 21 (initially 16, set by A/D)
Maximum speed / port	5.5 million pixels/sec
Clock drivers	Fully programmable (25.6V swing, 0.1V resolution, 2 A current)
Number of control bits	virtually unlimited
Analog biases	fully programmable with hardware limits for CCD protection
Gain settings	4
Filtering settings	4 (for different readout rates)
Offset levels	64k levels - each channel indiv. programmable
Data link	Gigabit/sec from detector head to computer (up to 400m long)
Telemetry	All voltages remotely readable
Test signals	Synthetic video generation
Software in the detector head	None - all in C, C++ in the embedded computer
Modularity	Separate modules allowing upgrades and expandability

***Table 1**. System specifications for the FIERA CCD controller*

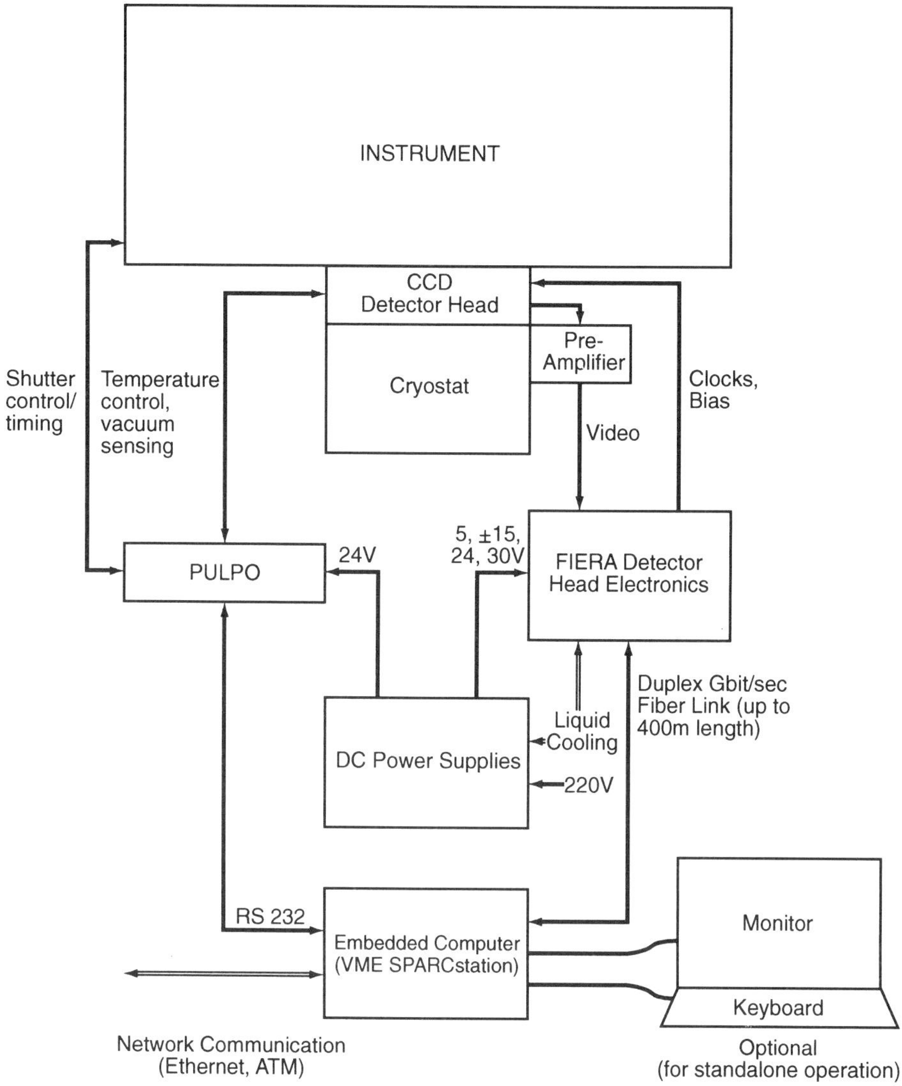

Figure. 1 Overview of the main components of the FIERA CCD controller

3.1 Embedded computer

Figure 2 shows the architecture of the embedded computer and the detector head electronics at the printed circuit board level. The CPU power of the embedded computer is provided by a SPARCstation which contains the control software, all written in C and C++. This embedded computer is a combination of commercially available components and a custom designed Detector Electronics Interface Board [3]. The interface board contains a Texas Instruments DSP, the TMS320C40, which provides the accurately timed generation of sequencer commands. The interface board includes a duplex gigabit fiber optic link based on the Finisar transceiver (same as used in the Messia system, see

[4]). We use a Hewlett-Packard HDMP module to serialize the digital data stream, providing 21 bit wide words at 50 million words per second.

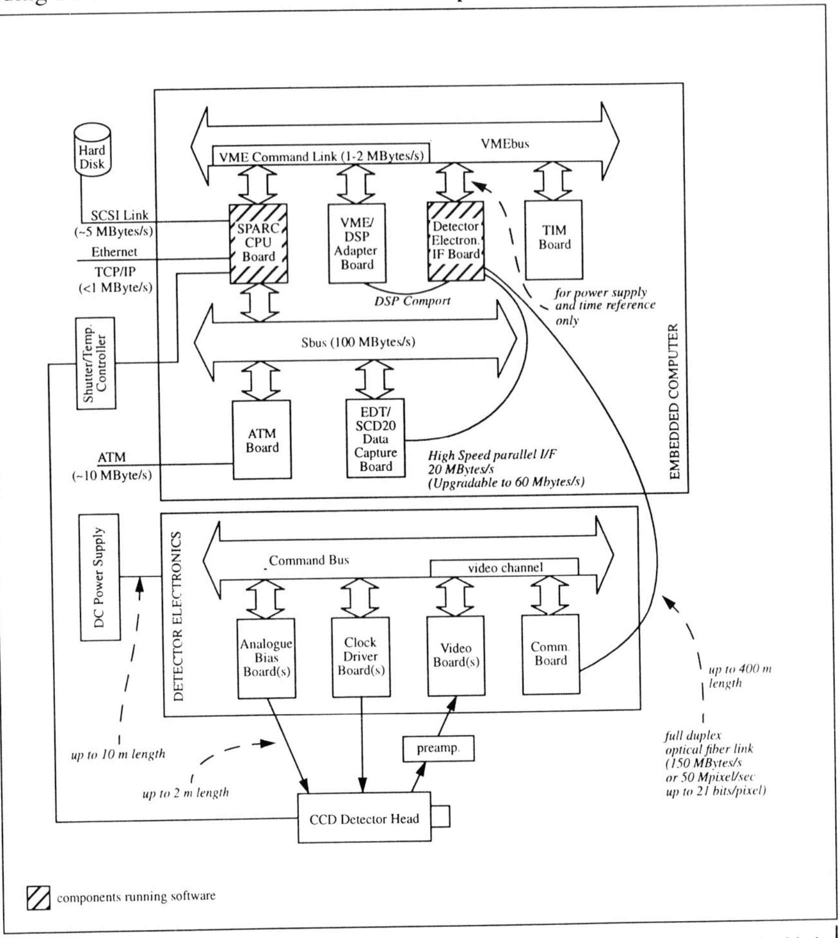

***Figure 2**: FIERA hardware architecture showing printed circuit boards of the embedded computer and the detector head electronics*

For the detector head control, two words are used to send commands "up" the data link to the detector head electronics (thus commands are sent at 40 nsec time boundaries, same as the global bus output of the C40). To transfer data from the detector head "down" to the embedded computer, there are 21 bits per pixel. At present, 16 bits are used for data and 5 bits for readout port number, but the system can be readily adapted to handle greater precision, up to 21 bits for data, if the video signal processing chain evolves to produce such data.

In general, the ultimate data rate limit of the FIERA system is established by the 50 million word limit of the fiber optic data link. However, it would be possible to pass up 75 million pixels per second, if the data was 14 bits/pixel and 3 pixels are packed into every pair of 21 bit words. This expansion is foreseen in the case of fast readout infrared detectors. At present, the data input into the SPARC is limited by the type of data capture board we are using, which provides a limit of 10 million words per second (16 bit data). This speed is adequate for all of ESO's needs through 1998, but there are commercially available data capture boards that can now receive 30 million (16 bit) words per second. FIERA also has the option of passing data directly from the interface board to an external computer; this link can run at the maximum 50 million words per second of the fiber optic link.

The TIM board shown in Fig. 2 is a Time Interface Module, a custom board developed by ESO to provide timing pulses with absolute accuracy of 10 microseconds, as derived from GPS satellite signals ([5], [6]). The embedded computer occupies 6 slots of a standard VME 6U chassis (boards are 233 mm high by 160 mm deep).

Within the VLT control system, the embedded computer of FIERA is called a SPARC-based local control unit (SLCU).

3.2 Detector Head Electronics

The detector head electronics is a combination of five custom developed printed circuit boards: backplane, communication board, analog bias board, clock driver board and video signal processing board. These boards are 6U high (233 mm) and 220 mm deep.

The 11 slot backplane uses VME standard P1 and P2 connectors but with ESO specific pin allocation. The 11 slot size was chosen as the maximum that could maintain the quality of 50 MHz command and data buses. This size is also adequate for all systems foreseen in the next few years at ESO, including those that utilize 16 readout ports (e.g. adaptive optics, wide field of view).

The communication board is the counterpart of the interface board which resides in the embedded computer, incorporating a Finisar transceiver and Hewlett-Packard HDMP modules. The communication board transfers commands from the C40 DSP sequencer to the command bus on the backplane and takes data from the backplane's data bus and sends it down the fiber link to the embedded computer.

The analog bias board produces 16 bias levels that are fully programmable over a range of -15 to +30 volts. There are hardwired protection switches that can be used to limit the output voltages of biases which could damage the CCD (e.g. output drain, output source and dump drain).

The clock driver board produces 14 clocks that can run at rates to the 50 MHz. The clock driver board is actually a mini-motherboard with SIMM modules inserted to produce the clock voltages. We have developed two types of SIMMs thus far for FIERA:

- A fully programmable bi-level SIMM that can have the two output voltage levels remotely set to an accuracy of 0.1 volt over the range -12.8 volts to +12.8 volts.
- A fully programmable multi-level SIMM that can output any voltage level between -12.8 volts and +12.8 volts, to an accuracy of 0.1 volt (256 levels), with an update rate of 50 MHz.

Both of these clock driver designs can drive large currents (2A); either for running CCDs very fast or for driving a large number of CCDs with a single clock driver board, as we plan to do with our eight CCD 8k × 8k mosaic.

To our knowledge, the fully programmable and very fast multi-level SIMM is a new advance in clock drivers. We look forward to using it to tailor clock edges to achieve any rise and fall curve desired. Our plan is to use this capability to maximize CTE, minimize spurious charge generation and possibly implement anti-blooming clocking in the best way possible.

Both the analog bias and clock driver boards have full telemetry built into the design; all voltages can be measured. There are relays between the boards and the CCD so that all voltages can be verified before sending power to the CCD.

The video signal processing board includes amplifiers, correlated double sampling (CDS) and analog-to-digital conversion for four channels at rates to 2000 kps (2 million pixels per channel per sec) with 16 bits per pixel. The limit at present is set solely by the A/D converter. Initially, we do not foresee any advantage for speeds faster than 1000 kps and are buying the less expensive 1 MHz A/D for most of ESO's systems. The systems that will use Tektronix or Loral CCDs will be outfitted 500 kHz A/D converters.

For the first implementations of FIERA, we wanted a system that could run at rates to 1000 kps. Thus, we have implemented the clamp-and-sample approach to CDS, rather than the lower noise dual-slope approach. We will try to overcome the $\sim\sqrt{2}$ disadvantage of clamp-and-sample by oversampling and digitally processing the signal. For instance, at 50 kps, we could take 20 samples per pixel period. This idea was discussed at the workshop, and it found support from the audience, but at the publication date of these proceedings (August 1997), we have not yet found time to test this approach. If we are not able to achieve the lowest noise at slow speeds with this video board, as is required for long exposure spectroscopy, we will design a new video board that implements dual-slope.

For CDS, there are a choice of four software selectable RC filter time constants placed just prior to the clamp circuitry. There is a synthetic video signal generator on the board, using the design of the multi-level clock driver, so that virtually any video signal can be generated. It will be tempting on cloudy nights to use this capability to generate synthetic images of galaxies!

The video board design and layout appears to be of sufficient quality. The rms noise of the board itself is 1.2 digital numbers and crosstalk is much less than 1 part in 100,000 when a highly saturated signal is put into one of the channels.

The FIERA detector head electronics is housed in an airtight box with fans and a heat exchanger with cooling liquid provided by an external source.

3.3 CCD Detector Head

All of ESO's "dewars" are now made from a design that has two parts, a detector head and a cryostat that provides the cooling function. There are three different cryostat designs in use at ESO, two "bath" cryostats designs and a continuous flow cryostat that can provide cooling to a detector head for one month via a LN_2 line from a 160 liter LN_2 tank. The same detector head can be used with any of the cryostats.

We are in the process of designing and testing the flex-rigid printed circuit boards for all of the CCDs that will be used in ESO detector heads. When this task is completed, we should be able to rapidly deploy a large number of CCD systems that are nearly identical.

Two different approaches are being used for the cryostat wiring. For Tektronix, SITe and Loral CCDs, a printed circuit board that is inside of the cryostat will be used to differentially amplify the video signal for transmission to the FIERA detector head electronics. The board inside the cryostat also provides passive filtering to the clock and bias lines. For EEV and MIT/LL CCDs, we have developed a new set of printed circuit boards with the pre-amplifier circuitry placed in a small box outside of the cryostat, as shown in Fig. 1. There is passive filtering on clock and bias lines placed on the printed circuit boards inside the cryostat.

For the pre-amplifier inside the cryostat, there is a single hardwired gain. In the pre-amplifier box placed outside of the cryostat, there is a choice of two gains, software selectable from the embedded computer. Our approach is to maximize gain as much as possible at the front end, since in some instruments we must allow cable length of 1 to 2 meters between the pre-amplifier and the video signal processing board.

3.4 PULPO Housekeeping Box

The PULPO module is used to measure and control temperature (via heaters) and for measuring the vacuum of the cryostat. It also can run shutters and transmit and receive optically isolated signals from shutter drivers. PULPO communicates to FIERA via the RS-232 port on the SPARC. PULPO is described in detail in another paper ([7]).

3.5 DC power supply module

As for any CCD controller, FIERA requires a number of DC voltages. Since size and weight are not significant constraints for ESO's optical detector systems, we have decided to take a conservative approach for DC power generation. We are using a switching power supply to generate the 24 volts needed by PULPO and linear power supplies to generate the 5, ±15, 24 and 30 volts needed by the FIERA detector head electronics. As time permits, or as is required by specified instruments, we will explore the use of switching power supplies.

The DC power supply module is housed in an airtight box with fans and a heat exchanger as done for the detector head electronics.

4. MICROSEQUENCER

A significant feature of the FIERA design is the implementation of a microsequencer that provides local generation of control bits. This design minimizes the number of high speed signals that are distributed across the detector head electronics backplane and it provides a virtually unlimited number of high speed control bits. The microsequencer is the key to realization of the high speed multi-level clock drivers and the 256 level synthetic video signal generation.

The main sequencer in FIERA is the Texas Instruments C40 DSP that is on the interface board within the embedded computer. This C40 puts out relatively high level commands to the detector head, such as "perform a parallel line shift" or "read out 1000 pixels".

These commands are distributed simultaneously to all of the microsequencers that are located in the detector head. The communication, clock driver and video boards all contain microsequencers (no microsequencer is needed on the analog bias board). After receiving a command, the microsequencers output the bit patterns that correspond to the operation requested. For instance reading out a pixel involves proper timing of the following signals:

- clock driver board - *serial phases 1-3, reset gate, summing well*
- video board - *clamp, sample, output data to data bus*
- communications board - *read in data, put data onto fiber link*

The microsequencer is simply a counter circuit combined with a SRAM (static RAM) memory chip. The counter cycles through address locations and clocks out data from the SRAM. The counter function of the microsequencer is performed by a programmable logic array, and it only needs 3 values to run a pattern (start address of pattern, length of the pattern and number of times to repeat a pattern).

For the microsequencer design to work properly, all microsequencers must work in lock step at all times, since there is no communication between microsequencers. Care must be taken with distribution of the commands and the microsequencer clock, but we have not had significant difficulty in getting this approach to work. The C40 DSP software keeps track of the length of the patterns "in the queue" and puts out the commands for successive patterns at exactly the correct clock transition. We have tested this operation thoroughly to be certain that successive clock patterns have seamless transitions. One attribute of the microsequencer that makes this task easier is that it has a 2 command buffer at its input.

In order to save memory locations in the SRAM, the clock frequency can be divided to utilize the minimal number of memory locations for storage of each pattern. The microsequencer clock division is performed by loading a register on the communication board. Of course, the microsequencers must finish any previously commanded patterns prior to changing the clock speed.

5. Control Software

The software architecture of FIERA is discussed at length in paper [8]. In short, nearly all of the control software of FIERA resides in the SPARC and is written in C and C++. There are a set of routines programmed for the C40 DSP which require DSP knowledge, but these are also written in C, not assembly language code. There will be a need for a few specialized C40 functions that are coded in assembler for the situations that have very tight timing requirements (e.g. adaptive optics). When operating in the VLT environment, there will be no monitor or keyboard attached to the FIERA SPARCstation and instead all commands will be sent to FIERA via the instrument network. In this implementation, the FIERA software extends to a module that resides in the instrument computer and the interface between FIERA and the instrument is located inside the instrument workstation.

The protocol between FIERA and instruments is that FIERA receives commands or queries from the instrument software and delivers replies or images (in FITS format) to the instrument workstation software. The FIERA software uses the real-time display (RTD) module developed for the VLT; this software enables virtually instantaneous display and analysis of the image data.

When FIERA is operated in standalone mode, the interface to instruments essentially migrates down to the FIERA SPARCstation and we operate FIERA as a standalone "instrument". All of the software developed for FIERA is thus used for either standalone mode or in conjunction with a VLT instrument. Since the FIERA software has a well defined interface, non-VLT software can access FIERA through a network interface (i.e. sockets). This will allow us to put FIERA into nearly any telescope environment, important for using FIERA at the telescopes of ESO's existing La Silla Observatory.

An important software module that has been developed for FIERA is the Waveform Editing Software (WES), which makes it relatively easy to graphically program the patterns of the microsequencers. In addition, WES is used to generate the sequences that are used by the C40 DSP code to schedule the transmission of microsequencer commands. The split between microsequencers and the main sequencer (C40) is transparent to the user, and it is easy to rapidly develop clock patterns.

6. Development, Production and Deployment Schedule

At the time of the ESO workshop, the first prototype of FIERA had just become operational. The results demonstrated that the architecture of FIERA was sound and that the microsequencer approach to system control would work as designed. However, as with any first prototype there were a number of errors made in layout and some basic improvements in the design were judged to be worth implementing. These changes were made to the printed circuit boards and five copies of a second prototype were integrated during May-August, 1997. These prototypes are functional and adequate for deployment on instruments, but a few bugs remain and thus a final set of relatively minor changes are being made to the design before producing 15 more systems. These changes, which will be carried out in the second half of 1997, will mostly provide for higher system reliability.

As of August, 1997 FIERA has been integrated with CCDs from several manufacturers, including: a Tektronix 2k × 2k, a Loral 2k × 2k, the EEV 2k × 4k, and the MIT/LL 420 × 420 frame transfer and 2k × 4k devices.

While we had achieved some respectable noise results on the EEV device (6 e^- rms noise at 300 kps), there was some lingering concern at to whether FIERA would have some fundamental problem with achieving lower noise. That question was resolved when FIERA was used in June, 1997 to achieve 2.3 e^- rms noise at 100 kps with the MIT/LL 2k × 4k device. Note that this result is with clamp-and-sample CDS.

In parallel with hardware development and production, the FIERA software is proceeding as scheduled. An engineering set of control software has been working well for months and the first release of software that communicates properly with VLT Instrumentation software is due for release in September, 1997. The software with all telemetry and data logging functions as required by the VLT environment will be released in January, 1998 - just in time for the 4k × 4k imager installation at the NTT 3.5-m telescope in February, 1998.

FIERA has already been integrated into the systems for two VLT instruments (FORS 1, VLTTC) and by the end of 1997 it will integrated into two more instruments. In addition, as part of ESO's procurement of 16 port adaptive optics devices from EEV, a 16 port system is being loaned to EEV for device testing.

Acknowledgements: Although the FIERA design was initiated by the three authors of this paper, its realization is the product of the hard work by all members of ESO's

Optical Detector Team, including Andrea Balestra, Claudio Cumani, Sebastian Deiries, Robert Donaldson, Reinhold Dorn, Christoph Geimer, Ali Mohammadzadeh, Holger Pieper, Gustavo Rahmer and Javier Reyes in Garching with critique and help from the La Silla ODT members - Jaime Anguita, Nicolas Haddad, Andres Rivera and Peter Sinclaire. The La Silla team developed the PULPO housekeeping module. The first prototype design and fabrication of the video board and the clock driver boards were done by SciMeasure Analytical Systems, led by Charlie Bleau. Design of the backplane was done by Helga Kasten. Gerhard Fischer has been invaluable in his assistance with prototype fabrication and construction, as has Eddy Pomaroli for finding electronics parts and specifying fabrication techniques. Fernando Pedichini of Osservatorio Astronomico di Roma helped design the analog bias board and visited Garching several times to work with our team. Johann Zipperer of Texas Instruments helped design the C40-based computer interface board. A "tiger team" review, which provided valuable midcourse critique in December 1995, was staffed by Preben Nørregaard and Jens Klougart of the U. Copenhagen.

7. References

[1] Beletic, J. (1997), "ESO's plan for optical detector systems", these proceedings, pp. 57-64.

[2] Yamashita, H., Y. Matsunaga, M. Iesaka, S. Manabe and N. Harada (1988), "A new high sensitivity photo-transistor for area image sensors", IEDM 88, IEEE, pp. 78-81.

[3] Beletic, J., R. Gerdes and J. Zipperer, 1996, "A DSP-based solution to remote control of detector head electronics for scientific CCD systems", DSP-96, Munich.

[4] Sekiguchi, X. (1997), "Messia CCD controller and data acquisition system for the Subaru telescope", these proceedings, pp. 157-164.

[5] M. Ravensbergen, K. Wirenstrand, 1995, "Time Reference System of the VLT Telescope Control Systems", SPIE vol. 2479.

[6] W.R. Lange, M. Ravensberger, 1995, "Time Reference System of the ESO Very Large Telescope", Ninth European Frequency and Time Forum, Besançon.

[7] Haddad, N. and P. Sinclaire (1997), "PULPO: Temperature, vacuum, shutter, LN2 level, all in one box", these proceedings, pp. 131-134 .

[8] Cumani, C. and R. Donaldson (1997), "The architecture for two generations of ESO VLT CCD controllers", these proceedings, pp. 115-122.

THE ARCHITECTURE FOR TWO GENERATIONS OF ESO VLT CCD CONTROLLERS

CLAUDIO CUMANI[1] *AND ROBERT DONALDSON*[1]

ABSTRACT

An explanation of the general VLT control system architecture is given. Two generations of ESO VLT CCD Controller systems (ACE and FIERA) are described, and a comparison between them is presented, considering both the architecture and the performances.

1. THE ESO VLT CONTROL SYSTEM.

The ESO Very Large Telescope (VLT) Control System is distributed on different platforms:

- high level operations and interface with the users are performed on Workstations (HP or Sun), running the Unix operating system,
- sub-systems hardware and time-critical operations are controlled from VMEbus based Local Control Units (LCUs), running the VxWorks real-time operating system,
- embedded software can run, in particular cases, in the low-level hardware.

All the computers are connected through Local Area Networks (LANs), and only standard components are used.

This applies also to all the Telescope Instruments, including those using CCD controllers. As an example, Fig. 1 presents an hypothesis for the Control System architecture of the VLT high resolution spectrograph called UVES (UltraViolet Echelle Spectrograph).

2. THE ACE-LCU CCD CONTROL SYSTEM.

Following the general architecture of the VLT Control System, the first generation VLT CCD System is based on three platforms: workstation, LCU and the Array Control Electronics (ACE), which is responsible for clocking the CCD and processing and digitizing the video signal. An overview of the system is shown in Fig. 2.

The CCD Controller LCU consists of:

- a MVME167 CPU Board, with a Motorola 68040 CPU,
- a MVME712/M LAN Interface Board,
- an IMSB016 Interface Board, for the communications with the ACE,
- an ESO B16LAB Link Adapter Board,
- an ESO Time Interface Module (TIM), optional.

---

[1] European Southern Observatory, Garching, München e-mail: **ccumani@eso.org**, url: **http://www.eso.org/~ccumani/**

J. W. Beletic and P. Amico (eds.), Optical Detectors for Astronomy, 115–122.

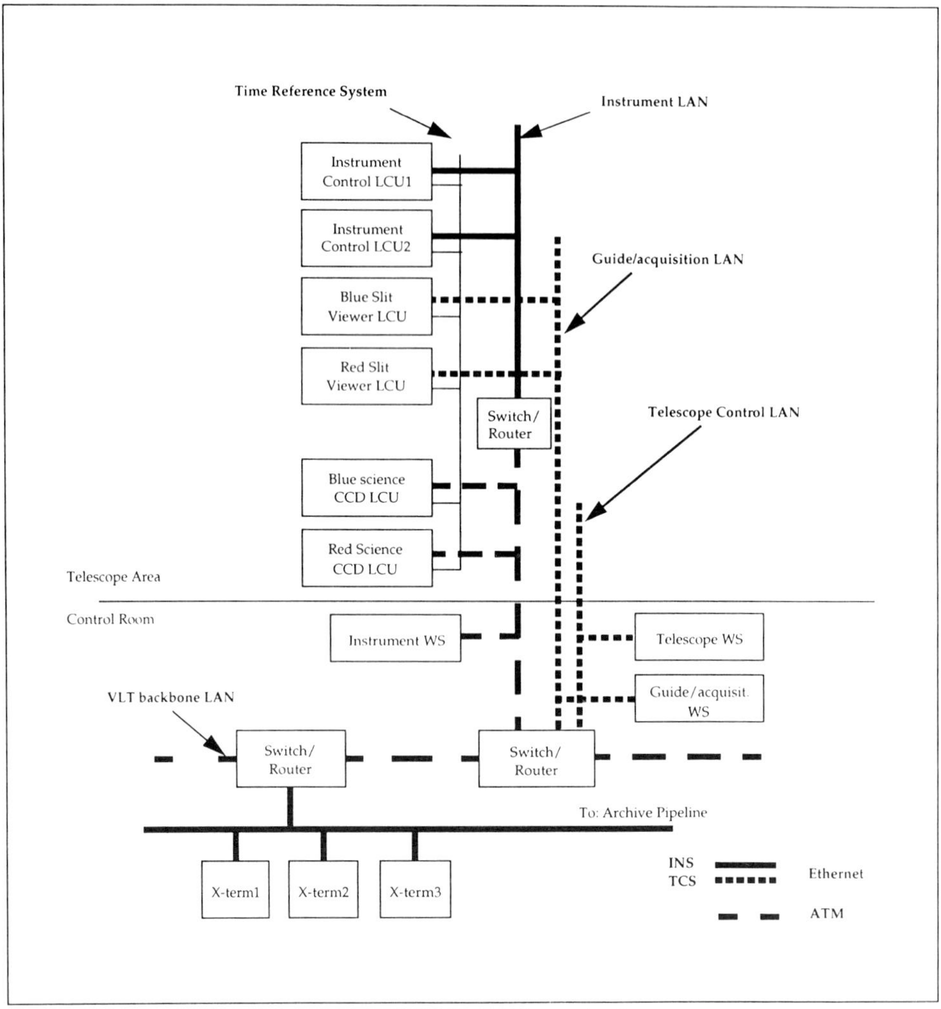

Figure. 1 - The UVES Control System Architecture

The CCD Controller ACE consists of a network of four transputers (one T805, two T225, one T222) and a DSP (Motorola 56001) distributed on four printed circuit boards (Sequencer, Video, Clock/Bias and Shutter/Temperature Boards) and responsible for the synchronization of the readout operations, the data reorganization and transmission.

The CCD Controller Software has been written in C (workstation, LCU and ACE DSP) and occam (ACE transputers), and Tcl/Tk scripts are used within the Graphic User Interface (GUI) panels on the workstation. The VxWorks version of the driver for the IMS B016 Board has been developed at ESO.

The CCD Controller Software has the following characteristics :

- it supports both scientific and technical CCD cameras for instrumentation and telescope sub-systems;

- it is by large extent (about 80% of the code) independent from the hardware used to control the CCD camera;
- it provides a programmatic interface to higher level software, such that it can work as a server for higher level software (e.g., Instrumentation Observation software or Telescope Autoguiding software);
- it provides a standalone mode, with a GUI, whereby the CCD camera can be considered as a simple instrument (see Fig. 2).

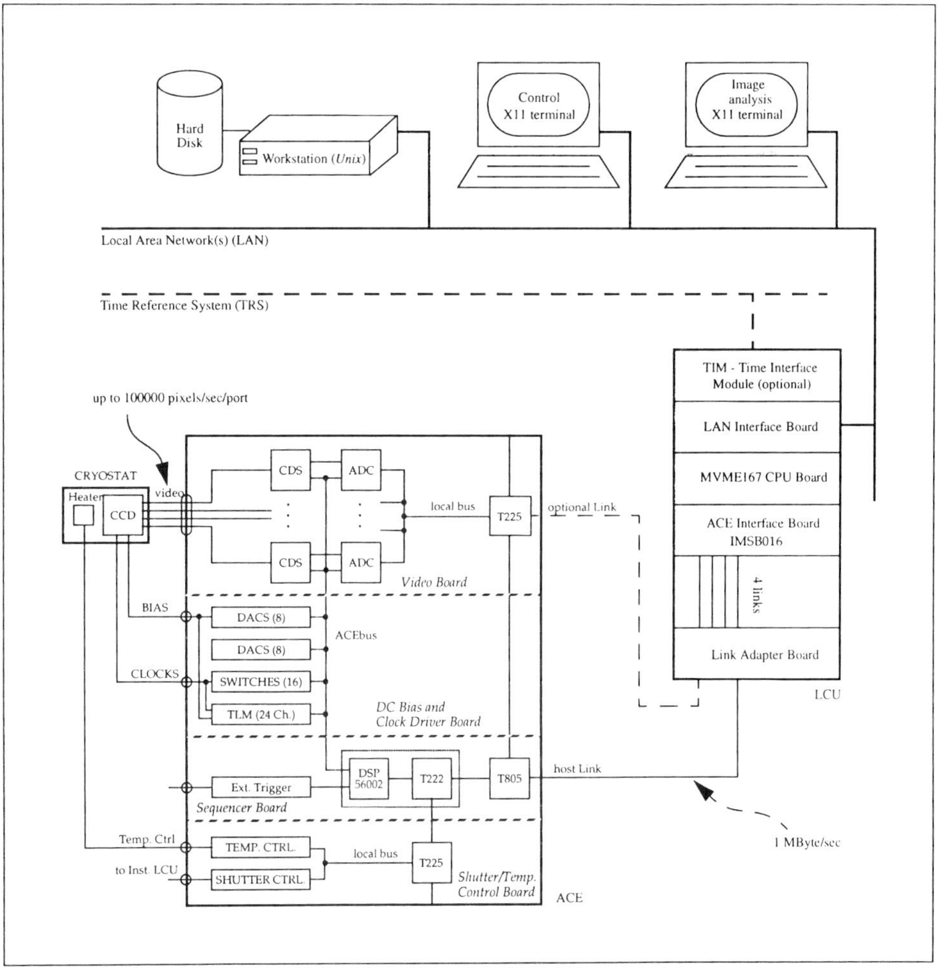

Figure. 2: The ACE-LCU CCD Controller (Standalone System)

The CCD Controller Software has the following functionality :

- single exposure execution (possibly repeated n times), with the possibility to define the readout parameters: binning, windowing, readout speed, gain, outputs used;

- images are provided both as FITS files and raw data, for quick look through the VLT Real-Time Display facility;
- a real-time image processing facility on LCU fulfills the requirements for technical CCDs (centroiding calculation, bias and flat-field correction, hook to user functions, e.g., for object recognition special algorithms);
- a telemetry information system is provided;
- a temperature control system is foreseen.

A first prototype of this system was successfully tested in January 17-20, 1995, on the New technology Telescope (NTT) in La Silla (Chile).

In November, 1996, three operational ACE/LCU systems were installed at the NTT, at the ESO Multi-Mode Instrument (EMMI) and the SUperb-Seeing Imager (SUSI) instruments.

The following table shows the readout performances measured on the EMMI-Blue Arm system (Tektronix 1124x1024 pixel full frame CCD), where the times shown indicate the seconds between the starting of the readout of the CCD and the end of the storage of the image on the workstation hard disk (in FITS format).

output(s) used	speed (*)	time (sec)
one	fast	27.61
two	fast	17.37
one	normal	32.24
two	normal	19.64
one	slow	41.43
two	slow	24.27

** fast = 2000 nsec for single slope of dual-slope correlated double sampling(~73 kilopixel/sec/port);*
normal = 4000 nsec for single slope of dual-slope correlated double sampling (~56 kilopixel/sec/port);
slow = 8000 nsec for single slope of dual-slope correlated double sampling (~39 kilopixel/sec/port)

3. THE FIERA CCD CONTROL SYSTEM.

In the recent years readout amplifiers for CCD devices have improved so that it is now possible to attain low readout noise even with fast readout speed (less than 2 electrons noise at 100 kilopixel/sec/port and 4-6 electrons noise at 1 megapixel/sec/port). Today 1 MHz readout rate is adequate for acquisition, focusing, flat-fielding, calibration, direct sky imaging, while the main need for very low noise is for high resolution spectroscopy.

Today it is possible to build electronics that can run any CCD detector or mosaic of detectors that we can envision for the next 10 years.

ESO must produce many CCD systems in the next few years in order to provide for the VLT instruments and to upgrade the telescopes in La Silla (Chile). Since the ACE-LCU system can not satisfy the readout requirements of the new generation CCD devices (primarily due to the ACE-LCU communication link, see Fig. 2), the most sensible approach for the ODT has been to produce a new single "universal" controller that can optimally operate all the scientific CCD chips or mosaics that ESO acquires for the next 10 years. Our philosophy is that the limitation of the entire detector system should be due solely to the CCD and the imagination of the user. This new advanced controller has been named FIERA (Fast Imager Electronic Readout Assembly).

The Architecture of FIERA differs slightly from the standard VLT Architecture. To achieve the readout requirements of the new generation CCD devices the LCU was replaced with a SPARC CPU, connected to the CCD Electronics through a 150 MBytes/sec full duplex optical fiber link. To avoid a major impact on the VLT environment, the FIERA system software will keep - as far as possible - the same interface to the VLT environment as the old CCD Controller. Therefore, FIERA should be able to replace the old controller in a clean and transparent way, communicating with the Instrument Software running on the workstation in the same way (see Fig. 3).

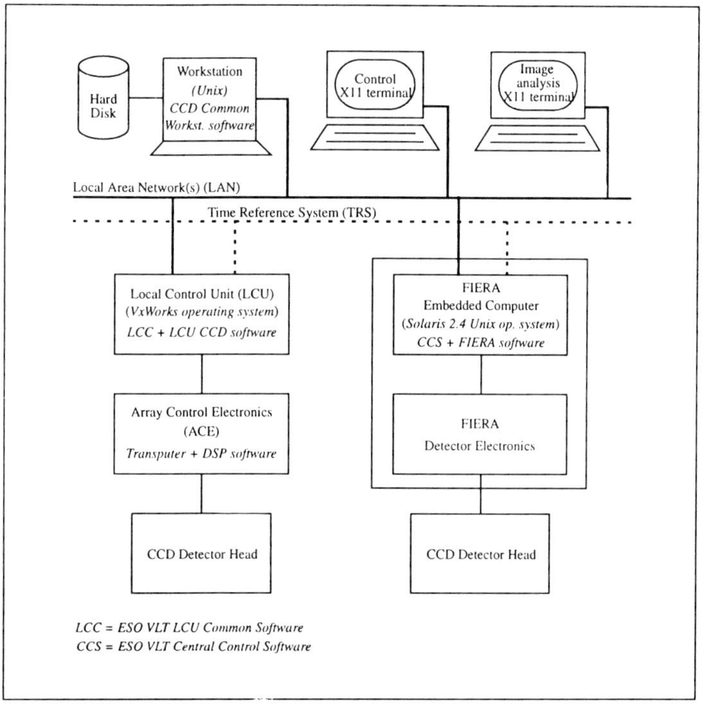

Figure 3: *A comparison of the ACE and the FIERA systems in the VLT Architecture*

The FIERA CCD Controller consists of two modules (see Fig. 4):

1. The Detector Electronics, consisting of
   - a Communication Board, for the communications with the rest of the system,
   - one or two Analogue Bias Boards,
   - up to four Clock Driver Boards (14 clock lines/board),
   - up to four Video Boards (4 channels/board)
2. The Embedded Computer, consisting of
   - a SPARC CPU Board

- a Detector Electronics Interface Board (with two DSP TMS320C40), for the communication with the Detector Electronics and real time operations
- a VME/DSP Adapter Board provides the interface between the SPARC CPU Board and the Detector Electronics Interface Board
- an EDT/SCD20 Data Capture Board, for fast data acquisition
- an ATM Board
- an ESO TIM Module (optional), for time reference.

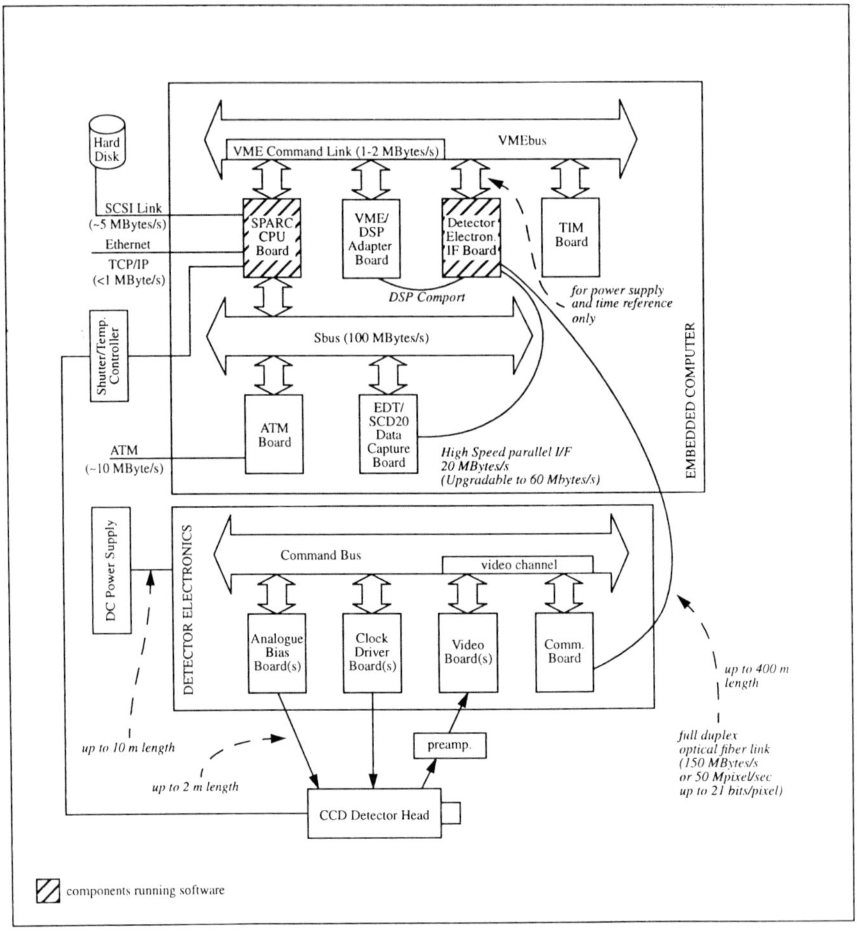

***Figure 4:** FIERA Hardware Architecture*

The Detector Electronics and the Embedded Computer are connected through a 150 MBytes/sec full duplex optical fiber link (due to the transmission protocol, the corresponding maximum video data rate is 50 Megapixels/sec, for pixels up to 21 bits wide).

All the software runs on the SPARC CPU (Solaris Operating System, C and C++ programming languages) and on the DSPs of the Detector Electronics Interface Board

(Parallel C). Analogue biases and high-current clock drivers are fully programmable, with hardware limits on potentially damaging bias voltages.

An *Oktoberfest '96* prototype of this system has been successfully assembled in October 1996, in the Optical Detector Team laboratories at ESO, Garching.

The following table shows the readout performances measured on a MIT/LL 420x420 pixel frame transfer CCD chip:

output used	1
readout time	0.20 seconds
readout speed	880 kilopixels/sec/port
system noise	1.3 electrons @ 1 megapixel/sec
crosstalk	Negligible

The system is still under test and redesign stage, but the general concept and architecture have been proven.

The following goals are foreseen:

- for the "1997 version" :
    - 2 megapixels/sec/port (limit set by 16 bits ADC)
    - 10 megapixels/sec total (limit set by Data Capture Board)
- and eventually :
    - 5 megapixels/sec/port
    - 50 megapixels/sec total

Acknowledgments: the authors would like to thank Antonio Longinotti and Philippe Duhoux, of the ESO VLT Software Group, who helped in the preparation of this article.

4. REFERENCES

[1] Beletic J., "The Plan for Optical Detectors at ESO", The Messenger, n. 83, March 1996

[2] Longinotti A., Cumani C., Duhoux P., "The VLT CCD Detectors Control Software", The Messenger, n. 82, December 1995.

[3] Reiss R., "ACE, ESO's next generation of CCD Controllers for the VLT", in "Instrumentation in Astronomy VIII, 13-14 March 1994, Kona, Hawaii", SPIE Proceedings, vol. 2198

The following documents are available under anonymous ftp on te1.hq.eso.org (directory /vlt/pub/doc) as compressed postscript files:

[4] VLT-SPE-ESO-17240-0227, 1.0, 08/04/93, CCD SW Functional Specification (file ccd_sfs1.0.ps.Z)

[5] VLT-SPE-ESO-17240-0601, 1.0, 28/09/95, CCD SW Design (file ccd_sdd1.0.ps.Z)

[6] VLT-SPE-ESO-13600-0568, 1.0, 08/12/95, ACE SW Design (file ccd_ace_sdd1.0.ps.Z)

[7] VLT-MAN-ESO-17240-0672, 1.4, 05/05/97, CCD SW User Manual (file ccd_sum1.4.ps.Z)

[8] VLT-MAN-ESO-17240-0917, 1.4, 05/05/97, CCD SW User Manual for Standalone mode (file ccd_um1.4.ps.Z)

[9] VLT-SPE-ESO-13640-1266, 1.0, 16/04/97, FIERA CCD Controller, Software Functional Specifications (file fs1.0.ps.gz)

Other information is available on the Optical Detector Team web pages (http://www.eso.org/odt/)

CRYMAC- CRYOSTAT MAINTENANCE CONTROLLER

REINHOLD J. DORN[1]

ABSTRACT

ESO has developed two types of cryostats for use with optical detector CCD systems: a bath cryostat, which has a tank of liquid nitrogen and a continuous flow cryostat that is fed from a 120 liter liquid nitrogen tank. Both types of cryostats have been designed to incorporate the same detector head, which is removable from the cryostat body. These cryostats can be split into two main parts:

(1)The detector head which houses the CCD itself and electronics such as preamplifiers, clock/bias filters and the temperature sensors and heating elements.

(2) The cryostat tank, which holds the liquid nitrogen for cooling the CCD to it's operating temperature. The cryostat tank is supplied with two sorption pumps, which are containers filled with zeolith . These sorption pumps absorb the gas produced by outgassing of the various warm surfaces and the electronic parts inside the dewar. The capacity of the sorption pumps is not unlimited and requires a regular regeneration to provide a clean vacuum system.

Therefore a vacuum & temperature control box was developed to provide the temperature sensing and control of the CCD head as well as providing enough electrical power to heat the sorption pumps for regeneration during maintenance and testing in the laboratory. This vacuum and temperature control box, which is designed for laboratory use, is named CRYMAC (CRYostat Maintenance Controller).

1. SYSTEM DESCRIPTION

Figure 1 shows a simplified block diagram of the system to be used with CRYMAC. In the detector head the electronics are split into a heater/temperature section for the CCD and a temperature section for the cryo electronics. The temperature is sensed via two PT 100 sensors, which are placed close to the CCD socket and up to seven Pt 100 sensors, which can be placed onto the cryo electronic boards and an auxiliary section for optional use. The temperature control is done by two 75 Ohm heating resistors, which can be used in parallel or serial mode.

In the cryostat body the temperature reading and heating system in maintained via three sets of PT 100 sensors and 3,5 Ohm heating wires. One set for the cold plate of the liquid nitrogen tank for rapid heating of the system and one for each sorption pump. The vacuum measurement is done by a single full range Blazers Gauge (PKR 250)

[1] Optical Detector Team, European Southern Observatory, Garching, Germany. Email: **rdorn@eso.org**

J. W. Beletic and P. Amico (eds.), Optical Detectors for Astronomy, 123–125.

2. CRYMAC FUNCTION AND SPECIFICATIONS

Although CRYMAC is a compact 19"/3HE single box, it is functionally split into two main units. A vacuum control unit and a temperature sensing and process unit.

The vacuum control unit is supplied with a Balzers Dual Gauge [1] and an RS232 serial interface. The temperature and process control unit is supplied with five Omega Microprocessor-Based Temperature/Process Controller CN76000 [2]. One for each sorption pump, one for the cold plate, one for the CCD section and one for sensing the cryo electronics temperature. These controllers read the temperature inside the cryostat and supply power, switched by a relay to the heating resistors, to maintain a programmed temperature. This on-off control is adjustable up to a limit ± 1°C in a range from -200°C to 200°C and can be programmed to the desired temperature.

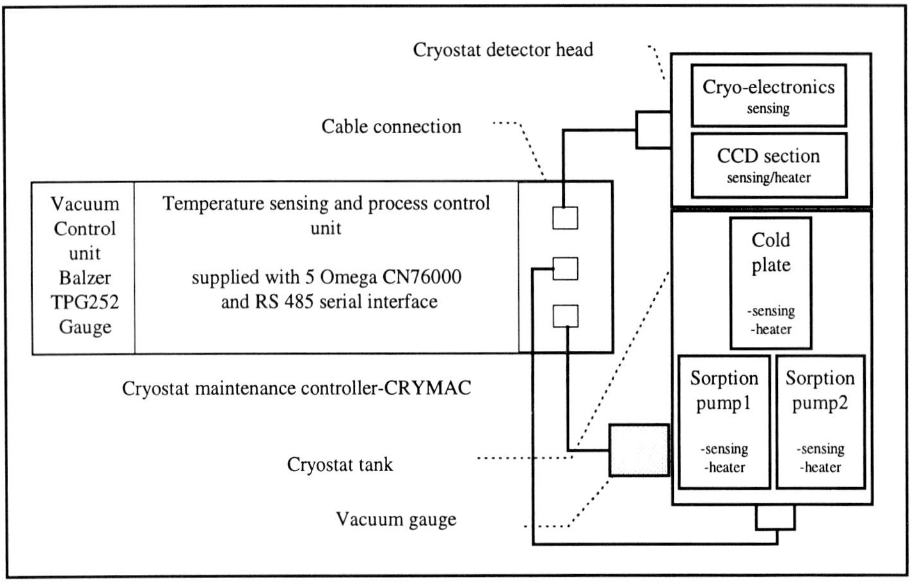

Figure 1: simplified Block diagram of CRYMAC for the vacuum/temperature control system

The sorption pumps are heated by a 5V/10A power supply and the cold plate and the CCD head with a 24V/10A power supply. With this power it is ensured that the sorption pumps can be baked to 450K for a longer period and a rapid warming up of the cryostat is possible. This is important if any technical problems require opening the dewar.

The PT100 sensors are connected using four wires to ensure an accurate measurement. The controller for the cryo electronics can accommodate up to seven PT100 sensors in the detector head that are wired to a rotary switch. CRYMAC also provides a connection for an additional sensor as well as a CN76000 controlled 5 and 24V output for optional use.

All temperature controllers in CRYMAC include a self-diagnosis with fault indication and a non-volatile memory to retain all process parameters when power is off, without a battery backup. Optional alarm function is provided to give an acoustic signal if the temperature exceeds present limits or if the sensors go-off line or are short circuited. The sensor break system deenergizes the control output to protect the CCD and cryostat.

Supplied with an RS-485 Serial communication line each instrument can be assigned with different addresses and baud rates. In this way a Computer is able to communicate with CRYMAC via data acquisition program or terminal software. Thus it is possible to modify the parameters of the instrument and to record all data during the maintenance and testing.

CRYMAC supports all requirements for an effective maintenance for the cryostats as a single compact box. It can be used during testing and integration in the laboratory as well as at the telescope.

Additional information on the Omega Temperature controllers and the Balzers Dual Gauge can be obtained either via the World Wide Web at http:/www.industry.net/omega, or direct from the companies.

3. REFERENCES

[1] U.S.A, OMEGA Engineering, Inc., Worldwide Headquarters, One Omega Drive, P.O. Box 4047, Stamford, CT 06907-0047

[2] Balzers Aktiengesellschaft, FL-9496 Balzers, Fuerstentum, Liechtenstein

SOME EXPERIMENTS WITH SIMPLE PREAMPLIFIER DESIGNS

JOHN C. GEARY[1]

ABSTRACT

A few likely candidates between commercially available amplifiers are investigated, in order to determine their suitability for use with present low noise CCD imagers. Tests and results are described and discussed.

1 INTRODUCTION

Anyone doing a search for recent literature on CCD preamplifiers will encounter a sparse field. Indeed, very little new information on this subject, and on the wider issue of optimal CCD signal processing, has been published in over ten years. Therefore, just what are other CCD systems designers doing for this critical first stage? There is a strong impression, reinforced by anecdotal information, that many systems builders have been relying on copies of electronic designs from long past. With the prospect of both very low noise CCDs at higher bandwidth and arrays of many such devices in a single system upon us, there is considerable motivation to start evaluating our signal processing systems in more detail. The following work is confined to an investigation of very simple preamplifier designs utilizing only commercially available IC operational amplifiers as active elements, evaluating their suitability for low-noise with the latest generation of CCD imagers. Our goal was to see what limiting performance can be obtained with the simplest structures, allowing very small, compact preamps to be developed for employment with massive arrays of CCDs.

2. AMPLIFIERS

A thorough reading of the specification sheets for the universe of commercially available opamps will reveal only a few likely candidates. There are a number of stringent requirements for such an amplifier, including input voltage noise, input current noise, slew rate, and settling time that will weed out most. The first two requirements essentially rule out anything except JFET-input opamps, while the latter two then serve to reduce the choices to a relatively few, but fortunately not zero! Two potential winners are the AD745 and the OPA627 (and perhaps also its faster sibling, the OPA637, for use above 100 kHz). Some of the salient parameters of these two opamps are summarized below:

[1] Smithsonian Astrophysical Observatory, 60 Garden St., Cambridge, MA 02138.
Email: **geary@cfa.harvard.edu**

J. W. Beletic and P. Amico (eds.), Optical Detectors for Astronomy, 127–130.

	AD745	OPA627
noise voltage (nV/√HZ)		
10 Hz	5.5	15
100 HZ	3.6	8
1kHz	3.2	5.2
10kHz	2.9	4.5
current noise@1kHZ (fA/√HZ)	6.9	1.6
slew rate (V/μs)	12.5	55
settling time to .01% (us)	5	0.45

One would be mistaken in believing that the above list can definitively pick a winner for all applications, however. Some subtleties in both the nature of the specifications and their relative importance to a particular application must first be understood. For example, the very rapid settling time of the 627 is only obtained for modest negative gains, while we will want to use positive gains in order to preserve high input impedance and not load the CCD output too much. Under these conditions, the 627 behaves similarly to the slower 745. For instance, the measured settling times to 0.1% (the best I could measure with simple equipment) for the 627 and 745 were about 700 and 800 nsec respectively for a 5 V step at a gain of +21 and with a critically-damped feedback time constant of about 100 nsec.

Similarly, one might worry that the higher current noise of the 745 might prove to be a problem, but this is not the case. Both of these current noise values are so low as to produce negligible input noise when coupled to a modern CCD. Referring to Fig. 1A, the AC input impedance is dominated by the source impedance of the CCD output FET, which is relatively low. For example, typical measured values for SITe imagers of recent manufacture are about 3 kΩ, resulting in an equivalent voltage noise much less than that of the opamp itself.

3. Test Configurations And Results

Two observing configurations of opamps were tested, shown in Figures 1A and 1B. The reason for trying the dual opamp version in Fig. 1B was to try to reduce the effect of input voltage noise by paralleling inputs. Such a configuration should reduce effective noise by √N. However, as implemented in this very simple scheme, it was evident that only the AD745 could be used in this mode, and then with no more than two amplifiers. Both the combination of 2 or more OPA627s and 3 or more AD745s produced severe slew rate and bandwidth degradation, making these configurations unusable without bulky isolation circuitry, which is undesirable for our application as a small, compact design for multichannel arrays. In our results below, 1-745 and 2-745 will refer to the single and dual AD745 configurations, while the 627 results will refer to a single OPA627.

These three configurations were tested at a gain of +21 with a low-noise SITe device of recent manufacture, calibrated with Fe^{55} for gain. All three produced approximately the same noise floor of about 3.0 electrons at a relatively slow 16 μsec integration time on the dual-slope integrator of our signal processor. This probably represents a close approximation to the noise floor for this type of device.

In order to more fully contrast the performance of these amplifier configurations, the wiring was modified to increase the closed-loop gain to 400 by dropping R2 from 200 to 10 ohms, and grounding the OS input.

The signals were fed through the normal signal processor and the resulting "image" could then be statistically analyzed just like normal CCD images. Knowing the overall gain of the system then allowed rms values for the equivalent input noise to be derived. In addition to the three opamp configurations discussed above, I also included the venerable JPL preamp, a NASA standard on many space missions and in many

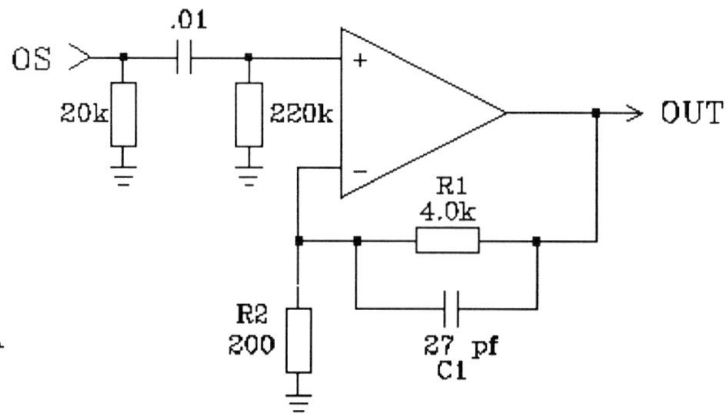

Fig. 1A

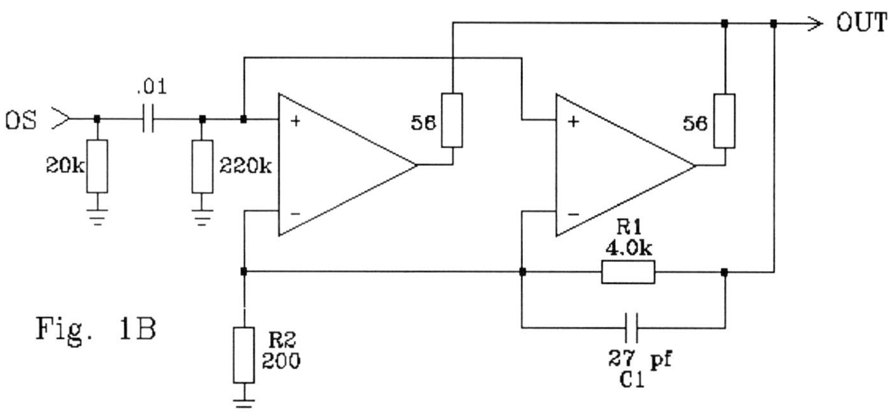

Fig. 1B

***Figure 1**. A. Single opamp used at a voltage gain of +21 (G = 1 + R1/R2). For both the AD 745 and OPA627, the feedback time constant C1R1 = 100 nsec seemed to produce critical damping for optimal settling time. B. Dual-amplifier version using the AD745, for decreased noise*

observatories but a very bulky and complex design rather unsuited to miniaturization. These results are summarized in Table 1 for 3 different values of the integration time on the dual-slope processor

	Integration Time		
	5 μsec	**8 μsec**	**12.5 μsec**
1-745	1.00 μV	0.89 μV	0.76 μV
2-745	0.78 μV	0.70 μV	0.65 μV
1-627	1.41 μV	1.30 μV	1.15 μV
JPL	0.78 μV	0.80 μV	0.80 μV

Table 1. *Total input noise, measured with three different integration times. Integrations were separated by 5 μsec to allow for transient settling needed with these fairly slow amplifiers.*

4. CONCLUSIONS

The low level of net noise demonstrated by all these amplifiers is cause for some celebration, as any of them could work well with the present generation of high gain, low noise CCD output structures. This can perhaps be more readily seen in Table 2, where we compare the computed net noise (in electrons) at 5 μsec integration time for three different CCD outputs, chosen to represent typical values for both older and newer designs.

	Computed net noise [e^-]		
	2-745	**1-745**	**1-627**
CCD output noise/gain			
5.0 e^- / 1.0 μV/e^-	5.06	5.10	5.20
3.0 e^- / 2.0 μV/e^-	3.02	3.04	3.08
2.0 e^- / 5.0 μV/e^-	2.01	2.01	2.02

Table 2. *Net noise for the three IC preamplifiers when added in quadrature with typical CCD outputs*

It is evident that all of these preamplifiers should work well with any of today's available CCDs. Even in the worst case, the somewhat noisier OPA627 used with a low-gain output from an earlier era, the net noise increase is only 4%. For typical high-gain outputs, the performance is even better, and any degradation caused by the pre-amplifier would be hard to measure. We can further see that the noise reduction obtained by use of the more complex dual-745 design (and the much more complex JPL amp) is probably not really needed for CCD outputs of the above type, thus allowing one to use the simpler scheme outlined in Fig. 1A.

The AD745 and OPA627 amplifiers thus seem well suited for most of our needs, as long as not too much bandwidth is required. Due to the relatively long settling times noted, a practical limit for our baseline preamp probably lies under 100 kHz pixel rate if dual-slope integration is used and perhaps a little above this for a pure clamp-sample processor. The latest CCD versions promise good noise performance at frequencies considerably higher than this, so the search goes on for candidate preamplifiers to meet these needs. In future experiments along these lines, I hope to try some new configurations to specifically address these requirements for increased bandwidth while preserving good noise performance.

PULPO:

TEMPERATURE, VACUUM, SHUTTER, LN_2 LEVEL, ALL IN ONE BOX.

NICOLAS HADDAD[1] & *PETER SINCLAIRE*[1]

ABSTRACT

We have developed a multifunction control and monitoring unit for CCD cryostats. The unit is able to read up to 7 Pt100 temperature sensors, control 3 heaters, drive IRIS type of shutters, control the timing for the shutter and read the detector head pressure when connected to Balzer full range vacuum gauge. One of the Pt100 inputs is reserved for implementing a LN_2 *exhaustion alarm.*

1. INTRODUCTION

PULPO, Spanish for octopus, is a CCD auxiliary functions monitor/controller unit under development at ESO since mid 1995.

The driving concept behind PULPO's design was to concentrate a multiplicity of CCD auxiliary functions such as detector temperature sensing and control, cryostat pressure, liquid nitrogen exhaustion alarm, shutter driving and timing, all inside one *intelligent* unit. Our goal is for PULPO to be able to communicate with any supervisor system having a standard RS-232 interface using a simple ASCII command protocol. It can also provide local information and manual control through an LCD and 3 push buttons.

2. PULPO FEATURES

- 7 inputs for 4 wire Pt100 temperature sensors (look-up table linearization) with 16 bits resolution. Other sensors such as 3 wire Pt100 or temperature diodes c also be used with some software changes.
- 1 mA current source for driving the sensors switched on under program control. The current is measured with 16 bits resolution for ultimate precision.
- 3 outputs for heaters with 12 bits resolution for 0 to 280 mA (6W at 75 ohm resistors)
- 1 current source output with 12 bits resolution (0-20mA) for driving an LED light source.
- 0-10V analog input and power supply for logarithmic vacuum gauge (Balzers model PKR 250 1 atm to 10-9 mBar) at 10 bits resolution.
- Shutter driver for electromagnetic type units (24V/100Ohm coil). Open pulse from 50ms to 200ms at 24V and keep open current from 50mA to 150mA, both adjustable with potentiometers.
- 3 opto-isolated TTL inputs for shutter status sensing (open/close/remote).
- 1 opto-isolated output for external shutter drivers (open/close)
- 1 normally closed relay output for telephone dialer alarm.

[1] European Southern Observatory - La Silla Cile. Email: **nhaddad@eso.org**

J. W. Beletic and P. Amico (eds.), Optical Detectors for Astronomy, 131–134.

- 1 RS-232 serial port for host/terminal communication at 9600 baud. (Software control of echo suppression)
- A buzzer that can be used as an audible alarm.

3. HARDWARE DESCRIPTION

All PULPO electronics (except the 24VDC power supply) are contained in a compact enclosure measuring 7. × 10.5 × 25 cm. This box has two Eurocard size boards: one for the DC/DC converters and one (main board) for the micro-controller and analog electronics. The DC/DC board generates all the other voltages needed by the main board (5 VDC @ 300 mA and +/- 15 VDC @ 150 mA). The main board has the necessary analog signal conditioning and digital circuitry to interface with all inputs and outputs.

The heart of the system is a credit card size micro-controller module (miniMOD535 by Phytec, Germany) based on an Intel 8051 family chip running at 12MHz (piggy backed over the main board). This micro-controller has 32K EPROM, 32K EEPROM and 32K RAM with battery backup, a Real Time Clock, a level adapter for the RS-232 serial port, and 4 of the 6 CPU's 8 bits input/output ports are accessible through the module pins. One port is used as 8 multiplexed analog inputs for the internal ADC (8 or 10 bits resolution).

4. SOFTWARE DESCRIPTION

The micro-controller is fully programmed in C using a compiler especially optimized for this CPU (C51 from KEIL Software, Germany). Using C, instead of assembly language, has significantly reduced the amount of time needed for software development.

The software is divided in three major components:

- The main task which runs in a permanent loop doing the PID control, vacuum gauge readout, display refreshing, serial link command parsing and the periodic execution of the State Machines. It also performs all required data logging.
- The 10 ms interrupt driver, which handles the periodic sensor readout, refreshing of the heater's power level and the push-buttons sensing. Other interrupt drivers handle the serial port and the shutter timing millisecond counter.
- The State Machines, which are routines that handle the push-button and LCD state, the alarm condition analysis and triggering, and the buzzer's different sound generation.

Additionally, there are various small drivers which handle the particular pieces of hardware in the system (ADCs, DACs, LCD display, real time clock, etc), therefore providing a high degree of software modularity isolating the major software components from the low level hardware.

The serial communication protocol, both for a remote host or simple terminal, is based on 2 ASCII characters commands plus numerical parameters (integer or float) separated with commas. The responses from PULPO will always be "OK" and the data requested, if any, or "ER" plus an error number. For use with a remote host, the character echo can be suppressed.

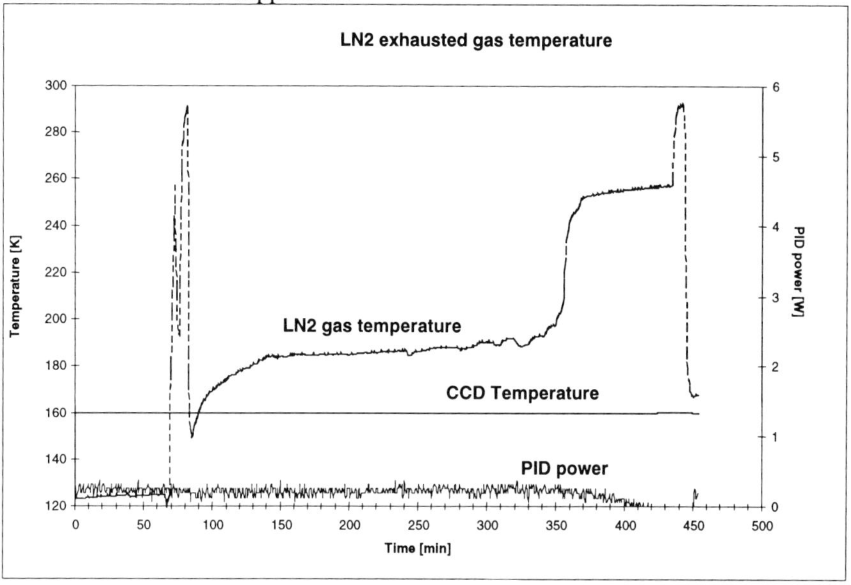

Figure 1: When last drop of LN_2 evaporates (~ 360min) the LN2 gas temperature starts to rise. The PID retains control over the CCD temperature almost 60 minutes after the LN_2 exhaustion. The peaks on LN2 gas temperature signal the cryostat refilling.

5. FUNCTIONS OF PULPO WITH ESO CCDS AND CRYOSTATS:

5.1 Temperature monitoring and control:

- On ESO dewars, the two temperature sensors are used for CCD temperature monitoring (at two different locations) and the other four Pt100 servers can be used for temperature monitoring on the pre-amplifier boards or even ambient temperature.

- To control the CCD temperature, one or two heaters are used with 75 Ohm resistors. Temperature regulation can be based on any of the two CCD temperature sensors after the PID parameters have been "auto-tuned" for the particular sensor.

5.2 LN2 exhaust alarm:

- One of the temperature sensors (number 7) is reserved for implementing the LN_2 exhausting alarm system. This Pt100 sensor is placed inside a holder assembly and inserted into the refilling tube (approx. 5 cm), to sense the temperature of the evaporating nitrogen gas. Once the dewar is empty, the cold gas stops flowing and the sensor temperature rises. When this happens, there is still about one hour period before the CCD starts warming up. An alarm will go off when this situation occurs.

- Additionally, it can also be used to measure exact holding time, as it records when refilling is done (the sensor has to be removed for that operation, thus there is a clear rising pulse in the temperature log).

5.3 Vacuum measurements:

- PULPO will provide vacuum readout for all cryostats that have been upgraded to the Balzers PKR250 gauges. An alarm will go off when pressure goes above a predefined threshold for a certain amount of time.

5.4 Shutter timing:

- Shutter timing is foreseen with 1 ms resolution and will be triggered through a simple one character command given by the host on the serial port to minimize overhead times.

VERY-HIGH FRAME-RATE BURST-IMAGE SENSOR

WALTER F. KOSONOCKY[1]

GUANG YANG[1], *CHAO YE*[1], *RAKESH K. KABRA*[1], *LIANSHENG XIE*[1], *JOHN L. LAWRENCE*[1], *VINCENT MASTROCOLLA*[2], *FRANK V. SHALLCROSS*[3], *VIPULKMAR PATEL*[3]

ABSTRACT

A 360 × 360-element very high frame rate (VHFR) burst image sensor captures images at maximum frame rate up to 10^6 frame/s. This is accomplished by continuously storing the last 30 image frames at the pixel locations. The 360 × 360 VHFR imager having a 2 × 2cm^2 chip is designed in the form of 4 quadrants each with 180 × 180 pixels. Each pixel occupies 50 × 50μm^2 and consists of a 337μm^2 photodetector with a fill factor of 13.5% and a 3-phase 30-stage (5 × 6) series-parallel type buried-channel CCD (BCCD) register for continuously storing the last 30 detected image frames. The chip uses 1.5μm design rules and 1.5 × 3μm^2 minimum-size BCCD storage elements.

The architecture of the VHFR imager is illustrated in Figure 1 for an array of 2 × 2 pixels. Each pixel consists of a photodiode with charge-collecting well under the G_1 gate, a blooming barrier gate G_2, a drain D, and the gate G_3 separating the charge collecting well from the 5-stage serial (S) register. The drain D is used for control of blooming during the optical (frame) integration time and for dumping the excess charge signals (excess frames) from the S register. However, the drain D could also facilitate the operation with subframe optical integration time. The 5-stage serial register coupled to a 5 × 5-stage parallel (P) register forms a 30-frame CCD storage at each pixel location. A block

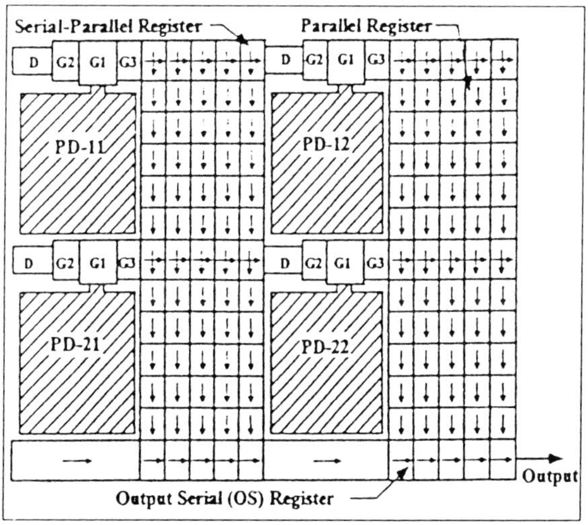

Figure 1:Architecture of 2 × 2 pixels

[1] Electronic Imaging Center, New Jersey Institute of Technology, Newark, NJ
[2] Princeton Scientific Instruments, Inc. Monmouth Junction, NJ
[3] David Sarnoff Research Center, Princeton, NJ

J. W. Beletic and P. Amico (eds.), Optical Detectors for Astronomy, 135–138.

diagram of the 360 × 360 VHFR imager is shown in Figure 2. This imager is organized into 4 quadrants to reduce transfer losses and to improve processing yield of usable quadrants. Note that all 360 × 360 photodetectors, PDs, have the same spacing, while the shape of the CCD pixel storage is different for upper and lower quadrants.

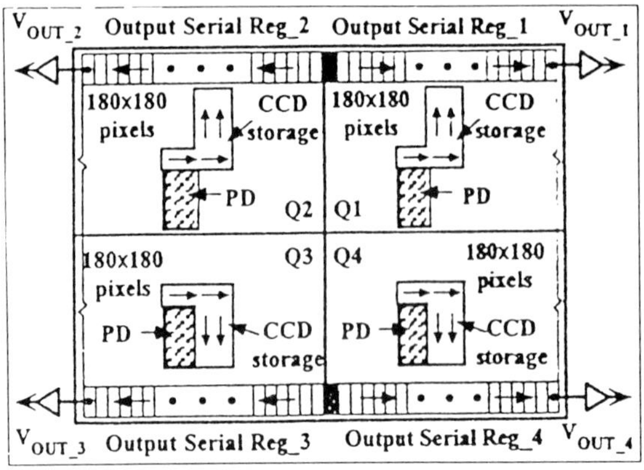

Figure 2: Imager block diagram

During the image acquisition cycle, the charge signal detected by the photodetector, PD, is transferred in series into the serial BCCD register of the pixel for detection of the successive frames. After the detection of 5 frames, the detected charge signals are transferred in parallel from the S register to the P register, providing a storage for the last 30 detected image frames. The continuous storage of the last 30 frames is obtained by preceding the loading of the S register from the photodetector by a parallel transfer into the S register of the charge signals from the last row of the P register of the pixel above. This last row of 5 charge signals at each pixel location is transferred to the dumping drain D while a new row of 5 charge signals corresponding to the next 5 frames is transferred from the photodetector into the S register. The readout of the last detected 30 frames can be initiated after the loading of the S register is completed. This can be at the end of any pixel row-time. At this time, all of the BCCD storage registers of each quadrant are converted (under control of the BCCD clocks) into a single large frame-transfer type CCD readout of 180 × (5+1) rows and 180 × 6 columns. For the 3-phase BCCD design, this charge readout involves up to 180 × 6 × 3 = 3,240 transfers in vertical direction by the parallel registers and up to 180 × (5+1) × 3 = 3,240 transfers in horizontal direction by the serial output register.

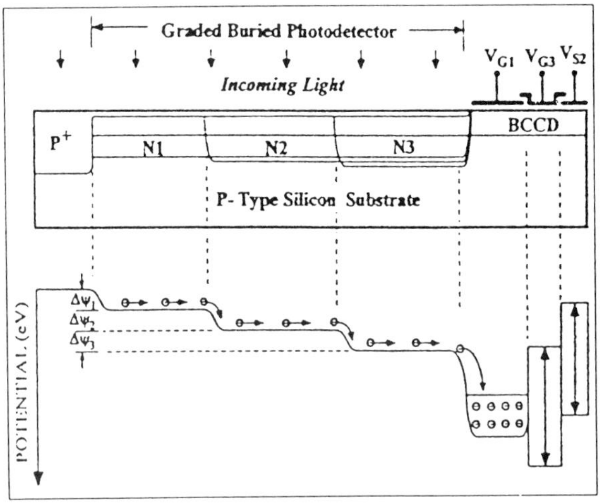

Figure 3: Photodetector construction and operation

To achieve a high-speed detection with essentially zero frame-to-frame lag, the photodetectors of VHFR imager are implemented as a pinned buried detector with three n-type detector implants in addition to a thin p and the regular buried-channel implants. The cross-sectional view of this photodetector and the associated potential diagram of the charge readout are shown in Figure 3. Figure 4 shows the comparison of the simulated charge readout from the photodetector with three implants and with a single implant after the illumination (corresponding to the intensity of 20,000 electrons/pixel in 0.5μs) is terminated.

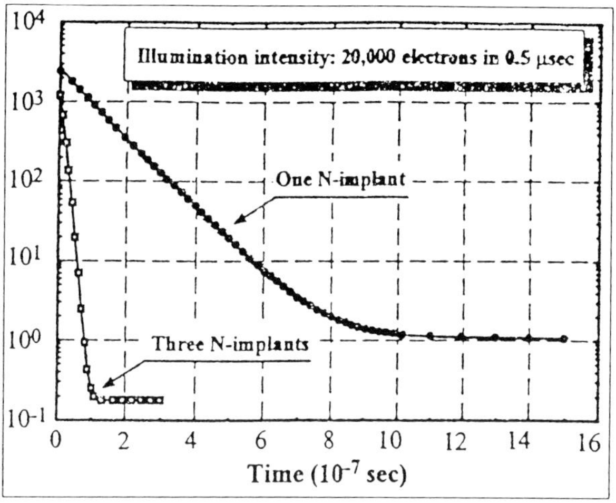

Figure 4: Charge decay in photodetector

The VHFR image sensor uses four levels of polysilicon, three levels of aluminium, and seven implants. To achieve high-density and low-transfer losses, the 3-phase buried-channel CCD storage register uses an additional level of polysilicon defining the channel stops for the serial registers at each pixel. Two levels of metallization are used for bussing the clocks to the pixels and the third-level metal is used for optical shield covering the BCCD readout registers. Arsenic and phosphorous implants are used as the BCCD implants to maximize the charge handling capacity and for providing high fringing fields to minimize charge transfer losses. The buried pinned photo-detectors use a thin p implant, the BCCD implants, and three additional n-type implants to achieve less than 0.1μs readout time. The imagers are processed using p (100) silicon substrate with a 7μm p-epi layer.

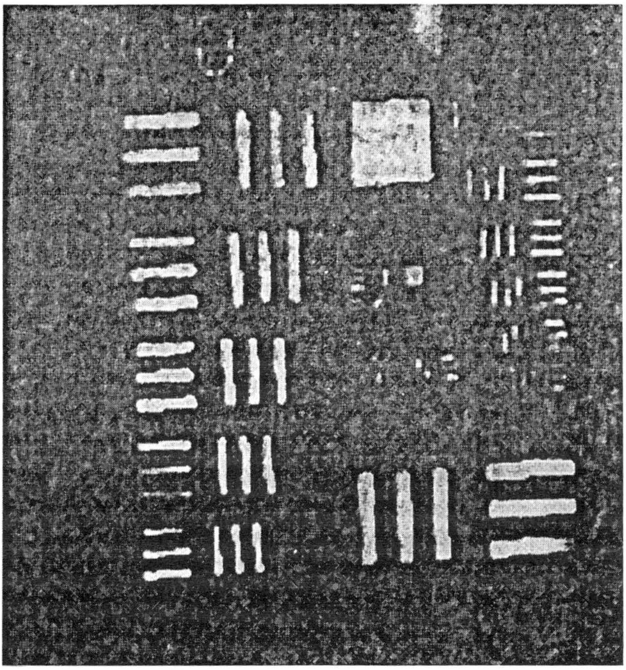

***Figure 5**: One frame image of test pattern illuminated by a 0.4 μs LED pulse*

The operation of the VHFR imager is demonstrated using one quadrant of 180 × 180 pixels. The chip is operated at -30° C to allow a readout time of about 6 seconds. Operation with frame-to-frame lag of less than 1% is demonstrated using LED array pulsed at a subframe time of as short as 0.4μs. Figure 5 illustrates an image of one detected frame for operation with 2μs frame time for the first frame and 1μs frame time for the next 4 frames of each 5-frame sequence. This demonstrates an effective frame rate of 8.33×10^5 frames per second. The highest level of signal in Figure 5 corresponds to 11,000 electrons/pixel and the dark current background is in the range of 100 to 200 electrons/pixel (from top to bottom). Observed ghost frame images due to transfer losses and lag represent less than 1% of the primary image. The characteristics of this VHFR imager are summarized in Table 1.

Number of pixels	4×180×180
Pixel size	50×50 μm^2
Fill factor	13.5%
CCD elements per pixel	30
Saturation signal, Q_{max}	11,000e^-/pixel
Responsivity	2.2μV/e^-
Readout noise	9 e^-/pixel rms
Dynamic range	1200:1
Maximum frame rate	8.33×10^5 frames/s

Table 1:*Characteristics of VHFR burst image sensor*

Acknowledgements: The authors acknowledge support by the Office of Naval Research under Contract No. N0014-91-C-0078 and by the New Jersey Commission on Science and Technology.

REFERENCE:

[1] Kosonocky, W.F., J.L. Lawrence, Very High Frame Rate CCD Imager, US patent Number 5.355.165, October 11, 1994

HIGH SPEED, WIDE DYNAMIC RANGE CCD SYSTEMS FOR APPLICATIONS IN ASTRONOMY

CRAIG D. MACKAY[1]

ABSTRACT

There are a great many applications that require high speed imaging. It is usually accepted that the image quality obtained from systems intended for high speed use is poorer than that obtained with precision slow scan scientific CCD cameras in terms of read out noise, low contrast capability and low light level sensitivity. The reason that this is the accepted wisdom is that it is virtually impossible using conventional techniques to build systems that use double correlated sampling signal processing chains that operate at pixel rates even approaching 1 MHz. This arises because the FET switches used amongst other components simply do not work quickly enough. By taking a completely new look at the way that controllers are designed it is possible to find alternate approaches to the considerable problems of designing fast read-out electronics that still allows precise, wide dynamic range operation to be achieved. The first fast CCD controllers became available from AstroCam Ltd. a few years ago that were able to offer very high levels of performance with specific CCDs. These have the potential to offer many researchers much better imaging system performance than has hitherto seemed possible and are now used widely in research fields outside astronomy. More recently astronomy has had need of much faster systems and this paper will describe two recent applications in astronomy that depend heavily on such systems and achieve very fast operation in very different ways.

1. ACQUISITION AND GUIDING OF THE TELESCOPES AT COAST

An exciting new telescope called COAST (Cambridge Optical Aperture Synthesis Telescope) is nearing completion in Cambridge University, England that is designed to give diffraction limited images with a resolution of up to 1 milliarcsec. It consists of four separate telescopes up to 100 meters apart feeding the light from the target star to a beam combining laboratory. The interference fringes are recorded and the data processed to remove the effects of the atmosphere using a method developed by radio astronomers called closure phase. A more detailed description of COAST may be found on the Web at: http://www.mrao.cam.ac.uk/telescopes/coast/index.html.

In order for the beams to interfere with the highest possible fringe visibility it is essential that each telescope is steered accurately and rapidly to remove the first order motions caused by the atmosphere. This must be done at high speed (over 100 Hz) but the need for really low read-out noise is acute since this effectively sets the limiting

[1] Institute of Astronomy, University of Cambridge, England; e-mail: cdm@ast.cam.ac.uk and AstroCam Ltd, Innovation Centre, Cambridge CB4 4GS, England; Email: **cdm@astrocam.co.uk**

J. W. Beletic and P. Amico (eds.), Optical Detectors for Astronomy, 139–142.

sensitivity of the whole telescope. Clearly if the star is too faint to allow the telescopes to be aligned, nothing will work.

We expect to get the lowest possible read out noise from a slow scan CCD system with read out times of several seconds per frame. However, COAST uses an intelligent controller from AstroCam to allows all four telescopes to be guided at over 100 Hertz with a read noise < 3 electrons rms.

The instrument science detectors work around 800 nm wavelength. The blue and visible light from each of the four telescopes falls on the corner of a cooled low-noise CCD near to its output amplifier. The CCD is used in the normal imaging (finding) mode to allow each telescope to be approximately aligned to the correct position. Then the controller is switched rapidly into a fast autoguider mode where pixels are fast binned to synthesize a quadrant autoguider geometry around each of the star images. The offsets are computed within the read-out cycle so that the error signals may be sent to the telescope control microprocessors at over 100 Hertz. By reading a tiny part of the CCD, the number of parallel and serial transfers are minimized, and as only 16 real pixels are read out each cycle, the aggregate pixel rate is very low and the read-noise is kept as low as possible.

The CCD is an EEV CCD 15-11 with 1024 × 256 pixels. Only a region of 192 × 48 pixels is read out at full resolution for acquisition, and then the central 12 rows are read out binned 6 × 6 pixels for the fast guiding mode. The signal levels in adjacent groups of 2 × 2 super pixels are used to compute the guiding errors.

2. ACTIVE AND ADAPTIVE OPTICS SYSTEMS FOR ASTRONOMY

Ground based astronomy is seriously hampered by turbulence in the atmosphere. The wavefronts coming from a star are distorted so that the images formed are smeared out to a resolution of typically one arc second in diameter. The telescope is capable of forming images perhaps 40 times smaller if no such phase errors were introduced by the atmosphere (for a 4 meter diameter telescope). Systems have been designed and built aimed at overcoming these limitations in the following way. The main requirement is to sense the phase errors by looking at the image distortions quickly so that a correcting element may be made to distort to compensate for these errors. The system has to work quickly (10 msec is the typical atmospheric correlation time in the visible), and it must work with the lowest possible signal levels.

Under typical conditions in the visible, the atmosphere shows relatively constant phase gradients across patches of 10-20 cm diameter. There will be typically a linear phase gradient across patches of up to 50 cm. On larger scales the phase errors become more complex. If the pupil of the telescope is re-imaged so that it falls on a lenslet array, then each lenslet forms its own image of the star with the light from the patch of the pupil it covers. By studying the relative motion of the multiple sub-images, we can establish how the atmospheric phase distortions are changing with time. This arrangement is often referred to as a Shack-Hartmann wavefront sensor.

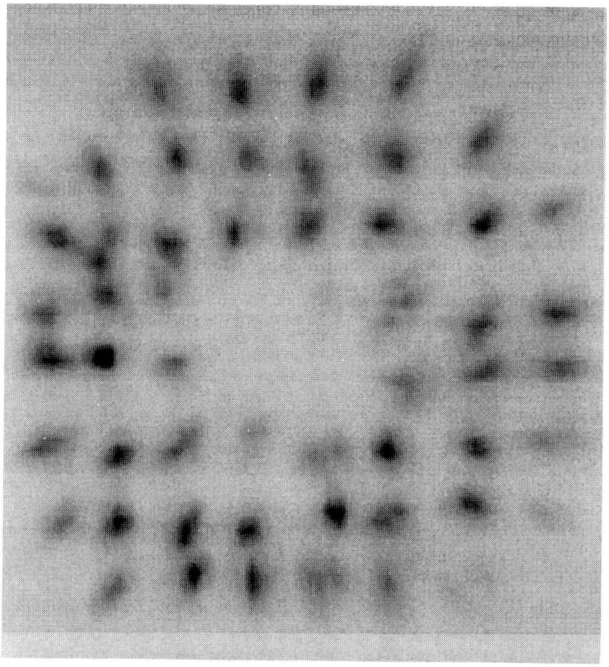

Figure 1: A typical image from the Shack-Hartmann wavefront sensor installed on the 4.2 meter William Herschel Telescope on La Palma. The distortions from a rectangular grid are partly atmospheric, partly due to telescope effects (particularly the mess at the 10 o'clock position), and partly due to optical geometric effects in the experimental set-up. The data were taken of the star Alpha Bootes in the red at 865 frames per second.

The atmosphere changes quickly, and the turbulence has different characteristics at different telescopes on different sites. If an effective AO system is to be built the designer needs to know a lot about the local conditions that have to be contended with. The UK AO programme has funded the development of a high speed imaging system that uses a lenslet array of 8 × 8 elements imaged onto a high speed Capella camera system built by AstroCam Ltd., Cambridge, England. It uses a custom 64 × 64 frame transfer CCD with 15 × 15 micron pixels designed by John Geary (Smithsonian, USA) and manufactured by Loral (USA). The CCD is unthinned (front illuminated). The Capella system reads out images at a maximum rate of 865 frames per second at full resolution into a SPARCstation computer. Data are taken at up to 5.5 MHz pixel rate with 12-bit accuracy (the Capella is capable of 14-bit accuracy at up to 5 MHz) at up to 32000 frames per run (set only by the memory capacity of the computer) before saving to disk. The data allow a detailed analysis of the motions of the star images from each of the 50 cm × 50 cm patches of the WHT 4.2 meter telescope on La Palma. The high maximum frame rate allows the movements from the atmosphere to be frozen and followed reliably. For even higher time resolution, the Capella may be programmed to read out a sub-array and give yet higher frame rates (such as 1200 Hz at 32 × 64 pixels).

The system is now in routine use as a seeing and site evaluation monitor on the 4.2 meter telescope. The scales (both temporal and spatial) of the turbulence increase as the working wavelength is increased, both by the power 1.2 of the wavelength. This makes it possible to work much more easily if the goal is for measurement and phase error correction to be effective only in the near infrared rather than the visible. The performance of the wavefront sensor is the most critical component in determining the limiting sensitivity of the system since if the target star is not bright enough to give a good signal in the correlation time over the mono-phase aperture then the system will not work. Telescopes are usually limited in sensitivity by other things and allow integrations of minutes or hours using the whole telescope rather than just a tiny sub-aperture.

RECENT RESULTS FROM THE EEV CCD42

A.P. OATES[1] *& P.R JORDEN*[1]

ABSTRACT

We present results of measurements from the latest EEV CCD42 devices. These are state-of-the art image sensors attaining very-low read-out noise and have several novel features in the design architecture. It is hoped that these devices will form the de facto detectors at the ING for spectroscopy and possibly in a mosaic configuration for the WHT. A suite of 6 devices will be delivered to the ING during the course of 1997.

1. INTRODUCTION

In December 1994, the RGO in collaboration with the Anglo-Australian Observatory (AAO) and the Italian Telescopio Nazionale Galileo (TNG) group placed a contract with the English Electric Valve Company (EEV Ltd.) to develop a new, large-format imaging sensor (CCD), known as the CCD42. The design specifications for these CCDs were determined during the period 1993-1994 by the above authors in collaboration with EEV; these devices are 2k × 4k, 13.5μm non-MPP devices with 3-edge buttability. The precise detail of the contract to provide the consortium with the required devices took many months of informed, interactive interchange between, RGO, and EEV. The details of this specification turned out to be a baseline for devices, which have now become the building block of many CCD cameras. Preliminary results from the first, front-side sample devices were presented in Ref. [1]. Since then, we have been undertaking test procedures and quantification of subsequent EEV thinned CCD42s and this paper discusses those results.

Although (as expected) the progress was at first slow, we have now received our first, thin, 2k × 4k CCDs: some potential science grade devices and 3 engineering devices. We hope to take delivery of the remaining science devices, 12 in total, in 1997. This paper describes test results we have obtained from the devices so far delivered.

2. DEVICE DETAILS

We have been operating a 2k × 4k un-graded device in the RGO detector laboratory since April 1996. Prior to this time we operated 2 front-side and one back-side 2k × 0.5k device housed in an interim package. These devices operated very well and permitted us to obtain experience in the operational characteristics before operating the 2k × 4k units.

The specification for these devices was determined during several visits to EEV during the period 1993-1995. It was clear, from a visit which we made to the U.S. prior

[1] Royal Greenwich Observatory, Madingley Road, CB30EZ Cambridge, UK. Email: **apo@ast.cam.ac.uk**

J. W. Beletic and P. Amico (eds.), Optical Detectors for Astronomy, 143–151.

to this time ([2]), that the $2k^2$ and $2k \times 4k$ pixel device would form the building blocks of CCD cameras of the future. In tandem with a wafer run at Loral-Fairchild (as it was then) for $2k^2$ devices thinned by Mike Lesser, we also negotiated a contract with EEV to provide a set of $2k \times 4k$ devices, in a collaborative venture with the Anglo-Australian Observatory and the Italian Galileo group. Figure 1 illustrates the basic architecture of this device.

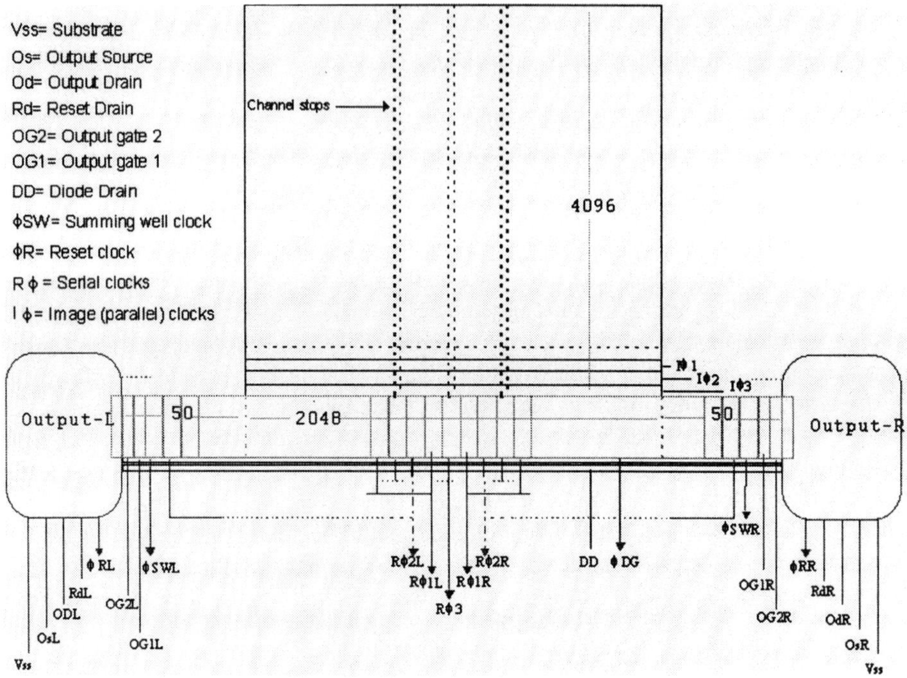

Figure 1. Architecture of the EEV42 CCD.

3. Test Results

3.1 Noise Versus Sampling

The CCD42 devices were designed to deliver low-noise, particularly at high read-out rates. Figure 2 illustrates 3 curves, 2 for devices already available (the CCD30 and CCD35) and the third, a prediction by EEV to demonstrate the performance of the output stage of the CCD42. As will be seen from theses curves, optimum performance is achieved for the CCD42 device over all sampling frequencies. The most obvious parameter to investigate with these devices therefore, is how the read-out noise of the on-chip FET performs with sampling speed. We are able to change this and pre-select speeds from a table and investigate how changing speed affects the read-noise.

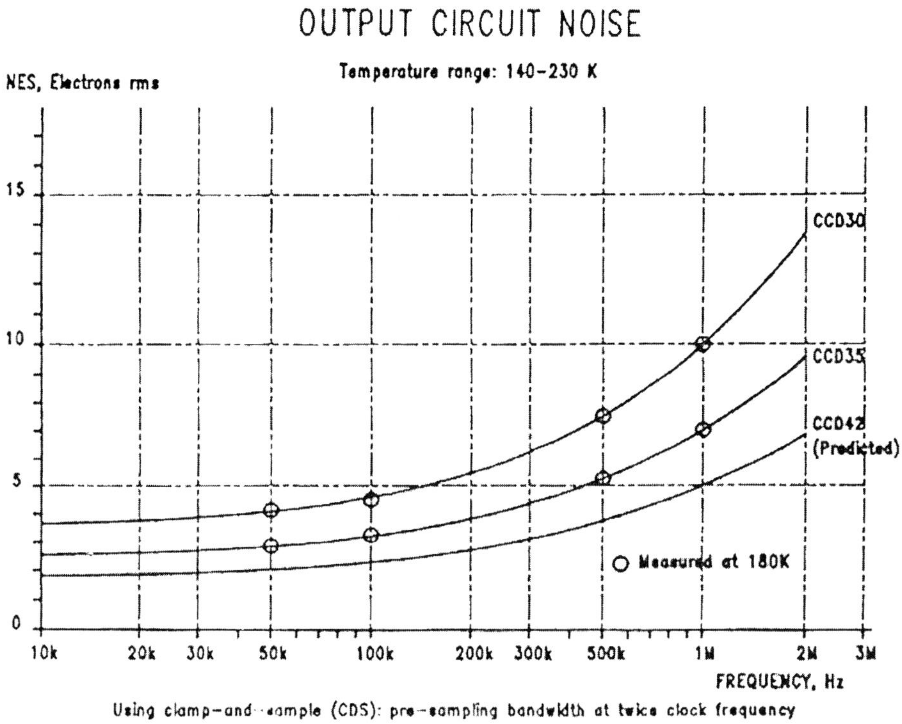

Figure 2. EEV noise performance as a function of sampling speed.

With our current hardware, the fastest pixel time we can achieve is about 5μs/pixel; though in the data taken here the fastest we tried was 10μs/pixel (with a CDS reference and sample of 0.5μs each). The speed of pixel timing was varied between 10μs/pixel and about 30μs/pixel with corresponding integrator sample times of 0.5, 1.0, 2.0, 5.0, 10.0μs. The system gain and read-noise were determined using photon transfer, utilizing 2 small LEDs in the cryostat as the illumination source. The gains were also checked against our Fe_{55} X-ray source. The expected noise levels EEV were predicting, just under 2 electrons at 20kHz using 'clamp-and-sample', indicated that we should expect ~1e^- using a dual-slope implementation of correlated-double sampling.

At the noise level quoted above, it is imperative that we measure the system noise and subtract this in quadrature from the total noise measured with the CCD present. Because of the levels involved and because of the environment, the determination of the system noise has been difficult to make. We are still looking into making more rigorous measurements in this area.

Table 1 lists the measured (and derived) noise parameters as a function of pixel sampling speed. From the EEV noise spectrum shown in figure 2 above we expect <2**e**$^-$ rms at 50kHz and 5**e**$^-$ at 1MHz sampling, in fact we are seeing slightly lower noise than this at the intermediate frequencies - a desirable feature. In addition, we also find that at the slower sampling rates, the noise isn't as low as the 1**e**$^-$ expected. This is being investigated.

CDS Speed	1	2	3	4
Parameter				
CDS sampling time, τ(us)	10+10	5+5	2+2	1+1
Gain, G(e/adu)	0.61	0.72	0.8	1.02
CCD + system noise variance, A(adu)	101.4	97.8	93.1	89.4
System noise variance, B(adu)	74.3	77.2	81.4	66.5
Net CCD noise = G.√(A-B) (e^{-})	3.2	3.3	2.8	4.9

Table 1. *Read-out noise as function of sampling speed.*

3.2 Gain Versus Output Node Size

One of the novel features which was deemed to be useful when the device specification was being drawn up, partly motivated by one of the members of the consortia requiring high signal handling capability in binning mode, was the addition of a gain switched output stage. The change in charge handling capacity is provided by an additional gate structure before the output node. There are therefore two output gates, the first performing the usual function of reducing clock feed-through to the output node from the serial clocks. The second (OG2), under normal operating conditions simply passes the charge through to the output and acts in a similar fashion to a summing well, which also exists at the end of the register. The voltage on the OG2 electrode is held at clock-low for this condition to operate. Putting OG2 HI, say at +3V (for historical reasons we operate all our devices with substrate at -7V), permits an increase in the node capacity by about a **factor of 3**. The following table indicates the changes in gain and noise measured by setting the gate voltage at -13V or +3V.

CDS Speed	1		2		3		4	
	OG2		OG2		OG2		OG2	
Node Gain(OG2V)	Hi (-13)	Lo (+3)	Hi	Lo	Hi	Lo	Hi	Lo
Gain (e/adu)	0.67	1.91	0.68	1.92	0.82	1.97	1.0	2.91
Total Noise(e)	6.3	16.3	5.9	16	6.4	15.6	6.7	18.6

Table 2. *Gain change with OG2.*

As can be seen from this table, the OG2 feature does indeed work and the measured and expected performance of this feature has been met.

3.3 Full Well

Reducing the system gain and implementing the low output gain feature of the output node allowed the full well of the chip to be measured. At a system gain of 4.2 e^{-} /adu using 0.5+0.5μs sampling, the full well, measured from the data in the graph below came out at 205,800e^{-} The minimum specification for the contract was 150,000e^{-} with a goal of 200,000e^{-}.

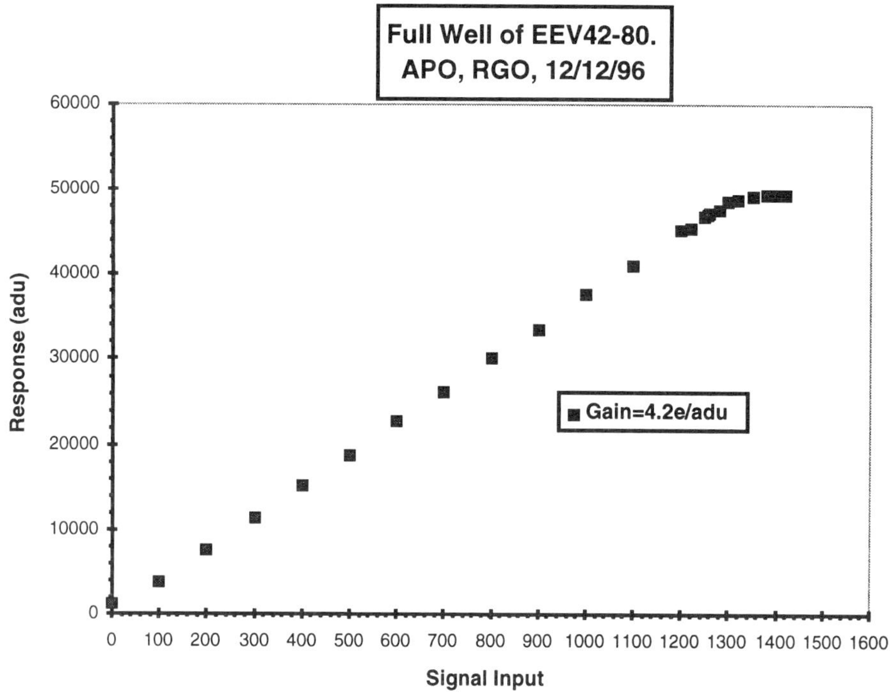

Figure 3. *CCD output response as a function of input signal*

3.4 Charge Transfer Efficiency

This was measured using residual signal in the vertical overscan on a low level (~100**e**$^{-}$) pre-flash and by utilising an Fe_{55} X-ray source. As mentioned briefly at the ESO meeting, there was evidence of low level vertical charge transfer in-efficiency on the device we were testing at that time and has been investigated further since the ESO meeting. This problem did not manifest itself in the global X-ray CTE measurement which produced the following results:

$$V_{cte} = 0.999994 \qquad H_{cte} = 0.999989$$

These levels are acceptable as far as the contract was concerned. On further inspection of the low-level pre-flash exposures, we did indeed find trailed charge in the vertical overscan at this low level. At low charge level, the vertical CTE was much worse than horizontal CTE:

$$Vcte = 0.9997 \text{ (@ ~100e}^{-}\text{ signal level)}$$

Indeed further inspection of the original X-ray frames revealed that the CTE problem did not appear in every column of the image and, in addition, appeared worse on the right-hand side of the frame than the left. This effect was also seen on an exposure which consisted of ten, co-added frames with around 15 e^{-}/pixel in each of the frames. It is therefore imperative that X-ray measurements are not relied on solely to determine the CTE, as this may give an erroneous indication of **low-level** charge transfer efficiency.

We believe, from discussions with EEV, that the devices are exhibiting a low level parallel-to-serial charge transfer problem and that this should be remedied by implementing process changes during the manufacture of subsequent batches.

3.5 Dark Current

The temperature at which we operate our camera heads is primarily dictated by the dark current performance. The following table shows the dark current measured at three operating temperatures for the first of the CCD42 devices.

Temp(°C)	-120	-110	-100
dark(e^-/pixel/hr)	**~0**	2-3	5

We therefore operate these detectors at a temperature of **-120°C.**

3.6 Defects

We have investigated defects using dark frames of 10 minute duration and processed them using IRAF. This involves using **imcombine** and setting the routine to detect the maximum and minimum signal on a pixel-by-pixel basis for the whole frame. 8 frames are used to produce reasonable statistics, the maximum and minimum pixels amongst the 8 frames are discarded and the other 6 form a mean for a new frame. In this fashion, the cosmic ray events are discriminated against, leaving the bright features due to the intrinsic chip properties. Using this frame, it is then possible to set a window for detection of bright pixels. For instance, on the 2k $\times$ 4k chip, a sigma=5 will result in random noise events amounting to 1 in 10^{+6} or about 8 events and this is in fact what we find. Modifying this value of sigma permits us to detect events which are any given value above the noise of the frame. Using this method we were able to detect (at a temperature of -120C) about a dozen events over the whole detector area, including, on the device discussed at the ESO workshop, 2 bright columns towards the top left hand corner of the array; these presumably being caused by single, saturated pixels blooming vertically to the top of the array.

Low level flat fields also revealed about as many single pixel or short, dark columns due: the former, we believe are due to single blemishes on the array, the latter case from pixel trapping sites.

3.7 Spectral Response

As with any detector, we aimed for the highest spectral response specification which we considered practical across the whole wavelength range. The table below summarizes what those figures were in terms of the absolute minimum and the goal the contract should seek to achieve. The figures are based on what we understood could be expected from the intrinsic thinning of the devices and from suitable selection of an anti-reflection (AR) coating. At the time of writing the details of the coating have not been finalized in terms of the peak wavelength so the UV response we are currently seeing on these devices, and indeed the peak, may change on the final devices.

Wavelength (nm)	Contract Spec. (%)		RGO (%)	EEV (%)
	Min	Goal		
350	15	50	28	38
400	20	70	55	57
650	70	85	80	85
900	25	40	25	31

Table 3. Spectral Response expectations for the CCD42 (EEV data c/o Paul Jerram (11/9/96))

The full RGO spectral response data for 2 devices, the CCD42-80 and CCD42-10 are illustrated in the following figure.

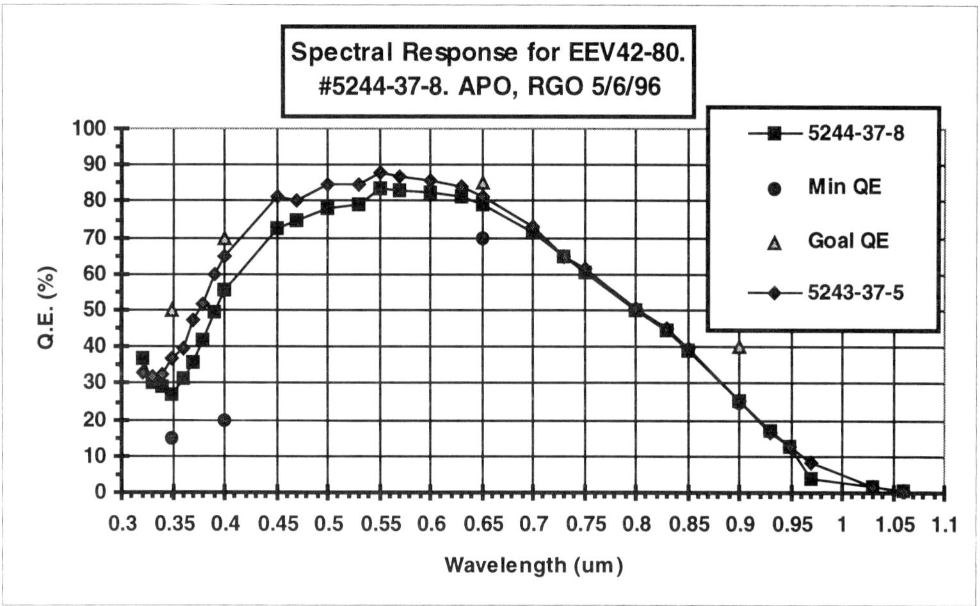

Figure 4. Spectral Response of 2 EEV devices

3.8 Flatness Profiles

Two dimensional surface profile measurements were taken, using the techniques described in [3], to determine any deviations from flatness. The data from these measurements indicated that the surface of the detector is flat to within ±5μm across the whole device but there was a 25μm deviation at one corner, opposite the read-out register.

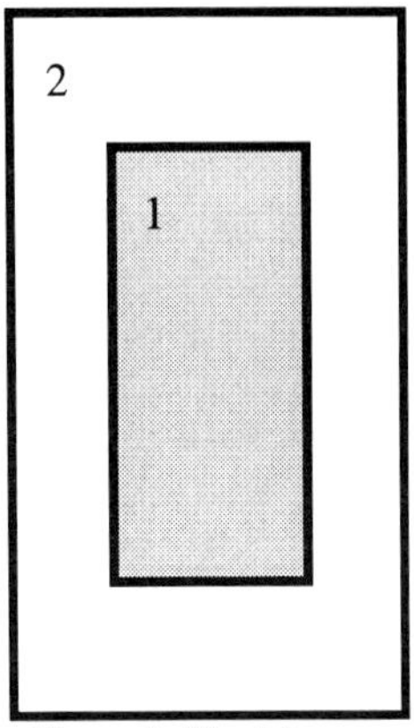

Figure 5. Blemish zones

The flatness specification for the devices is as follows:

±10μm, Zone 1 ±15μm, Zone 2

zone1 is 1200x3000 @ 424,548 i.e. the central 16x40mm, zone2 is the area outside zone1.

The surface profile data taken is illustrated in figure 6.

3.9 Fast Clearing

The EEV42 has a novel structure alongside the serial register - this is a **dump drain**. This feature makes it possible to clear the whole array in less than **0.5s.**

Traditionally we have implemented only one type of clearing operation but, in keeping with attempting to speed up the data retrieval process from devices which aren't getting any smaller, we introduced a faster clearing operation a while ago.

We have now implemented a 3rd option with the CCD42, permitting 3 options in all: **Standard, Quick & Flash Clear.** These refer to the means by which the chip is cleared and to the speed of operation.

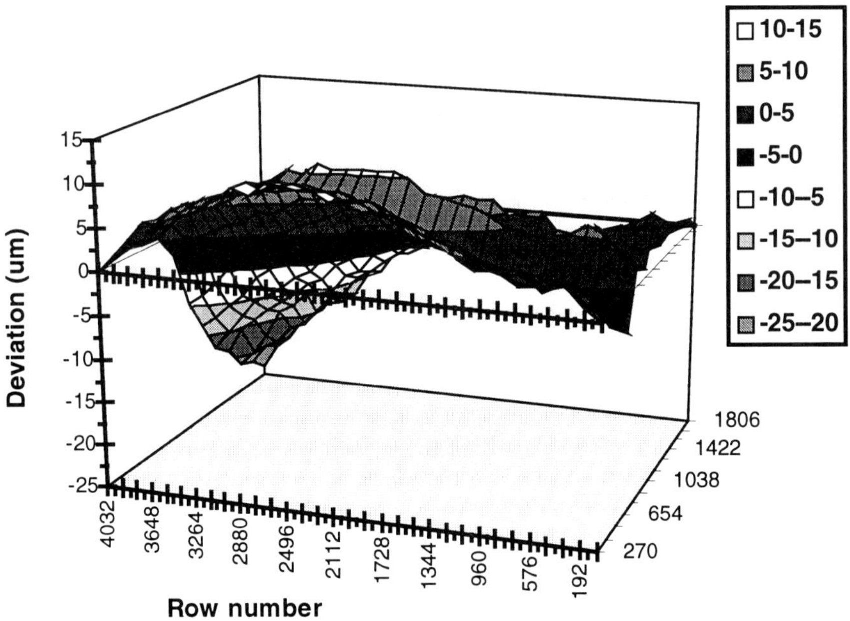

Figure 6. Surface Profile Measurement for the CCD42-80.

1. **Standard.** Each row of the CCD is clocked into the serial register which is then scanned with Reset gate HI. Repeated twice. t=**32s** (35μs Vclks)
2. **Quick.** Rows are clocked into the serial register and the register then scanned. The last 50 rows are not binned but read in separately and the register scanned. Repeated twice. **t=2s**

3. **Flash clear.** The Dump drain is enabled, each row is clocked into the serial register and at the end, the register is scanned to clear out remaining charge. Repeated twice. **t=0.4s**

NB: for **FLASH** clear the serial register must be scanned at the end OR some time allowed for charge to dissipate from the register. The **FLASH** clear routine can also be used during the skipping process when reading out windows, this speeds up this process as it doesn't matter how efficiently the unwanted charge is moved.

4. Final Remarks

It is a pleasure to thank work and the many useful discussions we have had with the team at EEV Ltd. amongst whom Ralph Holtom, Peter Poole, Paul Jerram and Dave Morris deserve special mention.

5. References

[1] Pool, P.J. Burt D.J., Oates A.P., Jorden P.R. **'A New Large-format, very Low-noise Scientific CCD.'** 1996 in press, Royal Society of Chemistry Special Publication #194. Proc. of 1995 ICSOI-3 Grand Cayman meeting.

[2] Oates A.P., Jorden P.R. '**New Large Format CCDs for the ING',** Technical Note #92, published by the RGO, Cambridge. UK. November 1993. Also available at **http://www.ast.cam.ac.uk/~apo/docs/ingccds/ingccds.html**

[3] Tulloch S.M. **'A Technique for the Determination of Co-planarity in Mosaic CCDs'., t**hese proceedings, pp.293.

THE AIP PHOTOMETRIC TESTBENCH

MARTIN M. ROTH[1]

ABSTRACT

A versatile laboratory test facility is described which is particularly useful for characterizing optical detector properties under well-controlled conditions. Moreover, as a generalized light source, it can be used for a large range of applications requiring calibration, stability, repeatability, and automatic control.

1. INTRODUCTION

Measuring CCDs and other optical detectors, in particular for the numerical characterization of overall detector performance, requires a test setup with a number of properties which are not easily realized on a standard optical table with standard lab components. The reason for this being mainly the need for a large dynamic range in illuminating light intensity, and for wavelength selection over the full range of the optical wavelength region, it is useful to have a dedicated test facility if one wants to routinely test and optimize detector performance. The Astrophysical Institute Potsdam, after having decided to start an effort in designing and building focal plane instrumentation for existing and future ground-based optical telescopes, has built such a testbench as a first step, not only to provide a tool for a the new optical and detector laboratories, but also as a test bed for acquiring experience with electro-mechanical drives and real-time instrument control using LAN-based embedded microcontrollers and workstations, i.e. the standard of modern telescopes. Some of these concepts and designs will enter a first instrumentation project (PMAS, [2]).

2. REQUIREMENTS

Many (although certainly not all) measurements of CCD performance involve the illumination of the chip in question with light of a certain intensity level, intensity profile, at a specific wavelength and with a certain spectral bandwidth. For example, the measurement of linearity and conversion factor using the photon transfer method [1] requires taking pairs of exposures over the full dynamic range of the detector. The light source must have a short-term stability below 10^{-3}. For RQE measurements, the light source must exhibit long-term spectral and intensity stability, and a calibrated optical detector must be available as a reference. In addition, for the extreme ends of the spectral range where the CCD quantum efficiency tends to drop, the light source must provide enough photons in order to keep exposure times reasonably short. For testing flat field properties as a function of wavelength, the CCD must be uniformly illuminated. When checking for low-level traps, the flat-fielding intensity must be controlled to a very small number of photons per pixel and exposure. Testing blooming and remanence

[1] Astrophysikalisches Institut Potsdam, An der Sternwarte 16, 14482 Potsdam, Germany
Email: **mmroth@aip.de**

J. W. Beletic and P. Amico (eds.), Optical Detectors for Astronomy, 153–156.

effects, however, requires illumination with a bright, concentrated spot. This may be driven even further; if sub-pixel uniformity is to be investigated [3]. When testing fringing properties, it is necessary to illuminate with a typical bandwidth as occurring in the respective instrument (spectrograph or Fabry-Perot, night-sky lines). Last not least, a seemingly trivial prerequisite which should not be overlooked is a light-tight enclosure and the suppression of unwanted light which may originate from spurious reflections from optical surfaces and the environment, indirect illumination through geometrical leaks, filter transmission curve leaks, unexpected secondary light sources like control lamps or displays of measuring instruments, etc.

In addition to the optical properties, we require the testbench to be remotely controlled from a workstation, and to offer a user-friendly operation with menus, macro-capability, and full access to data reduction software. Finally, the testbench has to be flexible enough to allow more measurements and tests with so far unforeseen equipment to be performed with very little effort and changes, and it should be capable to guide light out to a neighboring room in order to feed other experiments.

3. Technical Description

The testbench (see Fig. 1) basically consists of two optical tables, measuring 1.2 m × 1.5 m and 0.9 m × 3 m on the sides, which are put together to form an L-like arrangement.

Figure 1: *Photometric testbench, doors open. Rear left: illuminating optics, monochromator, filter wheels, and integrating sphere. Front right: CCD cryostat. The lamp cabinet is located in the background behind the wall to the right.*

The table tops are covered by a light-tight enclosure with doors at each side in order to give access to the interior. Using vertical walls, the whole volume of the box is divided into 3 parts for (1) the light sources, (2) the wavelength selection section, and (3) the measuring cabinet. The latter has at its full length of 3 m a rigid, 90 mm wide optical rail, defining the optical axis of the system (165 mm above the table surface).

The test objects under study may be mounted to sliding carriers, which are clamped at the appropriate position along the rail. For use with an integrating sphere (see below), the axis is parallel to the other, 380 mm above the table surface.

Opposite to the measuring cabinet, the light sources are located in cabinet (1), on the short side of the L. There are 2 alternative light sources available, namely a 100W tungsten filament lamp, and a 75W closed-loop stabilized Xe arc lamp in their respective housings, oriented at a 90-degree angle with respect to each other. Each light source has an UV grade, fused silica condenser lens creating a parallel beam. The desired source is selected by means of a 45° mirror on a linear stage. In addition, a spectral line lamp can be selected for wavelength calibration purposes. A light-tight fan, which is connected to the box by a flexible tube, allows for the removal of heat and the exchange of air in the lamp cabinet.

The beam coming from the light source section enters the wavelength selection cabinet through a small opening, connecting to a baffle, thus preventing any bright stray light from entering the following sections. A 45° mirror reflects the light into the optical axis of the testbench. A series of lenses which are all made of fused silica in order to give good efficiency in the UV, relays the light through the filter wheel unit, and optionally through a monochromator, which may be laterally inserted on a stage. Several stops and baffles are included in order to reduce stray light. The monochromator is an Oriel Model 77200, f = 250 mm, f/4.4 instrument with a motor-controlled grating drive. Two 1200 l/mm gratings with 500 nm and 280 nm blaze, respectively, cover the whole spectral range from 200 nm up to 1000 nm. The smallest possible slit width results in a spectral resolution of 1 Å. Four filter wheels, each with 20 openings of 25 mm diameter, carry various filters, including a set of interference filters (320 ...1100 nm, 10 nm FWHM), a set of neutral density filters (D = 0.1 ... 4), and Schott glass filters for order selection, red-leak suppression, and other purposes. The exit of the wavelength selection cabinet is formed by an electromechanical shutter, followed by a baffling tube towards the next section.

Filter Wheel Positions

	FW 1	FW2	FW3	FW4
0	open	open	Open	open
1	320	0.1	WG 335	UG 10
2	340	0.15	WG 345	UG 3
3	360	0.2	WG 360	UG 5
4	380	0.3	GG 375	UG 11
5	400	0.4	GG 385	BG 1
6	450	0.5	GG 395	BG 3
7	500	0.6	GG 475	BG 7
8	550	0.7	GG 495	BG 12
9	600	0.8	OG 515	BG 14
10	650	0.9	OG 550	BG 18
11	700	1.0	OG 590	BG 23
12	750	1.3	RG 610	BG 24
13	800	1.5	RG 630	BG 25
14	850	2.0	RG 645	BG 26
15	900	2.5	RG 665	BG 37
16	950	3.0	RG 715	BG 38

17	1000	4.0	RG 780	BG 40
18	1050	open	RG 850	RG 9
19	1100	open	KG 5	VG 6

Inside the measuring cabinet, the beam (now diverging at approx. f/8.5), is either propagating along the rail, or is guided into an integrating sphere by means of a small secondary „satellite“ sphere. The two modes of operation are selected by rotating the sphere in or out of the beam. The integrating sphere is a Labsphere Model US-120-SF and has an opening of 100 mm, thus accommodating most of the largest CCDs that are currently available. A small supplementary opening is used for simultaneous monitoring of the intensity produced by the sphere by means of a PIN diode (Hamamatsu S1337-1010BQ) whose current is read by an electronic picoamperometer with GPIB interface (Keithley Model 486). This feature is very useful for maintaining the photometric stability of the system, and, in particular, for absolute flux measurements. The latter is possible by calibrating the monitor against an identical diode with known responsivity which is placed on the rail exactly at the test object position (i.e. replacing the CCD). The target position can be varied along the rail up to a distance of 3000 mm within the testbench enclosure, and, as an extension into the next room, out to 8000 mm.

Instrument control is performed with a VME system under VxWorks, using a TSVME133 with a Motorola 68030 processor, TSVME409 IEEE-488 controller and MACCON MAC4-INC and MAC4-STP as four-channel DC and stepping motor controllers. This hardware is used to drive 4 motors of the filter wheel drive, the shutter, and the stepping motor of the monochromator grating, and to interface the picoamperometer. As a future upgrade, it is planned to replace the manually driven stage by a motorized one to switch the monochromator in and out automatically. The software driving these units is running independently on the VME system. A command line level user interface is provided through a TCP/IP socket interface to a SPARC workstation where operation of the CCD and control of the testbench is possible in combination with the desired data reduction tools (IDL, MIDAS, IRAF).

4. Summary

In conclusion, the testbench provides the following features:

- Diffuse uniform or directed illumination of targets with up to 100 mm diameter
- Wavelength selection with filters and/or monochromator
- Large dynamic range
- Diffuse illumination mode: simultaneous monitoring of intensity with PIN diode
- Manual or fully automatic operation of opto-mechanical functions (filters, monochromator, shutter, PIN diode)
- Remote control from SPARC workstation, integrated into LAN and access to MIDAS, IRAF, IDL,...
- Based on standard lab components, easy access for modification and upgrades

5. References

[1] Janesick, J.R., Klaasen, K.P., Elliot, T., 1987, Opt. Eng. 26, 972
[2] Roth, M.M., Seydack, M., Bauer, S. -M., 1996, SPIE 2871
[3] Schmidt, U., 1994, Proc. *OPTO94*, p.271

HIGH-SPEED DATA ACQUISITION SYSTEM *MESSIA* FOR SUBARU

MAKI SEKIGUCHI[1], *HIDEHIKO NAKAYA*[2], *HIROKAZU KATAZA*[3] *AND SATOSHI MIYAZAKI*[1]

ABSTRACT

We have developed a high-speed data acquisition system, Messia-III *(Modularized Extensible SyStem for Image Acquisition) for Subaru, Japanese 8.2m telescope.* Messia *has a direct 1-Gbps link to a host UNIX workstation.* Messia *is a VMEbus-based system but it does not have local OS nor CPU's in order to minimize software development and maintenance.*

1. INTRODUCTION

Messia-I, the first generation of the system, was developed in 1989 in order to serve the Japanese optical/infrared astronomical community to minimize the time that people spend setting up their data acquisition and CCD controller system for advanced instruments. *Messia* was designed to handle mosaic CCD cameras from the beginning. The basic philosophy of *Messia* architecture was laid out as follows and has not been changed:

1. Minimize software maintenance
2. No real-time Operating System
3. No online data processing in a VME crate
4. Direct connection between a VME crate and a host WS
5. Make use of computing power of a host WS
6. Any exception can be made using user VMEbus modules

Several *Messia-I* systems were manufactured and distributed. *Messia-II,* merely a debugged version of *Messia-I,* was manufactured in 1991 and distributed to more than 10 observatories and instrumental groups in Japan, one of which was 5Kx8K-pixel mosaic CCD group (http://indy.mtk.nao.ac.jp). *Messia-II* is still being used at these observatories and by the groups. In 1996 spring, *Messia-III,* which has much higher data transfer rate and real-time capabilities, has been manufactured and distributed to nearly 20 groups who are developing advanced instruments for Subaru 8.2 m telescope (http://www.subaru.nao.ac.jp). We are going to make *Messia-IV* system as the final version for Subaru in late 1997.

2. ARCHITECTURE

The data acquisition for optical & infrared observation is a very straightforward task. It is not event-driven and does not require real-time control, which involves interrupt handling. It has been often common to use some kind of CPU or processor with real-time operating system. Such a system is very nice and normally much more

1National Astronomical Observatory, 2-21-1 Ohsawa, Mitaka 181 Tokio, Japan.
Email: **maki@indy.mtk.nao.ac.jp,**
2 University of Nagoya.
3 University of Tokyo.

J. W. Beletic and P. Amico (eds.), Optical Detectors for Astronomy, 157–164.

powerful than we need. However, it is often forgotten how difficult long term maintenance of such system is. This is because the real-time system or operating system requires very special knowledge of hardware such as 68XXX, transputers, or DSPs, software such as OS9 or VxWorks, and programming languages such as machine-languages or FORTH as well as their device drivers. With such a system, an observatory has to spend considerable amount of human resources (i.e., very special programmers) for a quite long time normally ranging from a few years to as long as ten years. A typical problem situation is that an instrument operation can not be modified because an important software programmer is no longer employed.

Messia is designed to minimize these problems. There are no local CPUs or processors running an operating system.

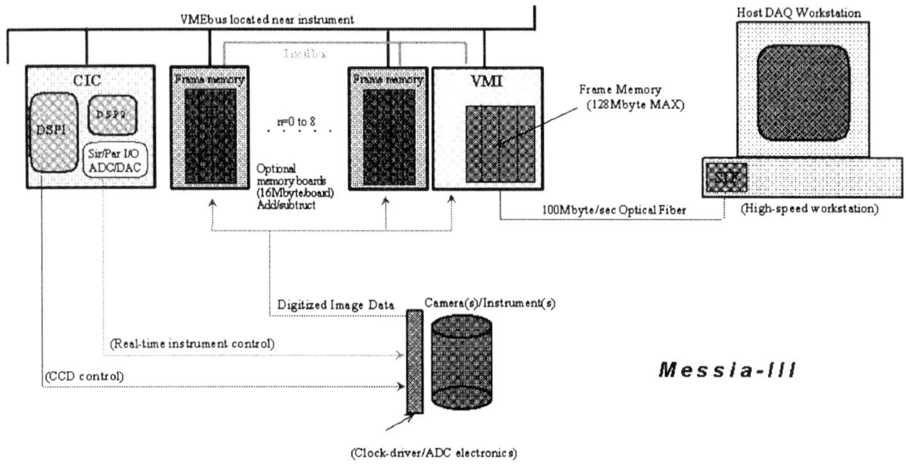

1. Three board minimul configuration : CIC (CCD & Instruments Controller), VMI (VME & Memory Interface), Optional FMM (Frame Memory Module)
2. CIC has two DSPs, one for CCD clock generation and another for real-time instrument control
3. ADC and clock-driver electronics are located near or in a dewar.
4. VMI has on-board frame memory upto 128Mbyte.
5. External memory modules, which have add/subtruct functions, can be connected through a high-speed local bus.
6. VMI and WS are connected by a fibre which is based on Fibre-Channel hardware.

Figure 1:*Block diagram of the Messia-III system*

Figure 1 shows the block diagram of *Messia-III* system. Everything and every I/O port can be directly manipulated by a host UNIX workstation, where a user can freely write his/her own C code using API functions, or Tcl/Tk scripts for their instrument without special knowledge. In *Messia-III*, I/O ports are shared by the access from the UNIX-host and on-board DSPs with an appropriate arbitration.

CCD clock patterns are generated by a hardware sequence (*Messia-I* and *II*) or by a DSP, Motorola DSP56156 (*Messia-III*). *Messia-III* has another DSP56156 for the instrument if a user requires real-time control. In this case, a user is not allowed to write raw DSP code but can only use official API (Application Program Interface) functions for the DSP. The API are well debugged, keeping the maintenance to a minimal level. All digitized data are stored in local memory, which acts as a buffer between a real-time world and UNIX world.

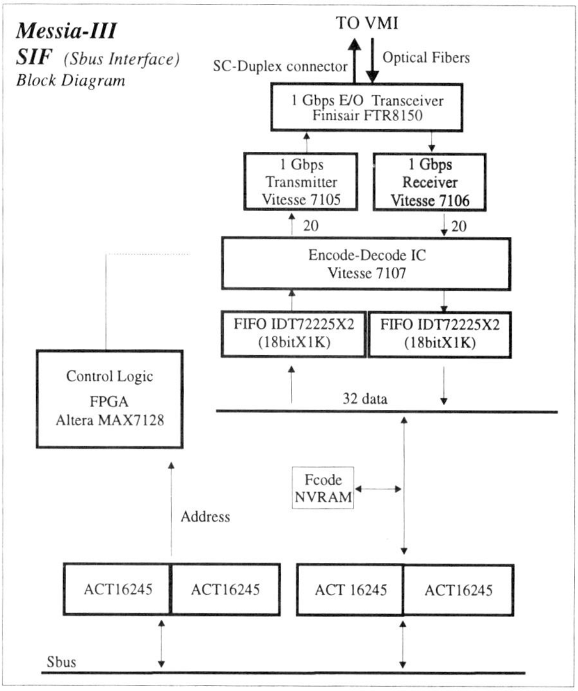

Figure 2: *block diagram of the Sbus Interface (SIF)*

Another difference between *Messia* and other CCD DAQ systems is that the *Messia* VMEbus is directly connected to a workstation bus. A user can make use of the latest high-speed workstation without additional effort. A user does not have to handle any TCP/IP hardware or software problems in order to acquire data. *Messia-I* and *Messia-II* uses 16-bit differential drivers between the VMEbus and SPARCstation Sbus. This type of electric differential connection has been proven to be very reliable up to a distance of 30m and a typical data rate of 6Mbyte/sec. *Messia-III* has a 1-Gbps fiber link between the VMEbus and Sbus. In this case, the data do not go through the VMEbus but go through an on-board bus or high-speed local bus.

We require a high data transfer rate because the mosaic camera used at Subaru has 10k × 8k pixels and the CCDs saturate quickly at the prime focus (see also at http://opal.mtk.nao.ac.jp/subaru/snl/snlv1n2/suprim.txt).

Messia does not include the front-end electronics such as pre-amps, video-ADCs, and clock drivers. These are not considered as parts of data-acquisition system and quite different for optical and infrared instruments. It is often a very good practice to separate the front-end electronics from other parts in order to keep the whole system modular and minimize the system noise especially for large mosaic cameras.

3. INDIVIDUAL MODULES

The minimum configuration of *Messia-III*, has three modules, (1) Sbus InterFace (SIF), (2) VME and Memory Interface (VMI), and (3) CIC (CCD and Instrument Controller). The optional module is Frame Memory Module (FMM).

3.1 Sbus InterFace (SIF)

The block diagram of SIF is shown on Figure 2 and its photograph is shown in Figure 3. SIF, a single slot wide Sbus card, is simply a FIFO buffer connected to the fiber link. A SPARCstation writes/reads the FIFO. The data written to the FIFO is sent to VMI through the fiber link and the data sent from VMI can be read from the FIFO. We use

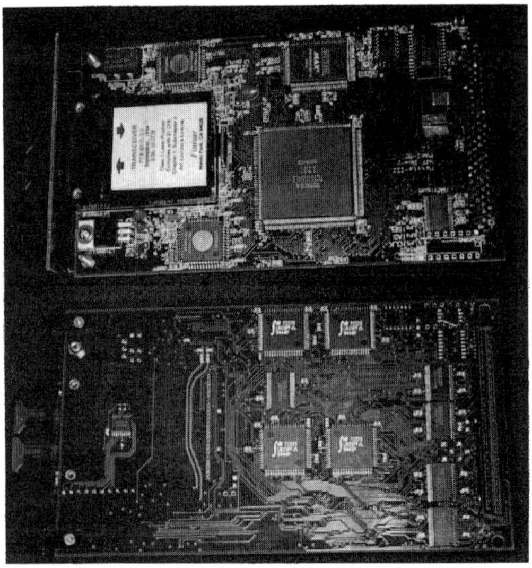

Figure 3: Photograph of the SIF (Sbus Interface) - Front and back

Vitesse VSC7107 (encode/decode), VSC7105 (transmitter), VSC7106 (receiver) and Finisar FTR8150 to convert electric signal from/to light. (see http://www.finisar.com/)

3.2 VME and Memory Interface (VMI)

The block diagram of VMI is shown on Figure 4 and its photograph is shown in Figure 5. VMI, a single slot wide master VME board uses the same Fibre - Channel hardware for its 1-Gpbs link to SIF. Based on the command(s) sent from SIF, VMI reads/writes various memories and buses such as (1) the on-board frame memory, (2) local-bus for external memory modules, and (3) VMEbus. The result and read-data are written into a buffer FIFO and sent back to SIF. VMI has opto-isolated inputs for CCD digitized data. The data can be either eight-channel 16 bit-serial or single-channel 16 bit parallel. The read/write speed for the on-board frame memory is about 20 Mbyte/sec and the maximum memory size is 128 Mbyte. This is good for typical large mosaic CCDs. The on-board frame memory is dual port configuration, which can be read from SIF during write.

3.3 CCD and Instrument Controller (CIC)

The block diagram of CIC is shown on Figure 6 and its photograph is shown in Figure 7. CIC, a single slave VME board, has many I/O channels in order to control both CCD(s) and instruments. CIC can generate 40 CCD clock lines (all differential output) which are controlled by a DSP56156. For instrumental control, it has 16+48 (optional) PIOs (differential output and opto-isolated input), eight-channel of 12 bit ADCs (differential output) and DACs (differential input), two RS232Cs, one serial channel as well as interrupt lines and a real-time clock. All I/Os can be controlled/read/written by a remote host workstation directly (i.e., by direct access from VME), and at the same time, by another DSP56156. A user is encouraged to use the direct control from the workstation, or to use API functions for the DSP only if real-time timing is absolutely necessary. We provide API (Application Program Interface) functions for all DSP operations so that a user does not have to know about DSPs or mess with DSP code.

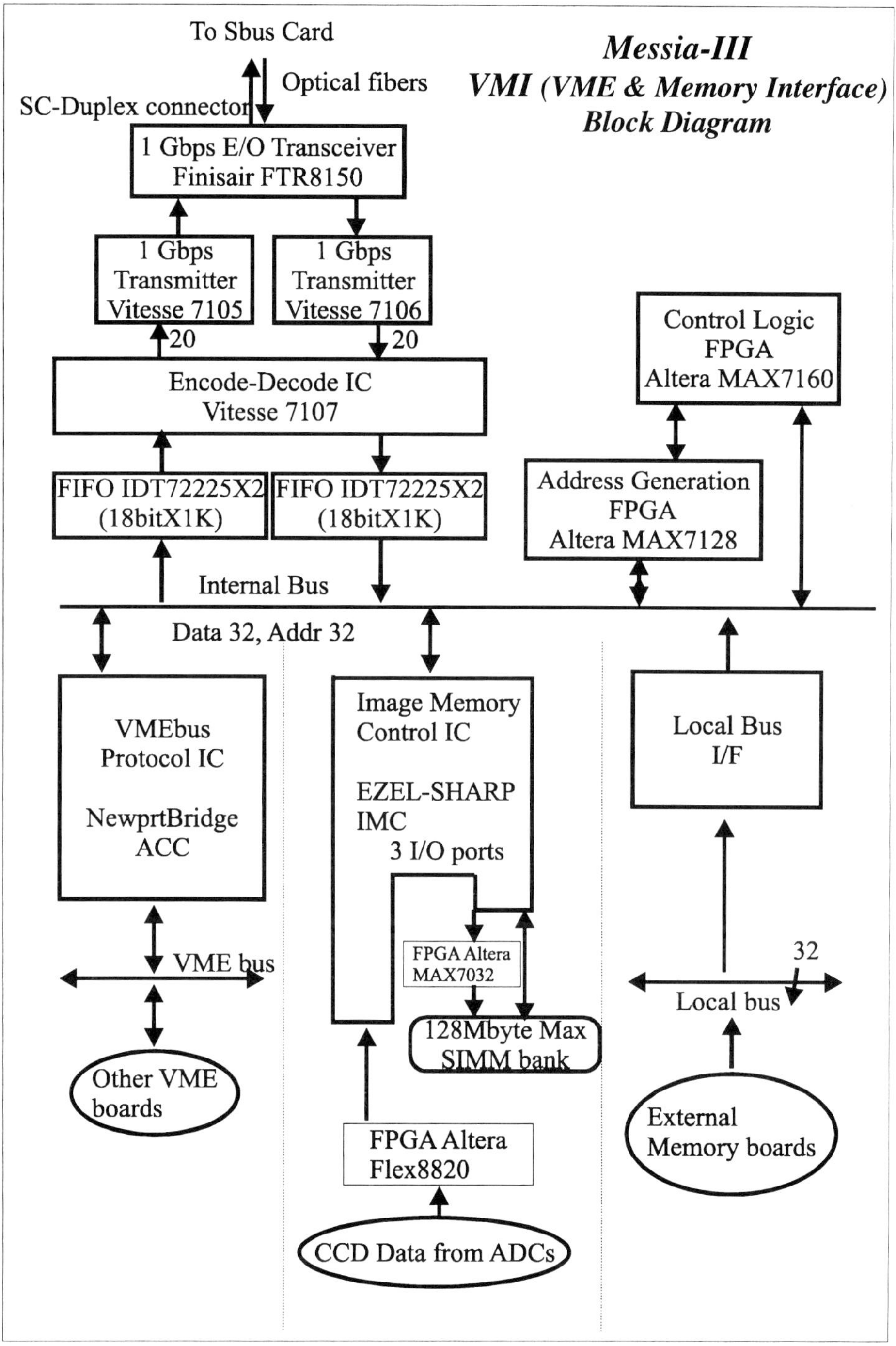

Figure 4: *Block diagram of the VME and Memory Interface (VMI)*

Figure 5: Photograph of the VMI board (VME and Memory Interface)

3.4 Frame Memory Module (FMM)

FMM, a single slot VME slave board, has an on-board ALU, which provides real-time add/subtract functions without any overheads. FMM accepts either 16 bit parallel data or eight-channel serial 16 bit-data. VMI and multiple FMMs are connected by a local bus. VMI can interleave up to eight FMMs on the bus. This would provide the readout data rate of 80 Mbyte/sec although each FMM has only 20 Mbyte data output rate.

4. TOP-DOWN DESIGNING

To make future maintenance of hardware as easy as possible, we adapted a complete top-down design of the system using VHDL (VLSI Hardware Description Language, http://www.acc-eda.com/h_intro.htm). Each entire board is designed and 100% described in VHDL and simulated. We wrote our own VHDL behavior or RTL (register-transfer-level) models of all integrated circuits on a board. We then compiled and synthesize VHDL code for FPGAs on the board. The gate level VHDL code was then fed back to the board level simulation for more accurate simulation. We used V-System (http://www.model.com/vsystem.htm) by Model Technology Inc. for VHDL simulator and Altera FPGA software (http://www.altera.com) for logic synthesis. It has been proven that such a top-down design minimizes:

(1) bugs on actual printed circuits,
(2) turn-around-time to remanufacture PCBs, and
(3) design cost.

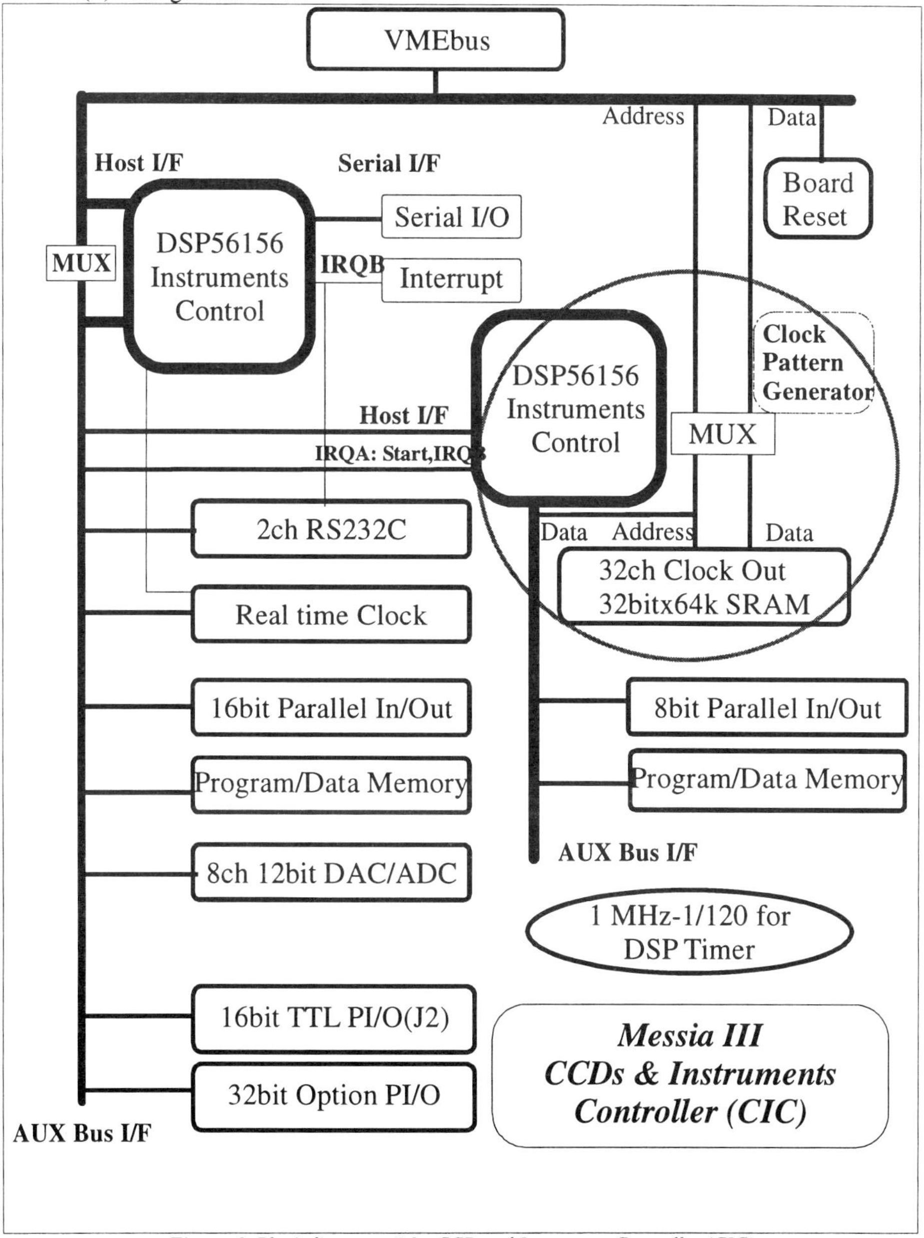

Figure 6: *Block diagram of the CCD and Instrument Controller (CIC)*

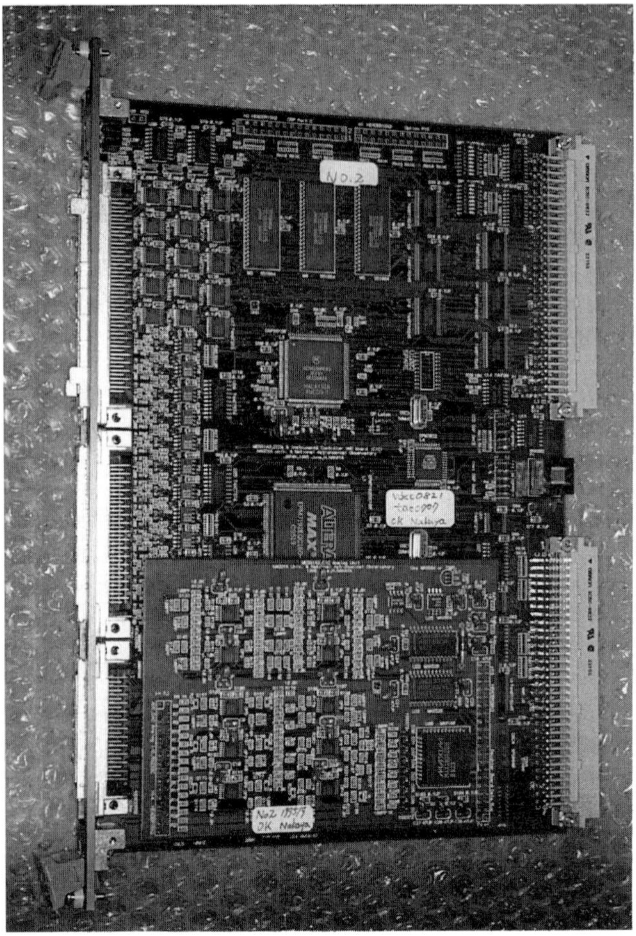

Figure 7: *Photograph of CIC (CCD and Instrument Controller)*

(*Messia-III* is best documented on WWW (http://www.messia.nao.ac.jp) in Japanese.)

READOUT SPEED OPTIMIZATION FOR CONVENTIONAL CCDS EMPLOYING DUAL SLOPE INTEGRATION FOR DOUBLE CORRELATED SAMPLING

ROGER M. SMITH[1]

ABSTRACT

This paper provides guidelines for improving the speed-noise trade-off for double correlated sampling of CCD video signals in astronomical applications. Only the dual slope integration method is considered since this is the most popular for CCDs with conventional output sensitivity(< 2.5 uV/e$^-$). Emphasis is placed on reducing readout time at little or no expense to noise in the 50 to 200 Kpixel/sec regime where a significant fraction of the pixel time can be taken up by processes other than integration on the signal.

1. READOUT SPEED REQUIREMENTS

One of the primary goals for CCD system design has always been to improve the trade-off between read time and read noise. A great deal of effort has been expended in the last 5 years constructing multi-channel CCD controllers to improve the readout time with no noise penalty. A new impetus for improving the readout speed *per channel* comes from the construction of large mosaics using 2k × 4k three side buttable CCDs having four times as many pixels per amplifier as the now common 2k × 2k quad readout CCD.

The most conspicuous benefit of fast readout is the improvement in observing efficiency for broadband imaging where exposures are short or observations of standard stars must be frequently interspersed with science targets. Rapid readout is essential for time resolved observations where the number of target and reference stars in the same frame is sufficient to make Regions-Of-Interest (ROI) an unsatisfactory means of increasing frame rate. For many programs the overriding importance of fast readout is the ability to acquire a complete set of high signal-to-noise flat-fields from the twilight sky, so that the quality of the scientific data is not limited by the instrumental calibration. At the latitudes of most major observatories (20°-30°), there are only about 15 minutes for flat fielding broadband filters, between the time when the sky saturates the CCD and when too many stars begin to appear. Short exposures must be long enough for the shutter time correction to be manageable. The problems encountered when using the twilight sky as a flat fielding source for astronomy are described by Tyson & Gal [3] and Chromey & Hasselbacher [1]. The total observing time required to reach a given signal-to-noise ratio can be reduced by optimizing dynamic range, through optimum gain selection, using *unsigned* 16 bit A/D conversion, and optimizing the CCD

---

[1] Cerro Tololo Inter-American Observatory, Casilla 603, La Serena, Chile. Email: rsmith@noao.edu
CTIO is a division of the national Optical Astronomy Observatories, operated by the Association of Universities for Research in Astronomy, Inc., under cooperative agreement with the National Science Foundation of the USA

J. W. Beletic and P. Amico (eds.), Optical Detectors for Astronomy, 165–184.

for linearity. Nonetheless, to eliminate the effects of cosmic rays it is desirable to take the median of at least three frames for each filter. To be able to acquire 3 exposures in each of 4 or 5 broadband filters, readout time must be less than 30 seconds, and preferably 20 seconds to provide some leeway in choosing exposure times (which vary rapidly), as well as time to change filters, and so on. The usefulness of fast read out for spectroscopy should not be overlooked either. Although exposure times are generally much longer, the total time saved per night on calibration exposures can be significant.

If readout time is reduced to 10 seconds, observing efficiency becomes limited by other overheads such as moving the telescope, changing filters and focus, so the 20 second readout appears to be a good goal. How does 20 to 30 second readout translate into pixel rates for popular CCD formats? Reading a $2k^2$ CCD in 30 seconds requires only 30 μs/pixel/channel through four channels. Most current CCD controllers can read at this speed, and some can do so at quite low read noise. However the 15 μs/pixel required in dual readout mode is beyond the capability of many controllers and the 2k × 4k three side buttable CCD which is becoming the new standard CCD format, requires the pixel time to be reduced further to 7.5 μs/pixel/channel.

The best CCD outputs now have sensitivity in the 5 to 10 $\mu V/e^-$ range and are capable of fairly low noise (−5 e^-) at ≤1 μs/pixel and state of the art noise (−1.5 e^-) at 5-10 μs/pixel. This pixel time cannot be achieved by most 16 bit astronomical CCD controllers in use today, due to signal settling time, or ADC and data link throughput limitations. Fortunately the high output sensitivity eases the task of the designer by amplifying the signal (and CCD noise) on the CCD so that less gain is required in the signal chain, and the available gain-bandwidth product can be allocated to faster settling rather than to amplification. The higher on-CCD gain and much lower output resistance (100-200 Ω, compared to 5 kΩ) make the controller input much more immune to interference. The lower output impedance also greatly reduces the settling time due to capacitive loading of the CCD output by the preamp and input wiring. The low source impedance allows the use of low noise video op amps with very high slew rate. However, these have very high input bias current and thus high current noise, prohibiting their use with conventional CCDs. The much shorter pixel time made possible by these CCDs obviously requires faster settling times in the signal chain and lower serial shift, CCD Reset and ADC sample time overheads.

As yet, few high sensitivity CCDs have been deployed for astronomy, and designers must still be concerned with optimizing the use of the existing inventory of CCDs. Furthermore, it is not yet clear whether the high sensitivity output amplifiers will be sufficiently common for the controller designer to rely on their benefits. Thus, it is the optimization of readout speed from *conventional* CCD outputs with sensitivity less than 2.5 $\mu V/e^-$, which is the theme of this paper.

Examples are taken from experience with Arcon, the CCD controller used at Cerro Tololo and Kitt Peak. It exhibits a very flat speed-noise curve with a variety of CCDs. (The noise degrades slowly as the pixel rate is increased.) The pixel rate of the Arcon is, for the time being, limited by the 10 μs/pixel/channel throughput of the Crystal CS5101A A/D Converter. 10.5 μs/pix/ch has been demonstrated in the lab and 13 μs/pix/ch is in regular use at the telescope. An upgrade is under development which will achieve 5 μs/pixel/channel at only slightly shorter dual slope integration times and thus comparable noise. The primary goal of this paper is not to report on Arcon performance, but to compile a comprehensive description of practical techniques for improving the speed-noise trade-off of any astronomical CCD system.

2. Overlap Dual Slope Integration Operations

I will only discuss the dual slope integrator implementation of the double correlated sampling video processor since this is the most popular in the speed range under consideration. Many of the techniques also apply to Clamp-Sample processing.

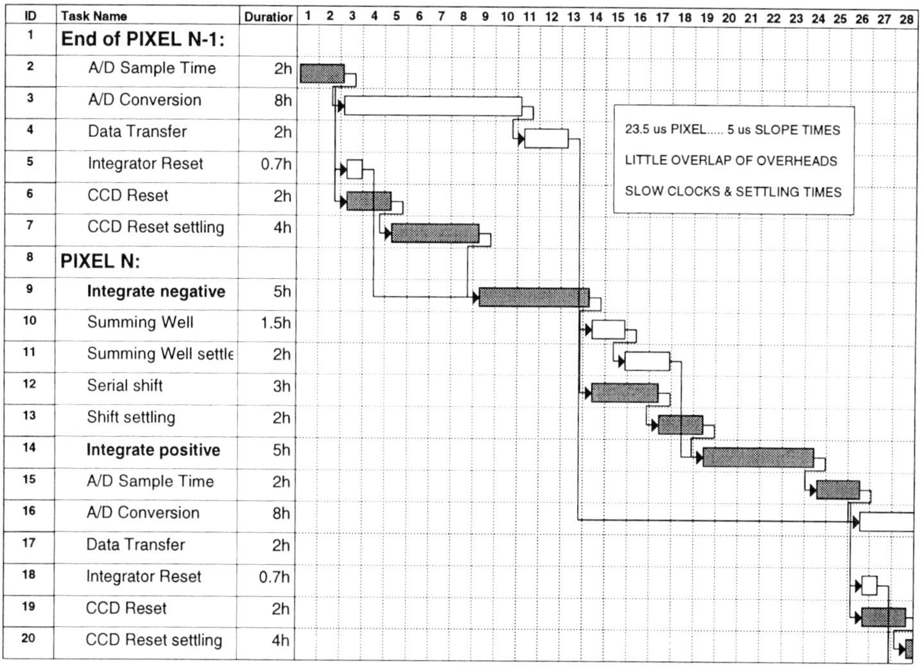

Figure 1: *Time line for a pixel, where operations are performed sequentially and settling times are conservative. Only the A/D Conversion time is overlapped with the subsequent pixel. Tasks on the critical path (solid bars) determine the pixel time. Each "hour" in the duration column represents one microsecond.*

The reader is assumed to be familiar with the dual slope integration method of double correlated sampling. The goal is to measure the charge dumped on the CCD output node in spite of uncertainty in the level after the CCD output node is reset, and the presence of white noise and 1/f noise sources. The baseline shifts are classically described as resulting from *KTC noise*, which is the thermal noise of the CCD, reset switch trapped on the output node capacitance when the switch opens. In practice the baseline subtraction performed by the double correlated sampler is also extremely important for rejecting other low frequency noise sources such as the CCD output FET's 1/f noise, noise coupled capacitively from the clock and bias voltages and any line frequency interference. Optimal noise filtering is achieved (at least for white noise) when the times spent filtering the reset and signal levels are equal. The two integration times must be matched anyway (to high precision) for maximum rejection of low frequency noise sources. The low frequency cutoff depends on the separation of the midpoints of the two slopes, which should be kept to a minimum. Harmonics of the integration time are also very strongly rejected.

The time line in **Figure 1** breaks the pixel into its components, showing the timing and dependencies for the most conservative implementation, where each step is performed sequentially and signals are allowed to settle completely before the next step begins. The goal is to maximize the fraction of the time spent integrating to filter the noise, but in this example only a small fraction of the time is allocated to dual slope integration. Solid bars indicate tasks which are on the critical path and thus extend the pixel time. Fortunately several tasks can be overlapped and thus removed from this critical path, and the remainder can be shortened.

2.1 Overlap A/D conversion with next pixel

The first optimizations to perform are those not requiring circuit changes. The biggest immediate gain is to overlap A/D conversion and data transmission with the subsequent pixel. (This has already been done in **Figure 1**.) The risk here is that the ADC and data links may couple digital interference into the video wiring or preamp. This is fairly easy to avoid with good layout, shielding and power supply filtering, but the most robust approach is to ensure that the digital signals are strictly synchronous with the pixel timing. I will return to this important point.

2.2 Overlap CCD Reset with A/D Sample Time

CCD reset can be brought forward to overlap with A/D sample time provided that this does not induce a transient at the ADC input. I find that the CCD Reset feedthrough pulse is blocked adequately by simply leaving the integrator input switches open and that it is not necessary to clamp the integrator input to ground. No transients are visible on the oscilloscope nor is any noise introduced into the data. If any feedthrough is occurring, it is stable and thus benign.

2.3 Overlap Serial Shift with A/D Sample Time

Thus far, Summing Well has been treated as though it was just another serial clock phase. However, it is possible to shift the charge from the serial register onto the Summing Well (SW), and from there onto the output node at different times. In particular, it is advantageous to perform the more time consuming serial shift, during the CCD Reset rather than during the charge dump. The SW pulse can be made very short since the capacitance being driven is so small and the charge generally moves from SW to output node in only a hundred nanoseconds or so. The SW feedthrough transient contains less energy and the signal chain settles faster if the SW pulse is made very short. (The settling of the signal itself is unaffected by the pulse width since charge dump occurs on the leading edge of SW). The width of the pulse may be limited by the degree of RFI filtering deemed necessary, given that SW couples significantly via about 1 fF of stray capacitance into the 100 fF output node capacitance.

The serial shift feedthrough is suppressed by overlapping it with CCD Reset, since the output node is shorted to the Reset Drain at this time and thus is not affected by small currents induced by capacitive coupling from the serial clocks.

One might consider performing only the low going edge of the Summing Well pulse between the integrator slopes, in order to dump the charge in the minimum time, postponing the rising edge until the deferred serial shift. However, this is inadvisable for several reasons. For all CCDs I have tested, with the exception of the Thomson 1K, the SW feedthrough is a large fraction of the maximum signal amplitude, so that applying a single edge would inject the equivalent of a large pseudo-signal consuming significant dynamic range. To make matters worse, the low frequency noise on the clock driver for SW would no longer be rejected by the double correlated sampling process since different sources would be coupled to the integrator during each slope.

2.4 Overlap A/D sample time with end of signal integration

The sample time of the Crystal CS5101A A/D Converter is divided into a fast-slew period, during which an internal current buffer drives the binary weighted capacitor array, followed by a fine settling time during which the capacitor array is coupled directly to the external ADC buffer to obtain optimum precision at the cost of slower settling. This strategy is designed to minimize settling time for an abrupt full scale input step. Since the dual slope integrator ramps smoothly to the final value, the coarse settling can be overlapped with the signal slope of the integrator. Some high slew rate sample time must still be reserved for the small step induced by charge injection when opening the input switch of the dual slope integrator. For a 1 nF integrator capacitor and 14 pC charge injection, this step is only about 7 mV (100 ADU) at the ADC input. (The Arcon ADC buffer has a gain of one half.)

2.5 Overlap SW and/or settle time with beginning of signal integration

In the "clamp-sample" method for double correlated sampling, an RC filter performs the noise filtering instead of an integrator. The signal filtering period is defined by the falling edge of SW and the clock feedthrough is seen by the filter along with the signal. The advantage of this is that there is no charge dump overhead. The disadvantage is that fluctuations in the clock feedthrough could affect the data.

This should work equally well for dual slope integration: it should be possible to start the 2nd slope shortly after the falling edge of SW, instead of waiting for the SW feedthrough transient to settle. The timing will be stable with respect to the dual slope integration, since they are both driven by the same Sequencer. So, the only side effect should be a change in the bias level. Making SW as short as possible reduces its effect on the bias level, however one might prefer to accept a bias level shift so that more RFI filtering can be applied to SW before entering the dewar.

2.6 Overlap serial clocks with integration time

If A/D sample time plus integrator reset time are shorter than serial clocking and its settling time, one needs to begin serial clocking before the signal integration ends or allow the settling time to occur during the subsequent reference level integration. Once again the only penalty should be a bias level shift that will be compensated by overscan subtraction and should not greatly impact dynamic range. Indeed, it may be possible to time the serial clock feedthrough to partly cancel the SW feedthrough.

3. IMPROVING SETTLING TIMES

3.1 Operational amplifier selection

Clearly, the bandwidth and slew rate of the video chain must be improved as the pixel time is decreased. The Precision Monolithics (now Analog Devices) OP27 bipolar op amp was very popular in astronomical CCD controller preamps due to its combination of low-noise, low input offset, tolerable input bias current, large common-mode input range and moderate power dissipation. Unfortunately, it has poor slew-rate and mediocre bandwidth. Its decompensated counterpart, the OP37 is of course significantly faster but is more difficult to use due to the requirement for higher gain to ensure stability. It must be operated at even higher gain to tame its propensity to overshoot. The FET-input Burr Brown OPA627 has emerged as a new favourite for medium speed applications, since it is guaranteed to settle to 0.01% in only 550 ns. Its voltage noise is remarkably low for a FET input op amp (typically 4.5 nV/√Hz at 10 KHz), but still greater than the OP27 (typically 3.0 nV/√Hz at 1 KHz). For a typical CCD output resistance of 10 KΩ (in parallel with a source follower load resistance of 20 KΩ), the greater voltage noise is largely compensated by the current noise being negligible.

Very high speed op-amps have been developed for video applications with very low voltage noise. For example the AD811 has voltage noise of only 1.9 nV/√Hz , however these require very low source impedance because they typically draw 2 μA bias current with commensurately large input current noise (20 pA/√Hz). Current noise will thus dominate voltage noise for source impedance greater than 95 ohm. This makes them inappropriate for preamplifying signals from conventional CCDs, since these have output resistances of 4 to 12 KΩ.

To reap the benefits of faster op amps, low impedance supply distribution and low stray capacitances are essential. Step response will be improved by using multilayer printed circuit boards with internal planar grounds, with oversized (100 nF) high frequency bypassing capacitors placed very close to the op amps. Attention must be paid to the width of traces to the bypass capacitors and placement of vias to supply and ground. The compactness of surface mount components is advantageous for reducing lead capacitance and inductance, both in the circuit board and within the IC package itself. Surface mounting components over planar grounds also improves conductive heat dissipation through the circuit board. Frequency compensation capacitors as small as a 2-3 picofarads are highly beneficial (or essential) for fastest settling even for amplifiers which are nominally stable at low gain. These should be adjusted empirically (±1 pF) since the optimum value will depend on parasitic capacitances at the summing node.

3.2 Reset Gate feedthrough

A 200 to 300 mV positive going transient is induced on V_{OS} by capacitive coupling between the gate of the CCD reset switch and the output node. This CCD Reset feedthrough appears to be opposite in sign but similar in amplitude to a full scale signal and thus carries a potentially serious settling time penalty. It is especially important to ensure that reset feedthrough does not cause op amp saturation, since some fast op-amps, notably the Burr Brown OPA627, can take microseconds or even tens of microseconds to recover (depending on the polarity of the saturating signal). Circuits to deal with this include diodes across gain setting resistors in the preamp (a linearity hazard), clamping switches to reduce gain during the transient, and wider supply rails to provide more headroom. The best solution is to attenuate the transient at its source.

Fortunately, the amplitude of the reset feedthrough can be moderated by increasing the low level of the Reset Gate clock. Changing RG_high has no effect since the CCD output node is connected to Reset Drain while the reset is active, and is thus determined solely by V_{RD}. The feedthrough begins to occur as Reset Gate (RG) falls below the threshold of the reset FET so that the reset switch opens. Thereafter, its gate-source capacitance (≈2 fF) can no longer be charged from Reset Drain, and charge is redistributed between it and the CCD output node capacitor ($C = Q/V = 1.6\times10^{-19} / 2\ \mu V/e^- \approx 80$ fF). Thus, the voltage change on the output node is proportional to the difference between the Reset FET's turn-off threshold and the low level of RG. Contrary to its appearance, the feedthrough pulse is really a high-duty cycle low-going pulse.

If RG_low is set too high then the reset switch does not turn off fully. The reset FET's threshold is about 7-9V below V_{RD}. For V_{RD}=13.5V, RG_low can be increased from 0 V (typical recommended) to about 3V, reducing feedthrough from about 250 mV to about 100 mV.

The settling time of RG may be governed by the shape of the driving pulse rather than the settling characteristics of the preamplifier, especially when RFI filtering is employed on all clocks entering the dewar. With this in mind, it pays not to drive the pulse any further positive than necessary so that the settling time of the trailing edge is minimized. The interference coupled via RG may need to be traded against settling time, by adjusting the RFI filter cut frequency, unless all sources of digital interference within the controller are strictly synchronous with the pixel timing and are stable in amplitude.

3.3 Integrator Reset

While the slew rate needed to follow the integrator slopes is not demanding, a fast op amp is required to minimize the integrator reset overhead, except in the case of "Interleaved Double Correlated Sampling" discussed below. This becomes important when the sum of A/D Sample and Integrator Reset times exceeds the CCD Reset Feedthrough plus settling time. The Integrator Reset settling time requirement is based on the desire to maximize contrast between a bright pixel and its successor. The OPA627 makes an excellent integrator op amp due to its combination of negligible input leakage current, low noise and fast settling: the manufacturer guarantees 450 ns to 0.1% and 550 ns to 0.01% for a 10V swing.

One must also ensure that the RC time-constant formed by the integrator capacitor and its reset switch does not limit the settling rate. High speed switches such as the HI-201HS or DG-271 with 50 ohm nominal impedance typically measure at ≈35 ohms in this application. Thus, to settle to 0.1% in 450 ns, the integrator capacitor must be kept below 1.9 nF.

3.4 Capacitive loading of CCD output

For conventional CCDs the FET output resistance (1/transconductance) can be as great as 10 KΩ. The transconductance (g_m) of the CCD output FET can be inferred from the gain of the source follower, determined by measuring the static response of V_{OS} to variation in the V_{RD} voltage:

$$\Delta V_{OS} / \Delta V_{RD} = g_m R_D / (1 + g_m R_D) \tag{1}$$

The output resistance of the FET is $1/g_m$, by definition, so that the equivalent source resistance of the entire source follower is,

$$(1/g_m) \parallel R_D \approx 10\ \text{K}\Omega \parallel 20\ \text{K}\Omega \approx 6.7\ \text{K}\Omega$$

The capacitance of the preamp input and its wiring must be kept quite small to achieve reasonable settling times. For example, in the Arcon, the capacitance of the 55cm (AWG#30) twisted pair from the CCD to preamp (plus the capacitance of the preamp itself) is 45 pF. Thus, in this fairly extreme case, $\tau \approx$ 290 ns and the settling time to 90% is 675 ns, and to 99% is 1.4 μs, requiring some overlap of the CCD Reset and Summing Well settling with the integrator slopes as discussed above.

The capacitance is dominated by wiring so there is scope for a 3 fold reduction in the time-constant by moving the preamplifier closer to the detector. The time constant can be dramatically reduced by adding a source follower FET such as a Siliconix U309 with much lower output resistance (100 Ω) as is done with many high output sensitivity CCDs. This implies that the 20 KΩ load resistor be moved into the dewar, which is a good idea anyway since its noise is reduced dramatically by cooling, and its ground is then referenced correctly to the CCD substrate. One must ensure that the noise of such a buffering FET (6 nV/√Hz for the U309) is significantly lower than that of the output FET of the CCD. These measures may not be required until the pixel time is pushed below 5 μs/pixel, and are certainly not required at 10 μs/pixel, although there may be other benefits such as immunity to crosstalk and interference.

4. TOTAL SYNCHRONISM

A recurring point in the previous discussion is that interference patterns can be converted to a constant offset by running all electronic signals perfectly synchronously with the pixel timing, provided that the timing jitter is low. The bias level shift resulting from strictly synchronous (constant) interference is then removed by overscan subtraction. However, perfect synchronism is not easy to achieve. One must pay careful attention to the execution of any program loops to ensure that the same instructions and data are executed at any given instant within successive pixels. The timing of a state machine is easier to control than that of a program loop.

The pixel time must be an exact multiple of the periods of all derived clocks. It is particularly important to synchronize the A/D sample time with the pixel timing. In the Arcon, pixel timing is in units of cycles of the 25 MHZ transputer clock. The ADC requires an approximately 8 MHZ clock, which is derived by dividing the transputer clock by 3. It is a common misconception that this is sufficient for synchronization. If the pixel time is not an exact number of the ADC clock periods (a multiple of 3 sequencer clocks) then the ADC clock will beat with the pixel control signals. This is inconsequential if the A/D sample time is long enough for the signal to entirely settle, however when squeezing the sample time to the absolute minimum a 3 pixel periodicity will appear in the data as the end of the sample time moves up and down the signal settling curve. This is easily avoided by padding the pixel time (and parallel transfers, serial skips, etc.) to be a multiple of the ADC clock (i.e. a multiple of 3 sequencer clocks). The 2 μs sampling time can then be cut approximately in half with the only effect being a slight change in the overscan level.

What happens if the ADC does not accept an external clock? It is conceivable that the start-convert strobe could start a train of ADC control pulses which remain in sync with the pixel timing, but it is more likely that the ADC has a free running internal clock to which the actual end of the sample time will be resynchronized. If the signal has not fully settled, then a pattern will appear in the data at a beat frequency formed between the ADC and pixel clocks, without the option of padding the pixel time to bring them into sync since they are no longer derived from the same master oscillator.

Fiber links encoders and decoder clocks should obey similar rules. Data should be transmitted synchronously with the pixel timing and not at a time determined by the host. Data FIFOs should be located at the host end of the link and packet or byte-wise handshaking should be avoided, since the timing of the acknowledge or handshake will depend on the host. Status updates, telemetry gathering and any other asynchronous activity should be disabled during readout.

As the pixel time is reduced, it becomes necessary to admit wider bandwidth into the dewar, with the risk of introducing low level patterns into the data. This is a different speed-noise trade-off to the one based on filtering the CCD noise. As the dual slope integration and A/D sample times are overlapped with the settling times of RG, SW and serial clocks, amplitude modulation of these transients has a greater effect on the data. All signals entering the dewar from the Arcon pass through surface mounted slope control filters. Their effectiveness as RFI filters depends on a low impedance path to the ground star-point located on the backplane. The filters are thus located on the backplane too (where the cable to the dewar is terminated) and the backplane is a printed circuit board with multiple planar ground planes to minimize the impedance to the ground star point. To enhance high frequency performance series inductors are used in the filters, and "feedthrough" type capacitors are selected for their negligible ground lead inductance. The series inductors also serve to produce a more ideal (trapezoidal) clock shape.

The extent to which synchronization is necessary will depend on the pixel rate required and the care taken to block the interference throughout the controller. In Arcon, the readout waveforms are padded to synchronize them to the ADC clock, so that complete settling of the A/D Sample is not required. All *direct* influences on the signal are strictly synchronous. However, strict synchronism is not enforced: the transputer which packetizes the data executes an asynchronous program and communicates over a fiber link employing byte-by-byte acknowledge, such that transmission timing depends on the host response. A FIFO is placed between the ADCs and this transputer violating another of the guidelines. This software based communication protocol allows the same fiber to be used for command, status, data and software download. The cost of this flexibility is the requirement for more care in PCB and wiring layout, grounding, shielding and filtering of both signals and power supplies, which has been successful but perhaps not the easiest solution. A very low level pattern is sometimes visible in bias frames, but it makes a negligible contribution to the noise statistics, and generally cannot be seen in line or column plots. We verify that the pattern is not significant by taking the FFT of each line of the image and then averaging the magnitude to obtain the power spectrum for all frequencies above the line rate. We see that the energy in the spikes is well below the random noise.

5. FLATTENING THE SPEED-NOISE CURVE

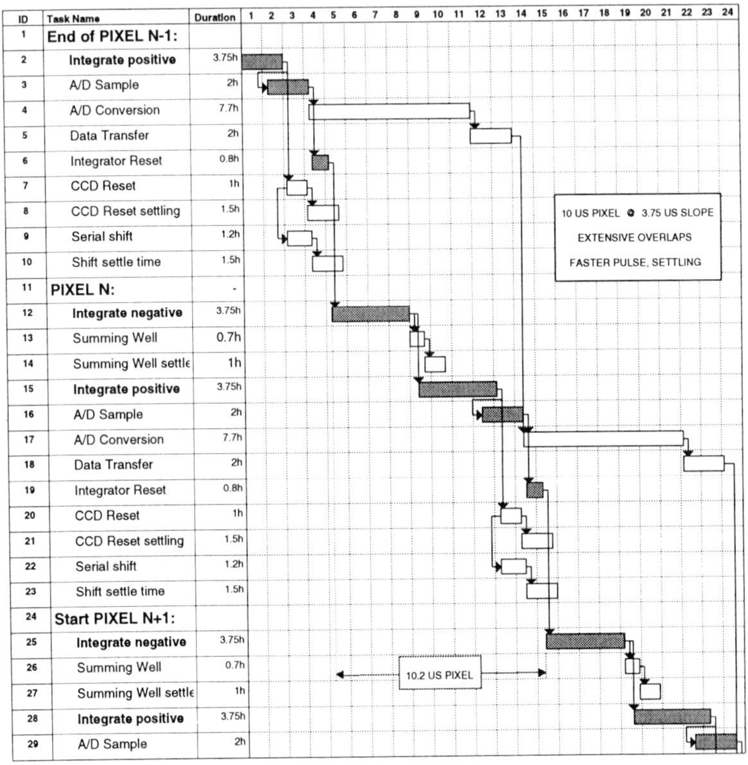

Figure 2: Time line for a pixel with heavily overlapped overheads and more aggressive pulse widths and settling times.

Figure 2 illustrates how the various timing overlaps enable the dual slope integration time to dominate the pixel time, as desired. The speed-noise curve is then the optimum permitted by the noise spectrum of the CCD and electronics. We have found that the speed-noise trade-off is not very steep for the CCDs in use at Cerro Tololo (Loral, Thomson, Tektronix & SITe). Thus a reduction in dual slope integration time results in only a modest noise penalty, provided that quantization and post-amp noise are not allowed to become significant. This requires switchable gain to compensate for the gain loss in the dual slope integrator as slope time is reduced so that postamp noise, ADC quantization and ADC noise do not become significant. Thus minimizing these noise sources and providing switchable gain are in fact tools for improving readout speed.

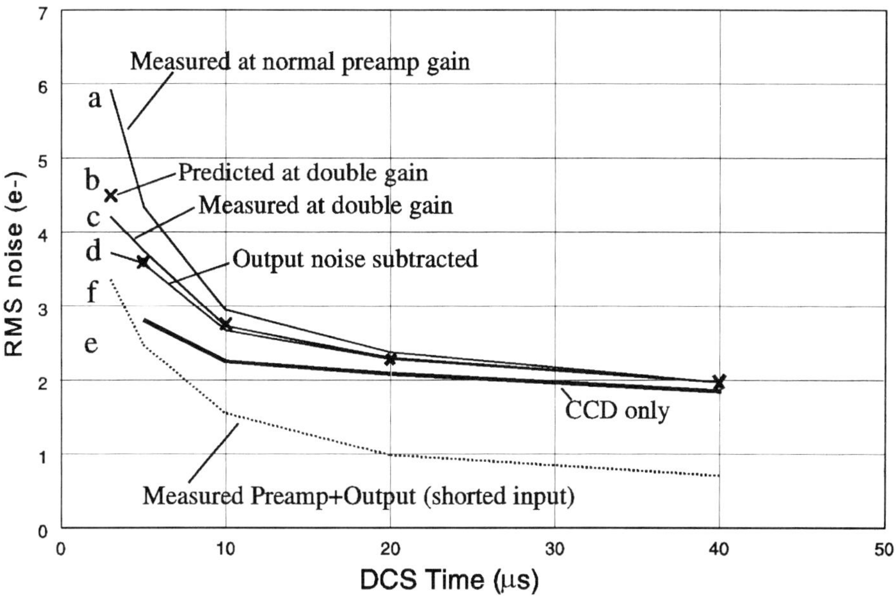

Figure 3: Noise versus double correlated sample time (one slope of integrator). CCD, preamp and output noise components can be deduced from these curves:
.a) measured in overscan = CCD + Preamp + ADC
b) measured noise for twofold preamp gain increase (solid line only)
c) predicted noise for twofold preamp gain increase (symbols only)
d) measured noise –minus output noise (quadr.) = CCD + Preamp
e) CCD noise only = $\sqrt{[(a)^2 - (f)^2]}$
f) Noise with shorted input (includes output noise)

Figure 3 shows the effects of gain switching on the speed-noise curve. It is widely believed that the decrease in noise with increasing slope time is primarily due to the band-limiting effect of the dual slope integration. However, in practice the noise variation is much weaker, due to a increasing low frequency noise just below the typical pixel rate (the so called "1/f noise" component). The steep increase in noise at short slope times is more often due to the increasing influence of the *output noise sources* as gain is reduced. This confusion arises from the widespread use of the dual slope integrator as a gain control element as well as a noise filter. The problem is most severe for CCDs with a large ratio of well capacity to noise floor.

A typical trade-off is to select a gain of 3 to 4 e^-/ADU at short integration times (say 5 μs each) so that the input range of a 16 bit ADC reaches full well (–200 Ke^-) for applications such as broadband imaging. When read noise has priority, 15 to 20 μs per integration is chosen so that gain is 1 e^-/ADU. This has 3 effects: to limit input noise bandwidth and thus achieve perhaps 3 e^- noise, to ensure that the CCD noise dominates post-amp noise, and to increase the noise to 3 ADU so that quantization and differential non-linearity (DNL) have negligible effect. Curve (d) in **Figure 3** shows the speed noise curve predicted for the case where the integrator capacitor is varied in proportion to the slope time so that the gain of the integrator is made independent of slope time. Curve (a) shows the normal case where gain varies in proportion to slope time.

To calculate the total noise as a function of integrator gain, noise sources must be classified and scaled to the equivalent input or output noise, depending on whether they occur before or after the dual slope integrator. The noise of the integrator itself is output referred since the integrator acts as a unity gain buffer for its own voltage noise, and only filters noise occurring earlier in the signal chain. For integrator slope time, τ, input resistance, R, and feedback capacitance, C, integrator gain is given by τ /RC.

$$\text{(Total noise)}^2\ [\text{ADU}] \propto (\tau/RC)^2 . \Sigma(\text{input referred noises})^2 + \Sigma\ (\text{output referred noises})^2$$

$$\text{(Total noise)}^2[e^-] \propto \Sigma\ (\text{input referred noises})^2 + (RC/\tau)^2\ \Sigma\ (\text{output referred noises})^2$$

Without gain switching the noise versus speed curve is dominated by the output noise at short pixel times where it is almost constant in ADU. This translates into a noise which varies inversely with slope time, when converted to electrons. The speed-noise curve can be quite steep even though most noise sources, including the CCD, tend to be fairly invariant with frequency. To compensate, the loss of gain in the integrator must be compensated electronically. I believe the best way to do this is to switch the integrator feedback capacitance.

The gain increase should be as small as possible for good noise in order to minimize the loss of dynamic range. The trade-off is improved by reducing output noise to the theoretical limit imposed by quantization. In the Arcon, the Crystal CS5101A ADC delivers 35 μV noise in unipolar mode as claimed in the specifications. This corresponds to the theoretical ½ ADU limit imposed by quantization! The anti-aliasing filter consists of a 50 Ω resistor and 1 nF capacitor. A passive filter is chosen for its low noise and predictable high frequency behavior (active filters suffer from poor power supply rejection at high frequencies). The capacitor is located very close to the analog input of the ADC and its ground is placed adjacent to the analog ground reference of the ADC. To obtain this performance, it is important to minimize the impedance to the central ground star point from the both the video circuitry's ground reference and ADC ground. Both the analog and *"digital ground"* pins of the ADC must be connected to analog ground! Incoming digital signals should be buffered by logic gates with clean ground and supply to block digital noise from entering the ADC via its control lines. These measures are beneficial to Differential Non-Linearity as well as ADC noise.

The equivalent noise bandwidth for the ADC buffer and integrator (set by the anti-aliasing filter) is 5 MHZ, and so their noise densities must be very much less than $35\mu V/(5\ MHZ)^{1/2} = 15.7\ nV/(Hz)^{1/2}$. This requires a low noise op amp such as the OPA627 or OP27, with fairly low resistor values (1-2 KΩ). Other FET input op amps which are candidates for the integrator tend to be marginally too noisy (e.g. 11 $nV/(Hz)^{1/2}$ for the classic low noise FET input op amp, the National LF356). The integrator noise and charge injection errors are halved by configuring the ADC buffer as a divide by two stage. (The 4.5 V input range of the CS5101A ADC then corresponds to 9V at the integrator output.) Configuring the ADC buffer as a differential receiver allows the integrator to be placed on a separate card without interference or crosstalk problems.

6. Variations on Double Correlated Sampling with Integrators

6.1 Digital Double Correlated Sampling

In the Arcon prototype, I used a variation of dual slope integration in which the subtraction was performed digitally. The post-CCD-Reset level was integrated (not its inverse) and digitized. The process was repeated identically after the charge dump, then the difference was taken digitally. This has the obvious speed *dis*advantage of an extra A/D conversion and integrator reset, but the advantage that all signals pass through the same path with the same timing so that no tuning is required to get excellent rejection of low frequency noise and interference, and very good bias stability. It is most suitable for the lowest noise applications where slope times are long and the A/D conversion overhead is inconsequential. At long slope times gain is sufficient to boost CCD noise above quantization noise, which is especially important since this is encountered twice per pixel when using digital subtraction. Digital subtraction requires the ADC to accept bipolar signals if the preamplifier is AC coupled. Since the coupling capacitor does not pass a DC component, the post-reset level equilibrates at minus half the average signal. This costs a factor of two in effective ADC resolution, unless a DC restoration circuit is employed. DC coupling does not suffer from this drawback, but at high gain it is necessary to ensure that input offset drifts do not drive the output below the unipolar ADC input range.

6. 2 Incremental Sampling Proposal

Digital Double Correlated Sampling suggests a method which could nearly double read rate without noise penalty, which I will call "Incremental Sampling". Nearly half the pixel time would be eliminated if the output node was reset once prior to a group of pixels instead of each pixel. The first pixel would be read by normal dual slope integration except that the CCD output would not be reset afterwards. Charge would be added to the output node (as for serial binning) with a single integration and A/D conversion after each serial shift: the signal would then be recovered by differencing the pixels binned on chip. Provided that gain is high enough for CCD noise (including the shot noise of the signal) to exceed quantization noise, there is no noise penalty. The 1/f noise of the CCD (and its bias levels), KTC noise, and shot noise of the previous pixels would be removed by the subtraction. Since the capacity of the output node of the CCD is generally designed to be about four times that of an imaging pixel, resetting after every 4th pixel might be a good choice. For low light imaging one might reset only once per line. This mode would complicate attempts to correct for non-linearity, but such correction is rarely necessary since gain variation due to CCD output non-linearity is typically only a fraction of a percent for a full scale signal change [2].

There may be some noise *advantage* to this method. Consider the simple case of aperture photometry, in an instance where (by chance) the aperture width lies entirely within one pixel group. When the pixels within the aperture are added, the total noise will be identical to standard dual slope integration if limited by shot noise or by 1/f noise, but the broadband (white) noise component of the read noise is only incurred once. The addition of pixels simply reverses the subtraction performed to extract the individual pixel values. The intermediate non-destructive readouts have been ignored and the result is the same as for on chip binning in the serial direction, but without the loss of spatial resolution The noise advantage is diminished of course if the aperture does not span a single "pixel group". I expect the noise advantage to be greatest at short pixel times where 1/f noise is less important. At short pixel times, the speed advantage will be less than a factor of two due to A/D Sample and Integrator Reset overheads. At long pixel times the noise advantage is diminished due to 1/f noise, while the speed advantage is closer to a factor of two.

6.3 Interleaved Double Correlated Sampling

How can one push the pixel rate significantly higher than 100 kpixel/s/ch with minimal performance penalty in order to take advantage of CCDs with high output node sensitivity (low noise at high speed)? It is desirable to design a video processor which can operate respectably at 400 Kpixel/s/ch *without sacrificing performance at lower pixel rates*, since some CCDs will be operated in both regimes and/or an observatory will operate both kinds of CCDs using a single circuit design. The most obvious approach is to employ faster op amps and ADCs. The fastest *monolithic* 16 bit ADCs currently manufactured are the Analog Devices AD976A and AD977. These are switched capacitor successive approximation converters with 200 KHz throughput. Their specifications are equivalent to the exemplary CS5101A converters, which have been so successful in the current generation of CCD controllers, except that the input noise is 1 ADU rather than 0.5 ADU. They dissipate 100 mW and occupy only 0.8 cm^2. Beyond this speed, hybrid bipolar technology is required and the power dissipation and size increase about ten fold.

The AD976A requires a sample time of 1.5 μs. The dual slope integrator op amp should be a FET input type to ensure low DC drift and low current noise. Op amps such as the OPA627 take 450 ns (or more) for integrator to reset to an adequate level to ensure that successive pixels are truly independent. Unfortunately, at 200 Kpixel/s at least 40% of the pixel time is consumed by activities other than filtering the noise with this Integrator/ADC combination.

The obvious solution is to use a much faster ADC with a sample time of only a few hundred nanosec. This has the drawback that the 10 fold increase in bandwidth into the sample-and-hold will pass $(10)^{1/2}$ times more noise from the preceding stages (dual slope integrator and ADC buffer). For example an OPA627 integrator op amp followed by an AD811 ADC buffer configured to divide by 2, with 250Ω feedback resistor will have a total noise of ~7 nV/(Hz)$^{1/2}$. With an equivalent noise bandwidth of about 18 MHZ (required for settling to 16 bits in 150 ns), the noise will be about ½ ADU for a 5V input range.

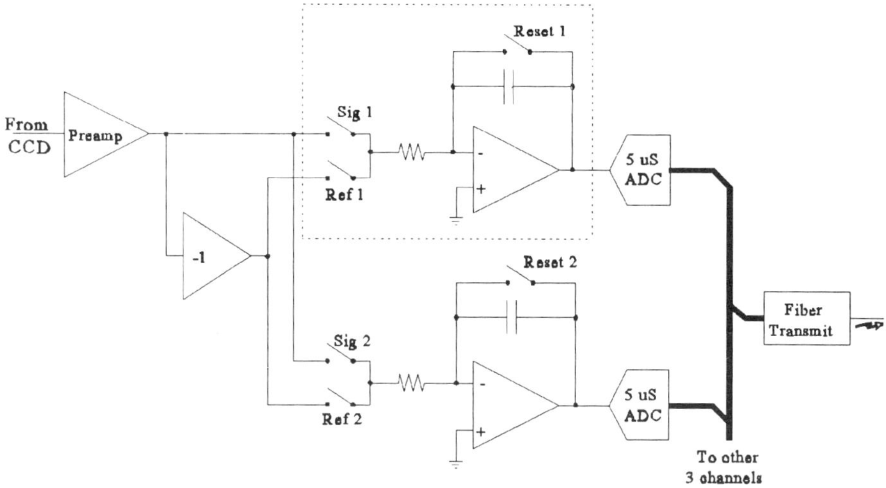

Figure 4: Block diagram of interleaved correlated double sampler (ADC buffer not shown). The standard dual slope integrator is enclosed in the dotted rectangle

I am currently testing an alternative approach using two dual slope integrators for each preamp, each with its own monolithic ADC. While one performs dual slope integration, the alternate integrator is sampled and reset as shown in **Figure 4** and **Figure 5**. In this way, the sample time and integrator reset time can be long, since they are hidden. AD converter throughput is doubled. A pair of AD796 ADCs will deliver 4 times the throughput of a single CS5101A while consuming less power and fitting into nearly the same space. The additional dual slope integrator requires few parts.

Each path will have slightly different gain and offset. The gain difference is removed by standard flat fielding, and to make the raw image look acceptable the gain can be trimmed electronically to better than 1%. Offset differences need to be removed by a modified overscan subtraction where odd and even columns are handled separately. I am using an output offset DAC to null the offset differences to about 0.1 ADU precision, so that raw images will look acceptable. While this will improve the appearance of raw images for Interleaved Double Correlated Sampling, it is primarily motivated by the larger offset differences between output channels in mosaics and CCDs with multiple amplifiers.

7. More CCD Outputs

At any given readout time, the noise performance can be improved by reading more slowly through more amplifiers. Another significant advantage is that a CCD system with redundant signal paths suffers only a speed penalty if some paths fail, since the image can be shifted towards an alternate output. This both reduces down-time in the event of a controller malfunction, and protects against CCD aging which can cause permanent serial register or output amplifier deterioration. Similarly, CCD manufacturing yield is increased in the sense that, rather than being rejected, a quad readout CCD can be downgraded to dual readout if one serial register is defective. Efforts to build mosaics from three side buttable 2k × 4k CCDs have led to construction of large CCDs with only one serial register, forfeiting the yield and performance advantages of quad readout.

It is worth examining whether a single mask set might be used to build both quad and dual readout 2k × 4k CCDs. The gap at the top of the three side buttable CCD cannot be reduced to zero since some clearance is needed between the CCD and the edge of the base plate to avoid chipping the edge of the CCD. I propose that the gap specification be relaxed in order to add a second serial register. This is preferable to segmenting a single serial register since column defects are not limited to half the CCD height, and charge can be read through any of the amplifiers. Exposures at several different positions are required to fill the image area (and to remove the effects of bad columns etc) anyway, so the size of the gap is not really critical provided that it remains a small fraction of the total area. This philosophy has been taken to the extreme by Tyson and Bernstein in the "BTC", a two by two mosaic of $2k^2$ CCDs separated by 800 pixels (http://www.astro.lsa.umich.edu/ btc/btc.html).

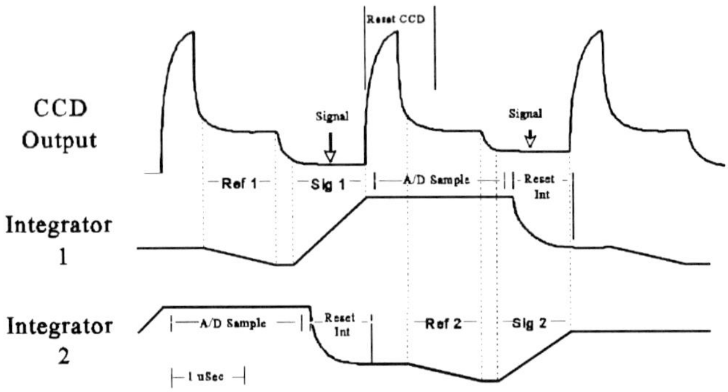

Figure 5:Timing waveforms for interleaved double correlated sampling. AD sample time and integrator reset are hidden within the subsequent pixel

Ideally all 2k × 4k CCDs would have two serial registers with amplifiers in each corner and then be packaged as 2-side or 3-side buttable (dual or quad readout) according to amplifier performance, trap locations, and so on. Increasing the gap by the serial register width would not drastically increase the CCD separation, especially if the bond pads and associated traces could be sawn off when the CCD wafer is diced, since these occupy more space than the serial register itself. A way to disable the unused serial register would need to be devised, and the parallel clock busses for the upper and lower halves of the CCD would need to be split or joined when packaged as 2 or 3 side buttable devices respectively. Wafer probing is needed to reap the yield advantage, but this is normal practice to avoid packaging rejects. An even more promising avenue would be to devise a method to package a CCD to be buttable on *all* sides by running traces to contacts lying within the profile of the imaging area (on the unilluminated side of course).

8. READOUT TIME FOR REGIONS OF INTEREST

A Region Of Interest (ROI) is a traditional method for obtaining higher frame rate when the full area of the CCD is not required. **Figure 6** compares the time spent shifting lines and skipping pixels to the time spent reading pixels for various sized square ROIs located on the centerline of a dual readout CCD with a single serial register. Neglecting preskips and serial register flushing after erasing a group of lines, the major components of readout time are:

ROI dimensions	$x = y = kX$	Fraction of width	$0 < k \le 1$
		CCD columns	$X = 2048$
Total line transfer time	$= YL$	CCD lines	$Y = 4096$
		Line shift time	$L = 270\ \mu s$
Total pixel skip time	$= \frac{1}{2}(X-x)yS$		
	$= \frac{1}{2}(1-k)kX^2S$	Pixel skip time	$S = 1.5\ \mu s$
Total pixel read time	$= (x/2+C)yr$	Overscan columns	$C = 30$
	$= (kX/2+C)kXR$	Pixel read time	$R = 10\ \mu s$

Parallel transfers dominate the readout time for small regions and pixel read time dominates for large regions. Pixel skips will limit readout time only for a small range of intermediate sizes, and often not at all (as in this example). Pixel skips can be more problematic for regions with high aspect ratio, as are common for spectroscopy. If the CCD orientation can be controlled, the long edge of the ROI (dispersion direction) should be made parallel to the rows, since the line transfer overhead is independent of region size or shape: each line is transferred into the serial register once whether it is being read, binned or skipped.

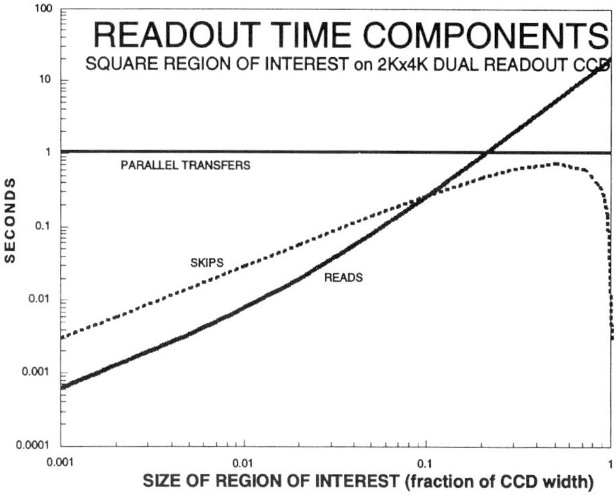

Figure 6: Contributions of line shifts, pixel skips and reads to the readout time of a square Region Of Interest on the centerline of a dual readout 2k × 4k CCD. 270 us per line transfer, 1.5 us per pixel skip, 10 us per pixel read, 30 overscan pixels per line, no preskips

The fastest way to erase the CCD or skip lines is to set all the serial clocks high, close the output reset switch, then transfer all lines to be skipped into the serial register at the normal parallel transfer rate. The serial register fills to the point that charge flows over the potential defined by the Last Gate electrode. Before transferring the first line of the ROI into the serial register, the residual charge in the serial register must be flushed out by normal serial shifting.

As pixel rates are increased in response to growing CCD size, the line transfer time becomes more significant especially for ROIs. As CCDs grow in width, the capacitance and resistance of each polysilicon electrode in the imaging area increases, so that the intrinsic time constant of the distributed RC transmission line can be expected to increase as the square of the width. Reducing the size of the pixels reduces capacitance while resistance is nominally unchanged. The details depend on electrode overlap, thickness, materials and processing. A 2k × 4k CCD with 15 μm square pixels might have C_{CCD}=100 nF per phase and R_{CCD}=150 Ω per phase so that the equivalent lumped RC circuit would have a time-constant of 15 μs. About 3 time constants are required between adjacent parallel clock transitions: 90% settling may seem excessive, but this is required to accommodate crosstalk through the inter-phase capacitance. For 6 transitions, the total parallel shift time is then 15*3*6 = 270 μs. In practice the rate may need to be as slow as 1 ms to minimize spurious charge injection or to achieve good full well capacity (e.g. if the resistance of the parallel bus up the sides of the CCD is too high). A 2k × 4k dual readout CCD requiring 1 ms parallel shifts suffers a 10% readout overhead at 10 μs/pixel, and readout time is dominated by line transfer time for ROIs smaller than about one third of the CCD width. At 270 μs, as in this example, the breakpoint is at one fifth of the CCD width, or 400 pixels square. Even if the parallel transfer time could be reduced to 100 μs, regions smaller than 200 pixels square will still be dominated by line transfers.

The average current drawn from the clock driver *at maximum frequency* is independent of the CCD capacitance. If there is a buffer between the clock switch and slope control filter, then the average current to charge or discharge the CCD capacitance is:

	I_{BUFFER}	$=C_{CCD}.\Delta V.f$	(per supply rail)
However,	f_{max}	$=(18R_{CCD}C_{CCD})^{-1}$	
so,	$I_{BUFFER}(\max)$	$=C_{CCD}.\ \Delta V/(18R_{CCD}C_{CCD})$	(per supply rail)
		$=\Delta V/(18R_{CCD})$	
		$=4$ mA	for $V_{high}=3V$, $V_{low}=-8V$, $R_{CCD}=150\ \Omega$

Buffer op amps always suffer from deterioration of their power supply rejection ratio (PSRR) at high frequencies and thus require the power supply to be very well filtered if clocks are not to be too noisy. By contrast, passive (unbuffered) slope control filters placed directly before the CCD are very quiet above the clocking frequency. The down side is that more drive current is required, since the filter capacitance must be about twice the CCD capacitance to suppress clock spikes induced by the inter-phase capacitance. This triples the total capacitance to be charged and discharged by the clock drivers. In this case separate op amps drive each clock level so that power dissipation per op amp is only increased by 50%.

$$I_{DRIVER}(max) = \Delta V/(6R_{CCD}) \quad \text{(per driver)}$$
$$= 12 \text{ mA} \quad \text{for } V_{high} = 3\Delta V,\ V_{low} = -8 \text{ V},\ R_{CCD} = 150\ \Omega$$

As CCD width and thus R_{CCD} increase, clock drive current decreases. As height increases more electrodes are placed in parallel, decreasing R_{CCD}. So, contrary to popular belief, CCD area does not affect the parallel clock driver current requirement, provided that the aspect ratio is unchanged. It is the maximum clock frequency which degrades in proportion to the increased area: R_{CCD} stays constant while C_{CCD} increases.

9. FASTER REGIONS OF INTEREST

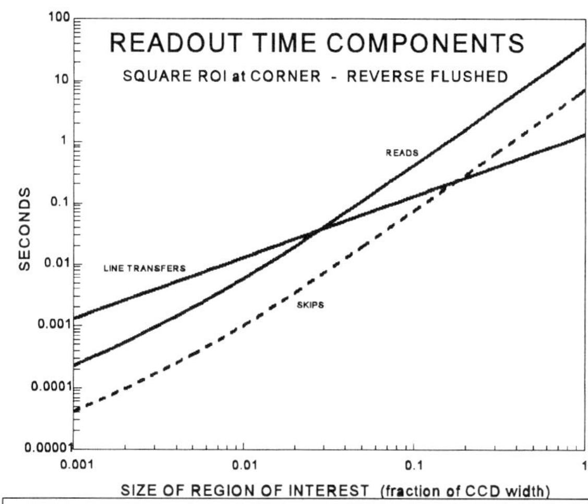

7:Readout time components for a square Region Of Interest at one corner of a quad readout CCD, reverse flushing serials by 120% and parallels by 150%. Same parameters as Figure 6, except that overscan size has been reduced from 30 to 10 columns. Note how the minimum ROI time has been reduced by 3 orders of magnitude without improving parallel and serial shift times

These ROI readout times are only a few seconds and are generally quite short compared to other overheads for single frame readout (filter and focus changes, telescope moves, typing, etc). The parallel transfer time is really only a problem for repetitive readout of a small region with very short exposure times. However, such a capability can be very useful for aligning a complex instrument which includes a large CCD. Fortunately, if the CCD has multiple output amplifiers, the line transfer overhead can be greatly diminished by placing the region of interest near the amplifier in use and employing the reverse flushing method described below.

After reading the data pixels on each line, the rest of the line would normally be skipped by serially clocking the charge to the output node. In this method the charge is clocked towards the opposite end of the serial register for only slightly more than the region-width (20% was used for **Figure 7**). This clears the portion of the serial register receiving charge in the next line of the ROI, and ensures that charge from columns outside the ROI does not accumulate in the serial register, without requiring the entire rest-of-the-line to be skipped.

Charge shifted into the ROI from the neighbouring image area must also be erased before the next exposure. For a quad readout CCD, the parallel clocks are driven in reverse to clear charge out of the region of interest. The image area is clocked towards the opposite serial register for slightly more than the ROI height so that on average the area beyond the ROI is driven into the far serial register which is configured as a dump-drain by taking all serial clocks high and closing the reset switches. (50% of the ROI height was chosen for the example in **Figure 7**)

A dual readout CCD with only one serial register is handled slightly differently: additional parallel clocking carries the image area behind the ROI *forwards* to dump charge into the serial register, which is configured temporarily as a dump drain. Prior to the next exposure, the parallel clocks are driven backwards by just the region width to clear out any charge there. The number of extra parallel transfers must be sufficient to prevent charge from bleeding into the ROI during the next exposure, but this may only need to be a few lines unless the area outside the ROI is brightly illuminated. If there is additional idle time between exposures, this can be used to continue shifting. As before, the serial register is clocked in reverse by slightly more than the region width just prior to the parallel shift of each line read. The number of shifts can be the same in each case. The quad readout CCD has the advantage that spurious charge generation is lower. **Figure 7** shows that parallel shift time still limits the ROI time, but much less than in **Figure 6** and not until the ROI is much smaller. The serial skip time remains well below the pixel read time even for small ROIs. Overscan pixels have been held constant and thus they dominate the reads for small ROIs. Where the zero level is not important, they can be eliminated so that the read and skip times fall more steeply (as the square instead of linearly), but this is of little benefit if line transfer time dominates. Narrower CCDs are less prone to this limitation due to the much shorter time constant of their parallel clock electrodes.

10. CONCLUSION

There is a great deal that can be done to reduce pixel time from the 20-30 μsec range to the 5-10 μs range while still achieving 3-5 e^- noise with CCDs having relatively low output amplifier sensitivity (2-3 μV/e^-). This places readout time in the critical 20 to 40 sec range for a dual readout 2k × 4k CCD even with conventional output amplifiers. If the number of pixels per amplifier is increased beyond 4 million, higher sensitivity CCD output amplifiers will be essential to achieve adequate read times. Until then, their most important scientific contribution will probably be in greatly reducing read noise for photon starved applications, while still reading at a respectable rate of 5 μs/pixel/channel. Quad readout should not be overlooked, even when high sensitivity amplifiers are available, since it not only improves the speed-noise trade-off, but also CCD yield and system robustness.

11. REFERENCES

[1] Chromey, F.R. & Hasselbacher,D.A, 1996, The Flat Sky: Calibration and Background Uniformity in Wide-Field Astronomical Images, *Publications of the Astronomical Society of the Pacific,* **Vol 108, 944-949**

[2] Smith, R.M, 1996, "How linear Are CCDs?", these proceedings, p. 251

[3] Tyson, N.D. & Gal, R. R., 1993, An Exposure Guide for Taking Twilight Flatfields with Large Format CCDs, *Astronomical Journal,* **v105, #3, 1206.**

SECTION IV: MOSAIC PROJECTS & SPECIAL TECHNIQUES

AN INTERFEROMETRIC METHOD FOR MEASUREMENT OF THE DETECTOR MTF

MICHAEL I. ANDERSEN[1] *AND ANTON NORUP SØRENSEN*[1]

ABSTRACT

We present an interferometric method by which the MTF of an array detector is measured directly. The method is not relying on high quality optics and precise alignment/focusing and it is therefore straightforward to set up.

1. INTRODUCTION

For most applications in astronomy, high spatial resolution is crucial. Measurements of the detector MTF is therefore just as important as any measurement of the quality of the optics in front of the detector. The MTF of a CCD may be divided into the following components;

- The pixel definition MTF as defined by the box car function corresponding to the pixel geometry.
- At longer wavelengths, the optical depth of silicon is several microns, implying that the focal plane is no longer well defined. In combination with an optical system with a low F-number, some degradation of the MTF will occur.
- If detection takes place in a field free region of the silicon, the photo-electron may not end up in the nearest pixel. This is usually termed *charge diffusion.*

By the appearance of Loral/Lesser thinned CCD's, notably the later problem has caught attention.

2. DESCRIPTION OF THE METHOD

Traditionally the line-spread function of CCD's has been measured in the *spatial domain* by scanning a small spot across the CCD [1]. The practical challenge of this approach lies in obtaining and calibrating the small spot. Our approach is to measure the MTF directly in the F*ourier domain.* The line-spread function may then be derived from a Fourier transform of the MTF. We generate a fringe pattern on the CCD by the test setup sketched in Fig. 1.

[1] Niels Bohr Institut for Astronomy, Physics and Geophysics, Copenhagen University Observatory.
Email: **mia@astro.ku.dk**

J. W. Beletic and P. Amico (eds.), Optical Detectors for Astronomy, 187–190.

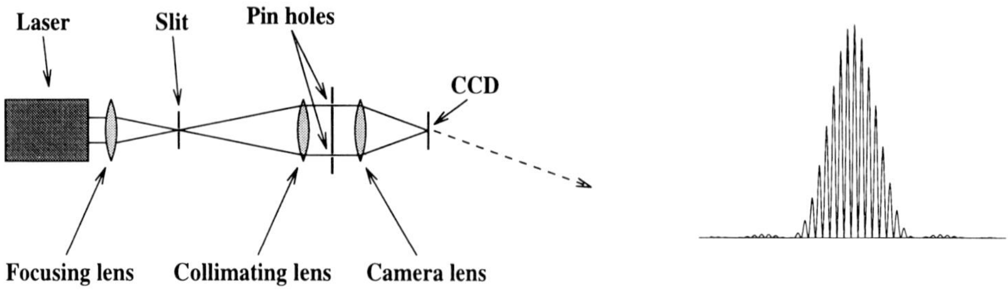

Figure 1: A laser beam is focused on a narrow slit. The light is then collimated, where after it passes two pinholes. The camera lens focuses the slit on the CCD. A cross section of the fringe pattern is shown to the right

Each of the pinholes form a diffraction spot on the CCD corresponding to the apertures defined by the pinholes. By focusing the slit on the CCD, the two diffraction spots will overlap and interference fringes will emerge. The line frequency is given by $\nu = D / f_{cam}\lambda = 1 / F\lambda$ where f_{cam} is the focal length of the camera lens, D is the separation of the pin holes and λ is the wavelength of the laser. The number of fringes across the combined diffraction spot is given by $2.44\ D / d_{pinhole}$. The fringe contrast as defined by $T = (I_{max} - I_{min}) / (I_{max} + I_{min})$ directly gives the MTF at the line frequency of the fringes. By adjusting D, we can change the line frequency and thereby obtain measurements at different line frequencies. Since we only use sections of the lenses, which are a small fraction of a mm across, aberrations in the lenses do not affect the measurement at all. Focal length, slit width etc. of our first test setup is given in Table 1.

Wavelength of laser, λ	670	nm
Slit width, Δ_{slit}	10	µm
Collimator focal length, f_{coll}	300	mm
Camera focal length, f_{cam}	180	mm
Projected slit width, δ_{slit}	6	µm
Diameter of pinhole, $d_{pinhole}$	30	µm
Diffraction spot size	9.2	mm
Range of pinhole separation	0.75 - 5.0	mm
Range of fringe frequencies	6 -40	cycles/mm

***Table 1**: Optical parameters for our interferometric setup*

The MTF as measured by this method has three components

$$T(\nu) = T_{pixel}(\nu) \times T_{slit}(\nu) \times T_{Intrinsic}(\nu) \tag{1}$$

where

T_{pixel} is the pixel definition MTF, given by the Fourier transform of the corresponding boxcar function of width Δ; $T_{pixel} = \mathrm{sinc}(\pi\nu\Delta)$

T_{slit} is the MTF of the slit, given by the Fourier transform of the box car function of width δ_{slit}; $T_{pixel} = \mathrm{sinc}(\pi\nu\delta_{slit})$, where δ_{slit} is the projected slit width given by the true slit width multiplied by the (de)magnification factor f_{cam}/f_{coll} for the collimator/camera lenses.

$T_{intrinsic}$ is the MTF of the CCD *before pixelization*. It can be considered to be the MTF of the silicon.

By selection of suitable lenses and a narrow slit, the projected slit width can be made sufficiently small. In our first setup, we used a 10 μm slit and lenses, which gave a (de)magnification of 0.6. By inverting the collimator/camera, the slit can be magnified and the width calibrated by measuring T_{slit}.

3. REDUCTION AND FIRST RESULTS

In practice the intensity of the two spots formed by each of the pinholes will not have the same intensity distribution, as the pinholes will not be identical. Accordingly, the fringe contrast will not be perfect. By taking images with the first pinhole closed, then the second pin hole closed and then with both open to get the fringe image, it is in principle possible to correct for this.

In practice, we found that the most reliable result is obtained from the highest fringe contrast in the image. This occurs in the zone of the image where the spot intensities are identical. Hence the fringe contrast in this zone will be perfect and we therefore get a direct measurement of the MTF by this approach. The image was divided into a large number of 64x64 pixel sub images. The fringe contrast in each of these images was then determined from the FFT, and the maximum fringe contrast found was adopted.

We have measured the MTF of two CCDs; A Tek 1024 × 1024 thinned CCD with 24 μm pixels and a Loral 2048 × 2048 CCD with 15 μm pixels. The Loral CCD was "overthinned" by M. Lesser to a thickness of ≈14μ. The measurements are shown in Fig. 2. In the two upper panels, we give the MTF relative to the pixel scale, and the two curves should therefore not be compared directly. In the lower panels the MTF of the two CCD's are compared. The total detector MTF as shown in panel *(c)* is more or less identical, with a small preference to the Tek CCD. This suggests that the effective pixel size of the Loral/Lesser CCD is around 26μ. From panel *(d),* it is evident that $T_{intrinsic}$ of the Tek CCD is superior to that of the Loral/Lesser CCD. Note the plateau in the MTF of the Loral/Lesser CCD at high frequencies at a value of 0.25. Our interpretation is that at 670 nm, 75% of the flux is detected in a largely field free region.

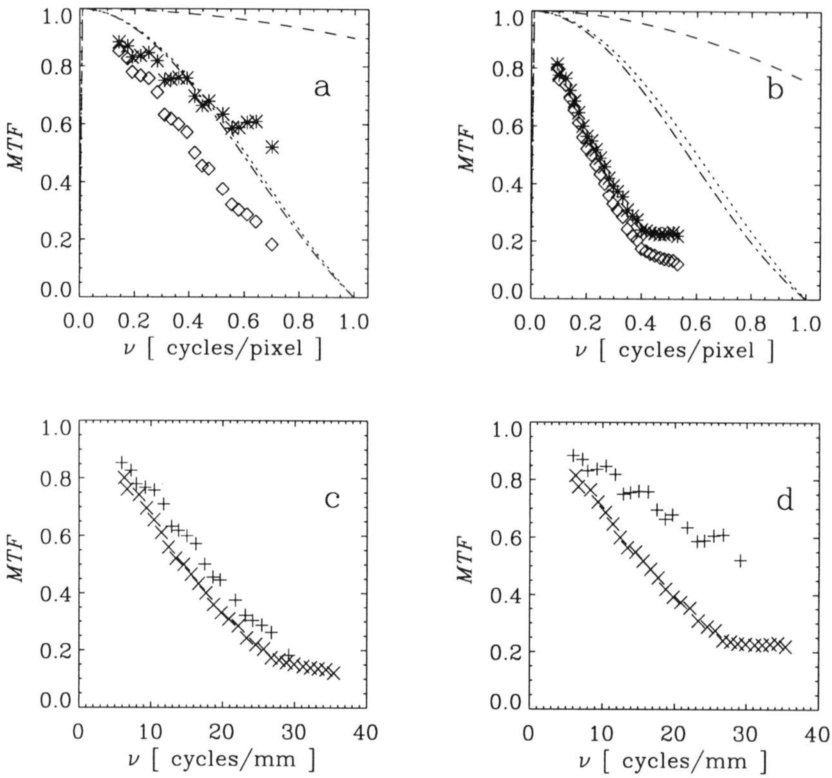

Figure 2: MTF measurements: a) Tek CCD, diamonds give measured TF. The dotted curve give pixel MTF, the dashed curve the slit MTF and the dot-dashed curve is the combination of pixel and slit MTF. Stars give the intrinsic MTF, i.e. corrected for pixel and slit MTF. b) Loral/Lesser CCD, symbols as in panel (a). c) The measured MTF for Tek CCD (plus sign) and Loral/Lesser CCD (crosses) at the same spatial scale. d) The intrinsic MTF. Symbols as in (c).

4. Conclusions

We have demonstrated a new and easy method by which it is possible to measure the MTF of a CCD. Improving our equipment, adding high quality pinholes and camera lenses of different focal lengths, should allow us to measure the MTF in a wider range of frequencies and at higher precision.

5. References

[1] Jorden, P.R., Deltorn, J-M. & Oates, A.P., 1994, in SPIE Proc. Vol. 2198, p. 836.

THE TWO EROS 4K X 8K CCD MOSAIC CAMERAS

F. BAUER[1] AND J. DE KAT[1]

FOR THE EROS 2 COLLABORATION:

C. Afonso[1] E. Aubourg[1], S. Aune[1], P. Bareyre[1], A. Bouere[1], X. Charlot[1], C. Coutures[1], N. Delabrouille[1], G. Dispau[1], D. Eppelle[1], O. Gachelin[1],C. Gaucherel[1], J.F. Glicenstein[1], B. Goldman[1], M. Gros[1], D. Hardin[1], C. Jeanney[1], , J.F. Lecointe[1], A. Le Coroller[1], E. Lesquoy[1], C. Magneville[1], A. Milsztajn[1], P. Perrin[1], M. Pinabiau[1], P. Rebourgeard[1], C. Renault[1], J. Rich[1], M. Spiro[1], R. Treillard[1], J. de Trogoff[1], N. Vignal[1], L. Vigroux[1], C. Walter[1], S. Zylberajch[1], J.N. Albert[2], R. Ansari[2], F. Couchot[2], C. Eder[2], O. Lodygensky[2], B. Mansoux[2], M. Moniez[2], O. Perdereau[2], G. Adrianzyk[3], J.P. Berger[3], J.C. Delclite[3], C. Figuiere[3], C. Guillaume[3], D. Kohler[3], D. Lacroix[3], A. Baranne[4], E. Maurice[4], J.P. Beaulieu[5], R.Ferlet[5], C. Loup[5], A. Vidal-Madjar[5], J. Andersen[6]

ABSTRACT

We present our new astronomical wide field imager, which consists of two 4k × 8k CCD mosaic cameras allowing simultaneous imaging in two focal planes with different color passbands. Each mosaic is maintained at 180K by using a fully automated original cryogenic system without LN2. A modular electronic system permits an individual computer controlled setting and a parallel readout of the CCDs. An acquisition system has been developed in order to process the data rate of 10-15 Gbyte per night. The described instrumentation will be used in a sky-survey and sky-patrol to search for gravitational microlenses, supernovae and red dwarfs

1. INTRODUCTION

The main scientific objective of the EROS 2 (Experience de Recherche d'Objets Sombres) collaboration is the search for dark matter in the form of massive compact objects such as Jupiters or brown dwarfs, which do not emit any detectable amount of light. Paczynski (1986) suggested that these could be found by the gravitational microlensing effect, which induces a temporary achromatic brightening of stars when the dark objects come close to their line of sight. As this transient effect is very rare, it is necessary to perform multicolor photometry of at least 10 million stars over a long time-scale. Fields in the Galactic bulge, the Large and Small Magellanic Clouds offer the required high surface density of resolvable stars. Our photometric detector should therefore be a wide field imaging system, taking images of the sky simultaneously in two

[1]CEA, DSM, DAPNIA, Centre d'Etudes de Saclay, 91191 Gif-sur-Yvette, France Email: florian@hep.saclay.cea.fr

[2] Laboratoire de l'Accèlèrateur Linèaire, Centre d'Orsay, 91405 Orsay Cedex, France

[3] Observatoire de Haute Provence, 04870 Saint Michel l'Observatoire, France

[4] Observatoire de Marseille, 2 Place Le Verrier, 13248 Marseille Cedex 04, France

[5] Institut d'Astrophysique de Paris, 98 bis Boulevard Arago, 75014 Paris, France

[6] Copenhagen University Observatory, Brorfeldeveij 23, 4340 Tollose Denmark

J. W. Beletic and P. Amico (eds.), Optical Detectors for Astronomy, 191–202.

color passbands on a dedicated telescope in the southern hemisphere. Electronics and data acquisition of the huge data flow must be rapid, in order to maximize the number of measurements per night. Pointing, dome rotation, and guiding should be done automatically.

2. OPTICS

The EROS 2 collaboration uses a dedicated 1m Ritchey-Chretien telescope mounted in the GPO dome at the European Southern Observatory, La Silla, Chile. Used until 1988 on the Mount Chiran near the Observatoire de Haute-Provence it gave full satisfaction, before being decommissioned for budgetary reasons. Its two mirrors are made of Cervit.

For the present survey the mechanics have been completely refurbished and is controlled by a pointing and autoguiding unit. The usable field has been increased by reducing the original focal ratio of F/8 to F/5,with three focal reductors and two corrector lenses in each color channel. The optical assembly is shown in Figure 1. Aberration, coma and astigmatism have been minimized to an acceptable level and spot diagrams show that 90 % of the light falls within a circle 30 μm in diameter. A dichroic beam-splitter permits simultaneous imaging in two focal planes with different color channels. The dichroic filter is sandwiched by two glass prisms in order to maintain his planarity and protection.

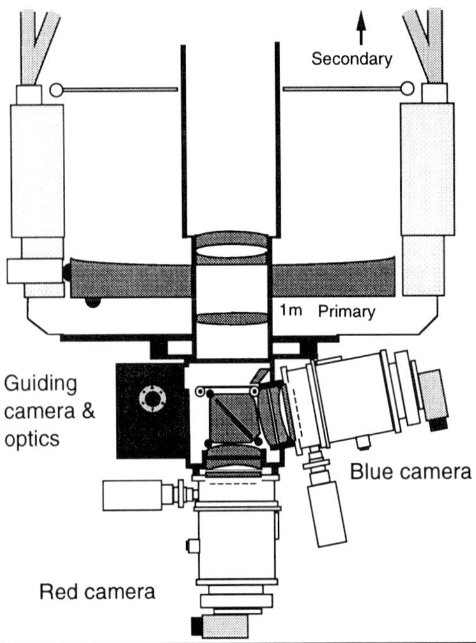

Figure 1: Technical view of the optical mounting

The measured passbands of the system are 450-650 nm and 650-1000 nm. The focal length for both colors is 5140 mm and the focal depth is about 200 μm depending on the wavelength. The mosaic used covers 0.7°x 1.4°. For applications other than the microlensing searches it is possible to insert manually filters in the two color beams. The available filters are V Gunn, i griz, r griz and a low pass filter at 900 nm.

As incoming light has traverses through 15 glass surfaces before getting on the detector, we had to make anti-reflection treatment on every surface. The total amount of reflected light after surface treatment should be around 10 %. A set of three diaphragms on each channel eliminates additional ghost generation. The average theoretical vignetting of the whole system is around 64 %, including the obscuration of the secondary mirror. A special wide field shutter system with an aperture of 12 × 24 cm^2 was developed. It is mounted above the dichroic filter and allows simultaneous and equal illumination of the two wide field cameras to exposure accuracy of 0.5 seconds. The shutter consists of a 98 cm long brass curtain with a rectangular aperture in its middle and is attached on two turnable cylinder blocks. A film of dark chemically deposited teflon coating covers the curtain and enables a smooth running. A small mirror mounted before the shutter west of the scientific field reflects light to an off-axis optical visualisation system with the autoguider camera. The available field is 11.3'× 11.3'. With a guiding limit of about the 15-th magnitude, the system can operate over the whole sky.

3. THE CCDS

We use Loral 2k×2k three edge buttable CCDs. The CCD was designed by J. Geary and manufactured by Loral Fairchild Imaging Sensor Company (Geary et al. 1990). The CCDs are built on 4 inch wafers, each wafer providing four 2k × 2k and eight 512 × 512 CCDs. Each foundry run produces 22 wafers. The cutting and packaging and the first tests at low temperature were made by M. Lesser (University of Arizona). Final tests were done in our laboratory. In a batch of 88 CCDs we found 32 dead on the wafer after production and 18 usable 2 years later. Six of them could be read correctly on both outputs. Each mosaic consists of eight thick Loral 2K3eb CCDs. The mosaic configuration is 2 × 4 CCDs and covers an area of 6.2 × 12.4 cm^2 (see figure 3).

CCD N°	Noise [e^-]	Dynamics [ke^-]	Cosmetics DC	BC	P	Usable outputs	Comments
R0	6.5	75	1			right	1LED
R1	8.7	105	3			left	2LED
R2	22.6	72	3			left	fault line of 360 pixels, amp has no more gain
R3	6.4	85	3			left	
R4	6.6	82	4			both	
R5	6.4	85	3	1	1	both	
R6	6.4	76	7			right	
R7	6.7	91	5			left	partial bad left CTE (7% lost) + 1 LED
B0	7.4	64	7			both	
B1	6.0	90	21			right	
B2	5.9	75	13			left	
B3	6.1	90	9			left	
B4	6.3	80	10			left	
B5	6.1	78	4			right	
B6	6.7	100	9			left	
B7	6.2	74	3		1	both	big cluster + ripples (7.7% lost)

***Table 1**: Characteristics of the 16 CCDs used for both mosaics. The number of dead columns are shown under DC, bright columns under BC, patches under P.*

Each CCD has one serial register with two low noise outputs at each end. A redundant design allows reading the whole CCD out of one or two outputs. This turned out to be very useful and allowed us to use even CCDs with an defective output. The three phases imaging zone can be used in MPP mode, in order to reduce the dark current level.

In addition to cosmetic defects like bad or bright columns, or clusters, we encountered many problems in our batch. The pressure used for the wedge bonding had to be very low (40 g), due to a local fragility of the chip. Many chips suffer from poor serial charge transfer efficiency (CTE). It was sometimes possible to use this device when the CTE in the other direction was sufficient. In other cases higher voltages on the serial clocks could improve the transfer at the cost of increased readout noise. For some devices a multi-parameter and time-consuming optimization was necessary. In a smaller fraction we observed parallel or partial CTE problems. Sometimes the CTE was deteriorated at high light flux and lowered the available dynamic range. We noticed an important correlation between a bad serial CTE on the left amplifier and the position of the device on the wafer.

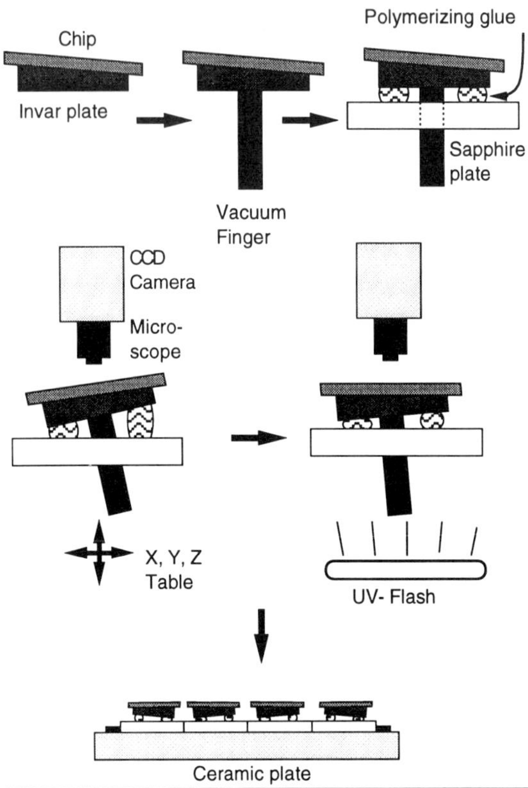

***Figure 2**: Mounting procedure of the CCD.*

Many devices had offset or gain problems on their amplifiers. We could again find a correlation with the wafer-position. In some cases the readout on only one output revealed rippled structure formed by abnormally noisy columns. This disappeared when the CCD was read with both outputs. No way to suppress this phenomenon could be found. Three CCDs presented a partial horizontal fault line of some hundred pixels. Behind this all the sensitive area was lost due to bad CTE. The length of the fault line can be reduced by an increased parallel clock voltage or a slower parallel shift. This phenomenon could be explained by an abnormally high local impedance on a single Φ_p electrode.

The most spectacular problem was an LED effect in the serial register of some CCDs. During long integrations in the dark, one pixel of the serial register emits photons in all directions. Local saturation can be reached after 20 minutes. Fortunately this phenomenon decreases exponentially with distance.

4. Mosaic mounting

As the delivered CCDs mounted on invar-plates have different thicknesses, we used the assembling technique proven in the EROS 1 experiment (Arnaud et al. 1994a), shown in figure 2.

Each CCD has to be glued on a different sapphire plate, slightly larger than the CCD itself. This allows us to control the planarity of the lower sapphire plate surface and the upper part of the CCD, to avoid touching once the CCDs are put together, and the replacement of a single CCD in case of failure. The gluing was made with a dedicated machine designed by the optical detector group of DAPNIA Saclay using UV polymerizing glue (Charlot et al. 1996).

***Figure3**: A view of the blue mosaic mounted in the cryostat*

The sapphire plate is fixed on a plane XY table. The CCD is fixed on a vacuum finger, that goes through a hole in the sapphire plate. UV polymerizing glue is applied between the sapphire and the invar plate. The vacuum finger can be rotated and translated in every direction in order to achieve the best parallelism between the CCD and the sapphire plate. The planarity of the CCD is controlled by focusing a microscope with a CCD camera mounted on its top. For this purpose special control software running on PC has been developed. Once the optimal position is achieved a UV discharge lamp is used to polymerize the glue. The main reason to use sapphire is its good UV transmission ability. All those plates are then bolted side by side on a ceramic cold plate. The sapphire and the ceramic material are delivered with a planarity better than 5 μm. A very low pollution is needed in order to avoid radioactive noise. The coplanarity of the mosaic has a σ of about ± 10 μm and the inter-chip alignment is better than 2 pixels. The average gaps between two adjacent CCDs is 0.5 mm.

5. Cryogenic mounting and controlling

Each mosaic has to be maintained at 180K or colder, in order to reduce thermal readout-noise. As a LN2 dewar would not fit between the telescope fork and due to the

LN2 filling problems in EROS 1, we designed a new cryogenic system which does not use LN2 (see figure 5).

The cooling system of each camera consists mainly of a 15 kg block of pure aluminium, a thermal resistance, a 1 kg block of copper which supports the CCD mosaic and a commercial one stage cold head. This system is packaged in a vacuum chamber of dewar. This system has a continuous two steps operating cycle. First, during the off-survey periods, the cold head is run and connected to the aluminium block to cool down. The final temperature is close to 50 K. Secondly, throughout the survey period, the cold head is shut off and disconnected from the aluminium block. The main benefits of this method are to avoid use of LN2 and also to remove induced vibrations of the cold head during the observations.

During the whole time, the thermal resistance controls the temperature of the copper block and the CCDs, close to 180K with an accuracy of ± 0.1K. This component of great importance is inserted between the copper block and the aluminium one. It directs the incoming heat-flows to the aluminium block. It is made of one circular plate fixed on the aluminium block by an epoxy knee joint and connected by copper flat spring-strips to the copper block. The inner part of the plate is covered with 900 flexible copper contacts. The thermal resistance value is controlled by varying the contact surface between the plate and the aluminium block according to the incoming flows and the state of charge of the aluminium block. The autonomy is 12 hours with a thermal stability around 0.1K. The aluminium block is cooled with a rate of 15K/hour.

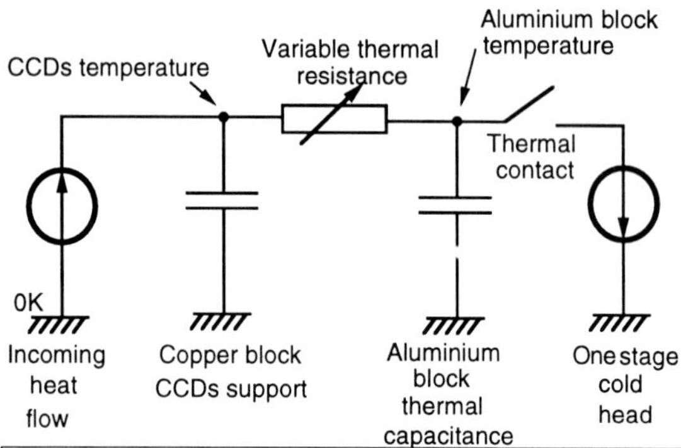

Figure 4: Electrical analogy of the cryogenic system. The grounding potential/temperature is set to 0 Kelvin. The incoming heat flow is mainly due to radiation through the entrance window

All electrical connections in the inside chamber are made of manganin wires which considerably reduce thermal losses by conduction effects. The inner walls of the dewar are covered with super-insulated aluminized mylar. The inside pressure is around 10^{-5} hPa. Three electric heaters inside the dewar are used when the vacuum pumping runs in order to improve the evacuation. A small electric heater fixed around the entrance window is designed to avoid condensation on the window surface.

Early in this project, a thermal model was developed in order to get a suitable compromise between the mass of the camera and the system time range and to simulate the feedback control laws with the purpose of checking the closed-loop system behaviour. The thermal model and terminology were established and drawn with electrical system's concept (see figure 4).

This model was used during the design and setting phases of the project. An automatic control system was designed and built to operate the wide field shutter and the cryogenic systems for the two cameras. The system is powered by an uninterruptible electric power supply. All the system is under the control of a Programmable Logic Controller (PLC) manufactured by Cegelec and General Electric-Fanuc (ALSPA 8000 series). This machine is fitted with axis positioning modules, communication modules and discrete and analog input-output modules.

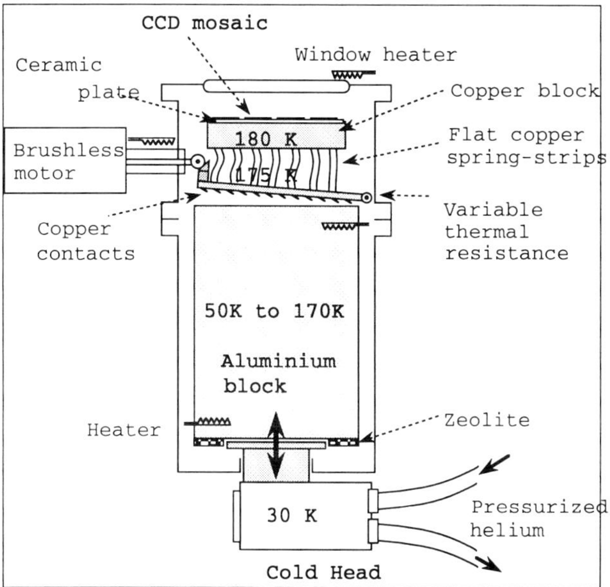

***Figure 5**: A schematic of the cryogenic system*

The PLC is connected with the VME acquisition crate by a serial RS422/RS485 link with RTU protocol. The PLC is also connected with a monitoring personal computer compatible PC. It works as a database which stores relevant information to the control system with a sampling period of 30 sec. It is connected to the Internet network as a server thus allowing remote control of the automatic processes. In this way, the engineering staff in France gets database files to analyze and repair the system in case of failure.

6. ELECTRONICS

The specifications for the EROS 2 electronic controller takes into account the experience achieved in EROS1 (Arnaud et al 1994b). The design should be more modular and flexible in order to be able to use any type and number of CCD with various output numbers. The readout of the 4k × 8k mosaic should be faster (the EROS1 controller would take 9 minutes) with a lower readout noise. Furthermore, great care should be devoted to reliability and CCD overvoltage-protection.

6.1. Organisation

Each camera has its own controller and VME acquisition system, linked together by a 50 m long pair of optical fibers. The controller is fully independent, having its own power supply and also a digital signal processor (DSP) and all necessary data for the CCDs sequences. In stand-by mode the controller continuously cleans continuously the CCD. The DSP controls the driver boards located in the same crate. Each driver board generates the polarization and clocking levels for every CCD. These levels are transferred to the cryostat by shielded cables. Inside the cryostat, small filter boards are mounted near the CCDs, in order to eliminate interference and overvoltages. The analog CCD output signal is transferred by twisted shielded pairs to the ADC crate located outside the cryostat, which contains an analog backplane, the ADC boards and an interface board to the DSP. Every ADC board contains two signal chains consisting of an amplifier, an ADC converter and an optocoupler interface. The pixel data are read by the DSP through the interface Xilinx board and transferred to the VME through one of the optical fibers. There, a custom board called MMB (Memory Management Board) descrambles the data flow in real time and put it in a RAM-board. The mosaic image is then available for visualization, archiving and processing.

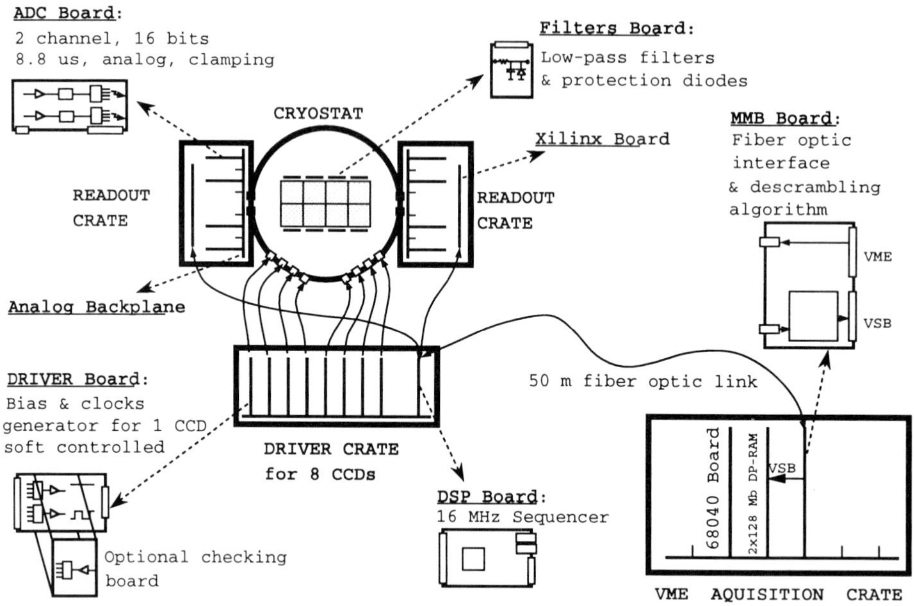

Figure 6: The cryogenic mounting

The separation of the functions in different boards allow a great modularity. Depending on the mosaic used, the system is assembled by modules and then described in the software of the acquisition. The theoretical limit is 64 CCDs with 4 outputs each, not taking into account problems due to power dissipation, power supply, weight, etc. At present the EROS2 controller is optimized for one DSP and eight CCDs with one output on each CCD.

6.2. CCD control

The DSP board does the sequencing of all the CCDs simultaneously. Its executable program can be loaded at any time from the acquisition VME station. It is possible to easily change the sequences but also to develop sequences, in order to repair unplanned CCD problems or to reduce contamination by blooming for example. This can be done without any modification in the hardware and without cutting off the power supply.

The driver boards are not designed for a special CCD type. They contain enough channels with large voltage-range to be adapted to any large scientific CCD array. With 32 8-bits DACs every driver board can generate up to 16 independent polarization voltages between −16 and 28 V and 16 clocks with levels between −16V and 16 V. This allows instant switching from MPP to non-MPP mode for example, or to easily optimize the control voltages for each CCD. This turned out to be very useful, since, for example, in the red camera only one of the eight CCD has the nominal voltages applied. Furthermore optional current buffer and slope control are available for the parallel clocks. All the voltage ranges are defined by resistances, in order to protect the CCDs against erroneous voltages requested by the user or software errors.

A daughter board can be plugged on the driver board. Thirty-two 8-bits ADCs redigitize the driver output levels, which can then be checked by the VME via the DSP.

In the cryostat each CCD is protected by a small filter board with Zener diodes against overvoltages induced by cabling errors, power supply or component failures.

The only parts of the system specific to a CCD type are the resistance supports on the driver boards, the filter boards, the connections inside the cryostat and of course all the files describing the system.

6.3. Analog part

Very early in the project we decided to place the whole analog readout chain outside the cryostat, in order to have easy access and to facilitate tests. As a consequence we had to give special attention to the transmission of the CCD output signal to the preamplifier. For this purpose we use a 15 cm long twisted shielded pair in the cryostat. The analog backplane, where the ADC boards are mounted, is directly plugged on the cryostat connectors. Even on the analog backplane the critical signal traces pass between ground traces and are sandwiched by grounded planes, in order to reproduce shielded pairs.

Each ADC board contains two channels. Each channel begins with a 10 kΩ CCD loading resistance followed by a coupling capacitance. Then a low noise preamplifier can be used in differential or single mode. Due to the clean connections to the CCD, the single mode turned out to be sufficient. The electronic gain is 30, with a bandpass limited to 2 MHz.

A clamping stage creates a new DC level after the amplification and does at once the subtraction of the CCD floating level. Special attention was given to this module, in order to avoid additional noise. The ADC is a true 16 bit converter (Crystal 5101) with a conversion time of 8.8 μsec. Our main reason for this choice, was its small dimension and power dissipation (300 mW), which are important parameters in a multichannel controller mounted in the Cassegrain focus. The serial output of the converters allow easy break of ground loops by means of optocouplers. Furthermore each ADC board has its own isolated DC/DC power supply.

Once the whole control system is assembled, only one electric reference potential exists and is connected to the metallic mass of the whole telescope on one point near the CCDs. Moreover the reference potential has to be as homogeneous as possible between the CCD and the ADC. It was also necessary to protect the critical analog signals from digital and other variable signals. As an example, we had to move the wire of the temperature sensors in the cryostat, because interference from them were crosstalking with the wires of the CCD signals.

The measured readout noise is 6-7 e^- for all the channels, CCDs included, reading at 83 kpixels/sec. The readout noise of our two multichannel cameras is comparable to other single CCD cameras, using Loral devices of this generation. No crosstalk could be seen down to 10^{-5}.

6.4. Readout

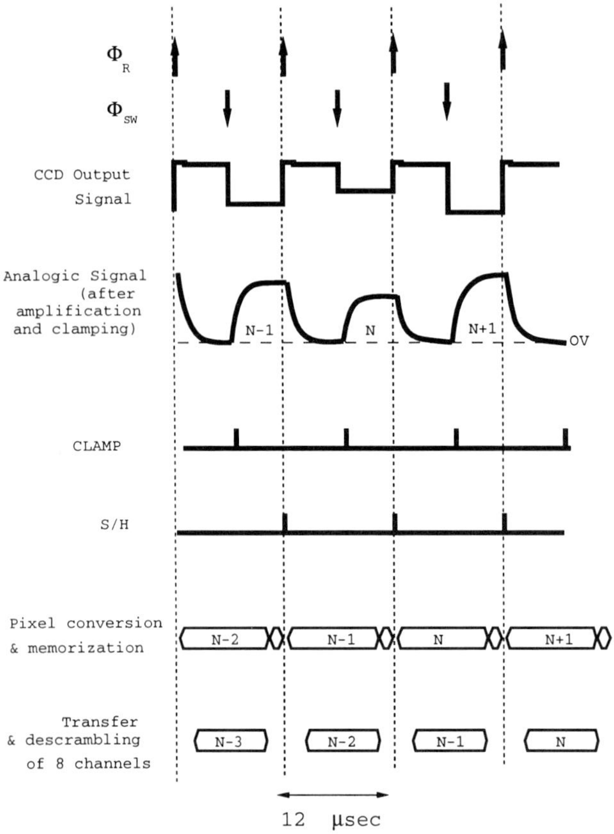

***Figure 7:** The pipeline principle*

Our goal was to read the mosaic as fast as possible, while sequencing the CCDs as slowly as possible. So we chose a pipeline method. In parallel, pixel n is shifted to the CCD output, pixel n-1 is preamplified, the ADC converter digitizes pixel n-2 and the DSP reads and sends pixel n-3 on the optical fiber link. The third operation gives the primary time limitation and takes 8.8 μsec for the conversion in the ADC plus 2μsec for latching on the Xilinx board and 1.2 μsec for varied operations in the DSP. Note that the analog subtraction realized by a clamping system allows a CDS operation with only one conversion. During these 12 μsec the DSP has just enough time for the broadcast sequencing and the transmission of the n-3 pixel data of the eight CCDs .This necessitates good optimisation of this critical loop in the DSP assembling software. Furthermore the CCDs are only clocked at the rate of 83 kpixels/sec and the noise contribution remains low.

The Xilinx custom board executes the simultaneous serial-parallel conversion of up to 16 output channels. A Memory Management Board (MMB), connected at the other end of the optical fibers in the VME crate, unscrambles the mosaic pixel data-flow and sends it to memory-boards via a VSB bus. The pixel addresses in image memory zone are computed in the MMB board by a hardware-designed unscrambling algorithm, which is programmable following the organisation of the mosaic, the number and type of CCD and outputs. This allows the writing of the incoming image directly in the correct order, at the speed of 300 ns per pixel. We read the 64 Mbytes of the mosaic consisting of eight CCDs (using one output only) in less than 50 seconds, unscrambling included.

6.5. Implementation

As the controller is mounted at the Cassegrain focus, we had to take into account the dimension, weight and power dissipation constraints. So we had to miniaturize all the boards and to use as much as possible surface mounted components. Every camera has three crates: one commercial single-euro crate hosts one DSP and driver boards plugged on a commercial backplane. The other two are custom crates which contain four ADCs and one Xilinx boards plugged on a custom analog backplane. For every camera the controller system weights 15 kg and dissipates 100 W. The power supply is deported under the telescope floor, where the dissipated heat is extracted by electric fans through a tube outside the dome.

7. DATA ACQUISITION

Due to the enormous data rate expected (5--8 Gbytes per night and per camera), we decided to divide the acquisition into two channels, one for each color. Every acquisition chain consists of one VME crate with a controlling and an imaging monitor and one Alpha DEC station with 20 GBytes of disc and a DLT-tape robot. They are located in a counting room 25 m from the telescope.

Each VME crate contains the same software, which permits full control of the whole experiment. Both crates may be independent or synchronized, depending on the data to be acquired. Each VME is connected by Ethernet to an Alpha DEC station running with Unix. Once the images are treated by the VME, they are sent in FITS format through the Ethernet. The actual rate is one 64 Mbyte image every 3 minutes. A switching bridge isolates the network traffic on each of the two channels. There second-level processing as well as data quality checks and flat-fielding are run. Optionally the raw data can be archived directly by means of a tape recorder mounted on the VME. The data is archived in FITS format on DLT tape in order to perform the final photometry in a computing center in France. The two acquisition chains communicate with a third Alpha DEC station with 30 Gbytes of disc. This station is used to do a raw photometry of images of the previous night in order

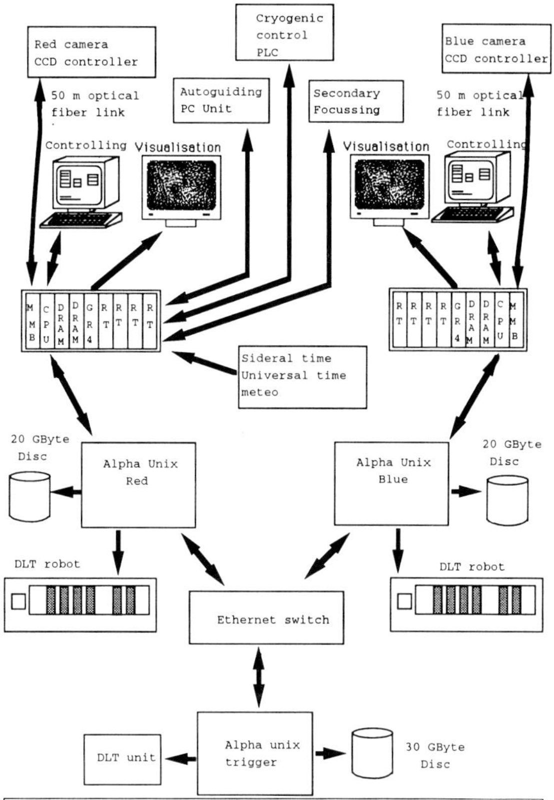

Figure 8: Data acquisition.

to trigger on microlensing or supernova events and to alert the community before the next night.

Each VME work-station consists of a standard screen for the dialog by means of windows with menus and a graphic screen 1200 × 1024 for display of the images. We built the acquisition system by assembling commercial boards in the VME/VSB crates. A Motorola M167A (68040 @33 MHz) controls our custom MMB, two 128 MBytes Double Port VME/VSB Memory boards, visualization, I/O interface and other development boards. A local disk contains the system and the custom acquisition software, running on the real time operating system OS9. Below the control-command process, ten other background processes have to run simultaneously in order to control directly the electronics, visualization, data transfer to the DEC alpha station and also the shutter, cryogenics, guiding and pointing of the telescope, via independent subsystems. Acquisition data like cryostat temperatures, telescope position and focussing and weather conditions are centralized at this level. A number of tools for the survey and interpretation of the images is also available in this system.

8. SUMMARY

The whole system saw first light in June 1996 and became fully operational on July 1996. This wide field imager consists of two 4k × 8k CCD mosaic cameras allowing simultaneous imaging in two focal planes with different color passbands. Each mosaic is made of eight 2k × 2k three edge buttable thick CCDs developed by Loral/U. Arizona. Special manufacturing techniques allowed us to reach a gap between two CCDs of about 0.5 mm and a planarity of the entire mosaic with a σ of 10 μm. Each mosaic is maintained at 180K by using a fully automated original cryogenic system without LN2. The cameras are mounted on a 1m Ritchey-Chretien telescope, at the La Silla ESO Observatory in Chile. The optics has a usable field of 0.7° × 1.4°.
A modular electronic system permits computer controlled setting and a parallel readout of the CCDs. The readout noise is around 6-7 e^-, the cross-talk is less than 10^{-5} when digitizing at 83 Kpixels/sec. The resulting 64 Mbytes per mosaic are transferred with optical fiber to a VME acquisition crate, which includes an user interface. The data rate of 10-15 Gbytes per night is then reduced and treated by a pool of three DEC alpha stations.
Acknowledgements : The authors wish to thank specially the ESO staff at the La Silla Observatory for assistance during the mounting.

9. REFERENCES

[1] Arnaud M., et al., 1994a, *Exp. Ast.*, **4**, 265

[2] Arnaud M., et al., 1994b, *Exp. Ast.*, **4**, 279

[3] Geary J.C., Robinson L.B., Sims G.R., Bredthauer R.A., 1990, *SPIE*, **1242**, 38

[4] Paczynski B., 1986, *Ap.J.*, **304** , 1

[5] Charlot X. et al., in preparation

WIDE FIELD IMAGING AT CFHT: THE MEGACAM PROJECT

O. BOULADE[1]

ABSTRACT

We present in this paper a project for a 1 square degree 16k × 16k CCD camera for the prime focus of the 3.6m CFH telescope on the Mauna Kea mountain. This camera will be used to make a deep multicolor survey of about 100 square degrees, with scientific applications ranging from studies of the solar neighborhood and the galactic structure to studies of the large scale structure of the universe or the weak lensing of galaxies by dark matter.

1. INTRODUCTION

All astronomers know that telescope time is a very rare resource. When using a classical telescope equipped with the current CCD cameras, one has to choose between observing the largest possible area of the sky, restricting the observations to the brightest objects in the field, or detecting the faintest objects the instrument can get, but on a very small field of view.

On the other hand, many astronomical problems require the observation, down to very low fluxes (B or V ≅ 26-27) of a significant (tens of square degrees) area of the sky. This is the case for example in studying the structure of our own galaxy: it is necessary to explore fields at different galactic latitudes, as well as fields in the directions of both galactic center and anti-center to derive the gradients in the distribution of the stars; and these observations must be deep enough to provide the statistics required to determine the faint end of the stellar luminosity function.

At the other end of the spectrum, the study of the large scale clustering of galaxies ([1], [2]) requires angular sizes of 50 to 100 degrees and B magnitudes down to 25-26 if one wants to put constraints on the nature of the structures at $z < 1$. It also requires a very good image quality in order to make a clear separation between stars and faint galaxies. Another very important cosmological problem, which can be addressed with large deep surveys, is the distribution of the dark matter. The weak lensing of the background galaxies by intervening masses can be used to determine the surface mass density ([3]), providing a direct estimate of the total matter distribution. As in the previous example, very good image quality is mandatory to accurately measure the shapes of the lensed background galaxies.

At present, there are no facilities which would allow the astronomer to make such observations in an efficient way: on the one hand, the Schmidt-type telescopes have a size too small to reach the desired sensitivities, and the large telescopes usually do not have the required field of view or image quality. On the other hand, the largest CCD cameras available to date do not cover fields larger than 30'. The only exception is the 3.6m Canada-France-Hawaii telescope, with a current 55' field of view at the prime

[1] DSM/DAPNIA/Service d'Astrophysique Centre d'Etudes de Saclay, 91191 Gif sur Yvette, France. Email: **boulade@hep.saclay.cea.fr**

J. W. Beletic and P. Amico (eds.), Optical Detectors for Astronomy, 203–207.

focus and seeing in the 0.5" to 0.8" range. This telescope is already equipped with UH8K, an 8k × 8k camera covering a 29' × 29' field of view ([4]), but this camera uses thick CCDs which make it almost blind at wavelengths shorter than 4500 Å.

2. DESCRIPTION OF THE MEGACAM PROJECT

2.1. Scientific specifications

The observations needed to address the astrophysical problems mentioned above all require a very good image quality over a wide field of view at the focus of a large telescope. In fact, the science consortium of the MEGACAM project, with scientists from most of the French and Canadian institutes, has compiled up to now more than 12 scientific proposals that could benefit from a deep large survey, only possible with a dedicated wide field imager. Table 1 summarizes the specifications common to all these proposals.

limiting magnitude (mag)	m_B = 24 to 26
lim. surf. brightness (mag/"^{2}) at S/N = 2	μ_B = 26.5 to 27.5
photometric accuracy (mag)	0.1 for m_B = 23
astrometric accuracy (")	0.01 to 0.2
spectral range (Å)	U to I + grism (15 Å/px)
image quality (")	≤ 0.8
sampling ("/px)	0.2
total field (deg^2)	25 to 100

Table 1: summary of scientific specifications

Moreover, the new generation of 8-10m telescopes will be able to perform redshifts measurements at an unprecedented sensitivity, giving us insight into the 3D distribution of faint objects. However, to do such measurements, one first needs to have catalogs of objects, with accurate positions and photometry. For such programs, there is then a *natural* separation between the 4m class telescopes, which can be used for imagery, and the 8-10m class telescopes, used for spectroscopy.

In this context, the Science and Administrative Council of CFHT recommended development of a 1 square degree imager. A new wide field corrector, giving a 1.4 degree field of view, will have to be built. To sample the seeing adequately, it will have to give a scale of about 0.2" per pixel, which in turn means that the imager will be based on a mosaic of CCDs having at least 16k × 16k pixels.

2.2. CCD mosaic and cryogenics

The mosaic will be based on 2k × 4k thinned CCDs, therefore 32 of these CCDs are needed (see figure 1). The CCDs are 3 sides buttable, and we will use a special packaging which will reduce the size of the connector area on the 4th side to less than 5 mm, by having all the connections made beneath the CCD itself. This packaging will give a filling factor of about 90% for the mosaic.

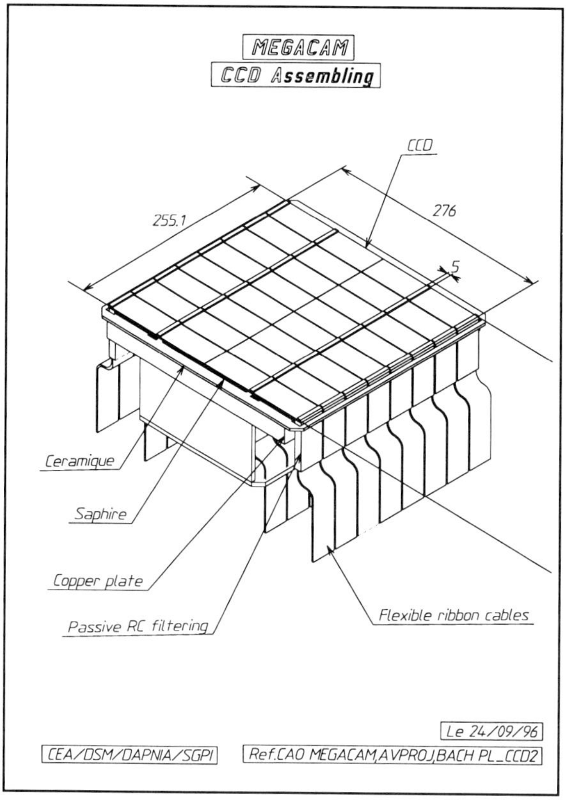

***Figure 1**: Layout of the CCD Mosaic*

With such a large field of view, there will be optical distortions, which will be calibrated by means of astrometric standard stars. Therefore, the distortions in the geometry of the mosaic itself are not a hard issue: they will be measured and corrected for with the optical distortions.

The most stringent requirement for the mosaic is its flatness, because of the large size of the focal plane. A rms flatness of 1 pixel is needed for each individual CCD, as well as from one CCD to the next one. A mechanism based on 3 suspension arms and 3 leveling screws, driven by stepper motors, will be used to align the mosaic axes with the telescope axises and the mosaic plane with the focal plane.

The mosaic will be cooled by a closed cycle cooler in order to automatize the operations of the camera as much as possible. The only problem with such a system is that it will induce vibrations at the prime focus, which will have to be damped properly.

2.3 Electronics and data acquisition

The main specification of the front-end electronics is that the total acquisition time for the whole CCD mosaic must be less than 20 seconds. This is necessary if one wants to do observations of the highest possible efficiency, knowing that the typical exposure time will be around 10 minutes.

The other main electronic specifications are readout noise less than 5 e^-, no crosstalk between channels. Because of the location of this electronics above the prime focus, the weight, the sizes and the heat dissipated must be kept to a minimum.

The chosen design of the controller is derived from the one used in the EROS2 experiment ([5]): for each group of 8 CCDs, one driver board and two readout boards, one per output, are used. The controller is hooked up onto the CCD mosaic, and connected to the outside world via fiber optics (figure 2).

Parallel readout of the channels is not enough to reach the required rate of 5μs per pixel. The system will use a full pipeline method: at the time the ADCs convert

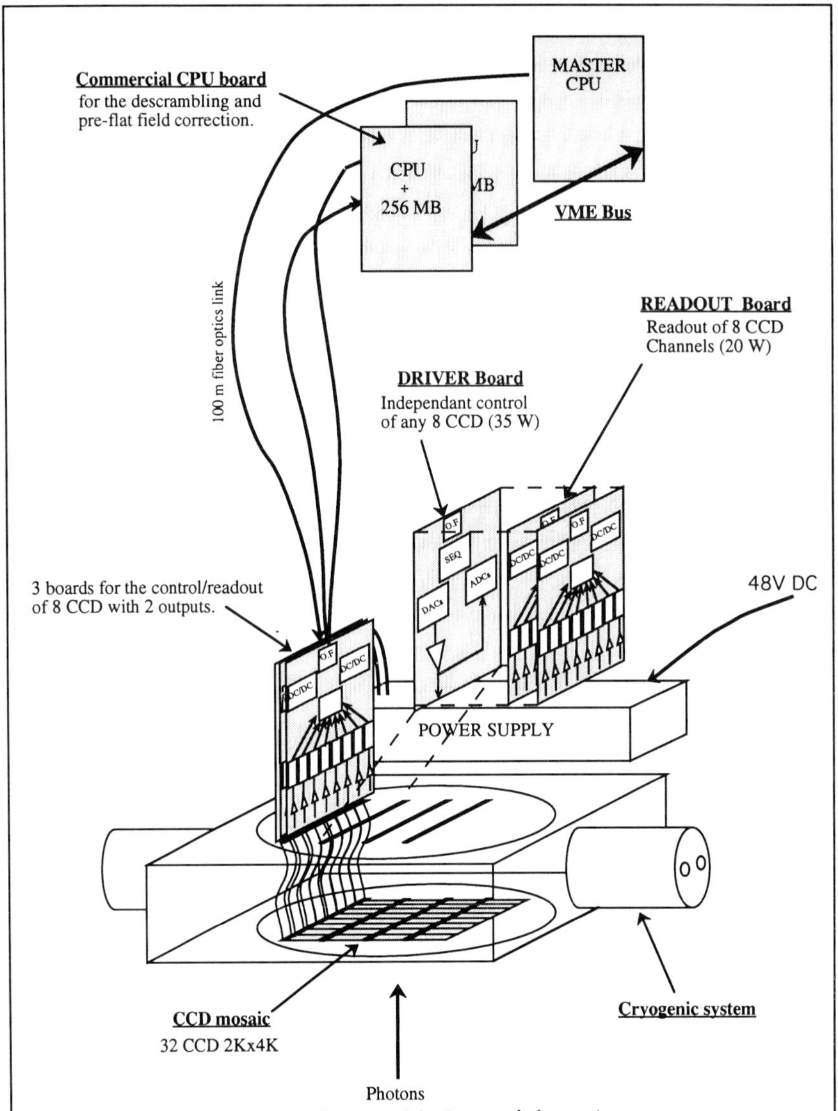

***Figure 2**: Concept of the front-end electronics*

pixels n, pixels n+1 are pre-amplified, pixels n+2 are shifted in the CCDs serial registers, and pixels n-1 are transferred in memory and unscrambled.

The estimated total power dissipation is 360 W, 300 W for the controller boards and 60 W for the power supply. This implies the use of a system to evacuate this heat away from the prime focus environment in order not to degrade the image quality.

At this stage of the project, the data acquisition has not been finalized: it strongly depends on the acquisition systems and software environment at CFHT, which are being reassessed for upgrade.

2.4. Optics

A new wide field corrector is needed for the prime focus of CFHT. It will give a 1.4 degree field of view, an output beam f-ratio around f/4, with image quality better than 0.3", edge vignetting better than 15% and low distortion. A tip-tilt correction is foreseen, as well as an atmospheric dispersion compensator.

This new wide field corrector, being much bigger than the current one, will not fit inside the current prime focus cage so most, if not all, of the prime focus top end will have to be rebuild.

2.5 Observing strategy

Most of the science proposals can be done with one single multicolor survey of about 100 square degrees. An ultra-deep survey will be done on a much smaller area (≅10 square degrees), and a few specific observations will be needed to complete some proposals (e.g.: additional fields of view for the study of the galactic structure, repeated observations of the same fields for detection of high redshift supernovae or proper motion measurements).

In order to improve the efficiency of the survey observations, the camera will be operated in a *pointed scan* mode derived from the drift scan mode: a given field of view is observed near the meridian for about 10 minutes in one filter, then the telescope is slewed back towards a contiguous field. At the end of the night, a stripe, 1 degree wide and about 50 degrees long, will have been observed in one filter. The same stripe is observed repeatedly during a 14 nights new moon run to achieve the required sensitivity. The total observing time needed to cover the whole survey field in all required filters is about 270 nights spread over 6 years.

3. PLANNING

The Scientific and Administrative Council made a recommendation to the Board of Directors of CFHT to approve and fund the MEGACAM project. If the project is funded at the December 1996 meeting of the Board, then the first thing to be done will be to start the procurement of the CCDs. The delivery of all 32 CCDs will take about 2 years, starting mid '97 and ending in the second half of '99. Most of the development of the camera head, the electronics and the software can be done in parallel with the CCD delivery; the integration of the CCD mosaic in the camera, and the commissioning phase on the telescope, will take place at the end of '99 and the observations should start in year 2000.

4. REFERENCES

[1] de Lapparent, V., et al., 1986, *Ap.J. Letters*, **302**, L1

[2] Arnouts, S., et al., 1996, *Astron. & Astrophys. Supp. Ser.*, in press

[3] Bonnet, H., et al., 1994, Ap.J. Letters, 427, L83

[4] Metzger, M., et al., 1996, submitted to PASP

[5] Bauer, F., et al., 1996, *this issue*

FRINGING - A USER'S PERSPECTIVE

RUSSELL CANNON[1]

ABSTRACT

I outline here of some of the key issues involving fringing as it affects observers in practice. Fringing is now the limiting factor in many applications of CCDs to astronomy, especially for spectroscopy at longer wavelengths.

1. ASTRONOMICAL CONSIDERATIONS

1.1. Imaging

Here generally the wavelength range is uniform over the field. Two sub-cases arise, broad (~ 100 nm) and narrow (≤ 10 nm) band-pass imaging. In narrow band work, the fringes are stronger and sharper, while the wavelength range may not be strictly uniform across the field due to instrumental effects. Note that the "narrow band" case arises also in broad-band imaging when there are strong emission lines within the band-pass, either in the astronomical target itself or in the night sky. In these cases, the source, not the filter, defines the narrow band-pass that matters.

1.2. Simple spectroscopy

Here the spectra cover a few rows on the CCD and each small set of pixels perpendicular to the direction of dispersion receives light of a different wavelength. The severity of the problem depends on the relative length scales of the fringing and the astronomical features of interest. For ordinary stellar spectra, absorption line features are typically sharp and on a scale (in wavelength) smaller than the fringes, while sometimes the opposite is true. In such cases, it is not too difficult to correct for the fringing. However, there are situations where the fringing and astronomical length scales are comparable, particularly in very low dispersion spectroscopy, or e.g. the emission lines in SN 1987A, whose bumps and dips gave important information on the lumpy structure of the supernova ejecta. When fringing *is* a problem it hurts in two ways: it gives rise to spurious features in the actual spectrum and, since it varies from pixel to pixel even at the same wavelength, it limits the accuracy of sky subtraction.

1.3. Two-dimensional spectra

These can be obtained with long slits or with multi-object fibre spectrographs. The full CCD array is used, with wavelength being constant (more or less) along columns (or rows). In practice, this is only a small extrapolation from case (1.2) above.

[1] Anglo-Australian Observatory, PO Box 296, Epping, Australia and ESO. Email: **rdc@aaoepp.aao.gov.au**

J. W. Beletic and P. Amico (eds.), Optical Detectors for Astronomy, 209–212.

1.4. Full two-dimensional cross-dispersed echelle spectra.

In this application the wavelength varies continuously across the whole image, since the echelle disperses the light in one direction and the cross-disperser (prism or grating) spreads the orders out perpendicularly. Every pixel receives light of a different wavelength.

Note that in 2-D spectroscopy the fringing pattern is a complicated convolution of each monochromatic fringing pattern with the wavelength distribution across the CCD. It is *not* the same as the pattern seen in any more-or-less monochromatic test exposure. In addition, the effect on a given pixel may change *drastically* with only very small changes in wavelength registration on the CCD, making correction very difficult.

2. FLAT-FIELDING

The usual way of dealing with fringing is to obtain "flat fields", which can be used to correct what amounts to a sensitivity variation across the detector.

2.1. Direct imaging

The problem is at least tractable in this case, although there is no ideal way of getting flat fields. Three methods are commonly used:

- *Twilight flats.* These are generally easiest, but there are practical problems. They are time-critical, in that there are only periods of a few minutes at dusk and dawn each night when the sky is dark enough to get good unsaturated exposures (high S/N) but not so dark that star images begin to appear; second, they can be affected by the small non-linearities present even with CCDs; third, for large arrays of CCDs it is no longer safe to assume that the sky brightness will be intrinsically constant across the field. And of course, if the instrumental set-up is changed during the night there is no solution.
- *Dome flats.* These use an artificially illuminated screen and are reasonable, but the light does not follow exactly the same path as starlight and it is hard to get uniformity on the largest scales.
- *Deep sky flats.* By combining the background sky from many deep exposures, after first removing the star and galaxy images from each frame, one can achieve sufficiently high S/N. This is OK if the same set-up is used all night, and at high galactic latitudes where there is no nebulosity and the sky is intrinsically flat. For broad-band imaging with the same filter and CCD combination one can actually use archival deep flat fields of this type over very long periods, even several years.

Note however that an essentially uncorrectable problem arises if the fringes are due to night sky emission lines, since these are intrinsically variable on short timescales.

2.2. Spectroscopy

Here flat fielding can be done in principle, although *only if the format is fixed.* However, much astronomical spectroscopy involves changing grating angles to observe different wavelength regions. In such cases, virtually every observation requires a separate flat field. Even when the set-up is fixed, flexure or temperature changes often lead to small shifts of the spectrum on the CCD and hence to significant changes in the fringe pattern.

There are a number of severe technical problems in the spectroscopic case, some of them of a rather fundamental nature:

- Most light sources are not spectroscopically flat, i.e. they are not grey. Scattered sky light is particularly useless, having the very complex spectrum of a metal-rich G-type dwarf star, with the added complication of atmospheric night sky emission lines. Incandescent lamps are better but do not give uniform output over wide wavelength ranges, so often a series of exposures has to be taken.
- For artificial light sources, it is hard to ensure that the light path is the same as for celestial targets.

In the case of 2-D spectra (multi-object or echelle), a further problem arises if there is any mechanical flexure in the instrument. Then the pixels illuminated by the flat field may not be the same as those exposed to the sky.

3. General conclusions

There has been quite a lot of discussion about fringing and what should be done about it, but astronomers and CCD manufacturers have to get together to decide how to tackle the problem. A few key points follow:

- Fringing is not a single-parameter quantity; as well as being an obvious function of wavelength, what matters to the user are both its strength and its length scale.
- It is of only limited use to specify acceptable levels of fringing, unless the precise cause of the fringing is known so that manufacturers have some control over it.
- Is it true that we are always better off with high sensitivity, anti-reflection coated, thinned CCDs, as has been claimed? Compare two CCDs, one with 90% QE and 20% fringing, the other with 45% QE and only 2% fringing. Given sufficient information about the fringing pattern at high S/N, then the first device must be better. However, if determining the fringe pattern costs as much telescope time as the main observation, then the second device is at least as good. Where the fringe pattern is unknowable or determinable only to limited accuracy, then the lower sensitivity device will actually do better astronomy. At the AAT, observers sometimes still use the unthinned Thomson CCD for spectroscopic observations beyond 900 nm, even though its QE is only about half that of the thinned Tek 1024.

Finally, lest all this look too negative and gloomy, it is worth remembering that some of these fringing problems have only become significant because CCD detectors are so good. Only now, when astronomers are getting used to working with S/N per pixel of many hundreds (or even a million, in the case of spectropolarimetry), rather than with tens of counts per pixel, are these fringing problems really becoming serious. Blueward of 900 nm, fringing is generally below the 1% level and can be corrected to better than 20% of the peak value, so it is a tractable problem. For very faint objects spectroscopists still typically work with S/N of order 10 per pixel, at which level fringing is irrelevant, at least out to 950 nm. In addition, it is in the previously almost unexplored far red and near-infrared regions of the spectrum, opened up now by CCDs, that some of the most exciting new astrophysics can be done. So we certainly do want these red-sensitive CCDs; the only questions are how best to specify what we want, and how best to utilise what we get.

Acknowledgement: I thank Raylee Stathakis at the AAO for helpful comments and input on some of these issues.

DEIMOS DEWAR SYSTEM

DAVID COWLEY[1]

ABSTRACT

Design work resumed on the DEIMOS detector system in September 1996. The major change from the initial description is the inclusion of the focus and one axis of the flexure control system inside the vacuum head of the dewar. Traditionally Lick has not included any active stages inside the vacuum vessel on visible light instruments. However the appeal of having these stages inside the vacuum is that the moving mass is greatly reduced, reducing the gravity induced motions and the power requirements to move them. The motion stages inside the vacuum are flexures to avoid the need for lubrication, the drives are outside of the vacuum for the same reason and also to reduce the need to open the vacuum for servicing of components.

The mosaic is in an f/1.3 beam, and is being used spectrographically, thus it must be very flat and stable both with changes in gravity and over time. The design currently being developed includes a mosaic system that has materials matched for their thermal coefficient of expansion.

1. INTRODUCTION

This report is intended to gather together and present the early conceptual design of the detector and dewar system. Following this report the concept will be further refined and prototypes fabricated of the mosaic, the focus and FC mechanisms. This second stage is planned to be complete for the detector review planned for February 11, 1997. Current plans show the dewar detector system be complete and ready for testing by July 1997.

2. FUNCTIONALITY/SPECIFICATIONS

The DEIMOS dewar is planned to have considerably more functionality than is presently included in UCO/Lick designed dewars.

DEIMOS requires a mosaic of eight 2Kx4K CCDs with 15μm pixels. The mosaic is in an f/1.3 beam, therefore needs to be very flat as compared to mosaics used in direct imagine mode on a large telescope. The mosaic specifications are:

Flatness	10 μm peak to valley
4K edge gap (detector/detector)	1 mm ± 15 μm
2K edge gap	0.1mm ± 7.5 μm

[1] UCO/Lick Observatory, University Of California, Santa Cruz. Email: **cowley@ucolick.org**

J. W. Beletic and P. Amico (eds.), Optical Detectors for Astronomy, 213–216.

3. MOSAIC

3.1 Clips

The CCD that we plan to use in this instrument are being mounted on aluminum nitride (AlN) packages that have an indentation for clamping the CCD to a baseplate. It is proposed to hold the AlN package to the mosaic back plane using clips made of Invar 36. Invar moves very little in relation to AlN when cooled to -120°C. An Invar plate has been made that holds three packages. The proposal is to cool this plate and determine if there is any detectable change in either elevation of the chips or changes in their translational relationship to each other. Exactly how the assembly is cooled and how the measurements are to be made is yet to be determined.

3.2 Back Plane

The intent is to make a back plane out of a material that is similar enough in coefficient of thermal expansion (CTE), that the CCDs can be aligned warm, and then to -120°C while staying within the error budget.

One possible choice of material would be Invar. However Invar is heavy and does not conduct heat well. Another material is a Carbon/Carbon (C/C) composite. This material is normally used in high temperature vacuum furnace applications, but shows no signs of out-gassing when pumped down with our leak detector. The material is about 10% the weight of Invar and is nearly as thermally conductive as AlN. The thermal expansion of various materials in shown in Figure 1.

3.3 Alignment

The proposal is to machine/polish the clips bases attached to particular packages to an elevation accuracy of ± 2.5 μm. The feasibility of this has been demonstrated with the prototype clips using the grinding machine that can take 2.5 μm cuts from a clip. We need to demonstrate that we can do this with a mosaic set of clips.

The transverse placement of the chips may be done by developing a transitional placement stage. It may be possible to do this with spacers, and using the test plate we will test this alternative.

A method to measure the position of the chips cold has yet to be developed.

3.4 Focus Motion

The focus motion system proposed, uses four sets of flexures, and is driven externally by a commercial vacuum feed though linear motion mechanism. This mechanism is manually driven as supplied, and a concept of a servo driven system will be prototyped.

The system would have a resolution of about 0.5 μm and a expected repeatability of 1 μm. The focus stroke is planned to be ± 1.5 mm.

This flexure is designed so that the base is connected by an aluminum plate, the height to the point at the apex does not change with temperature fluctuations. This concept was developed by Peter Hastings of Royal Observatory of Edinburgh and has been adapted to our requirements. The focus stage will be balanced and expected to have a gravity deflection of less than 0.5 μm.

A prototype of this flexure and drive system is planned.

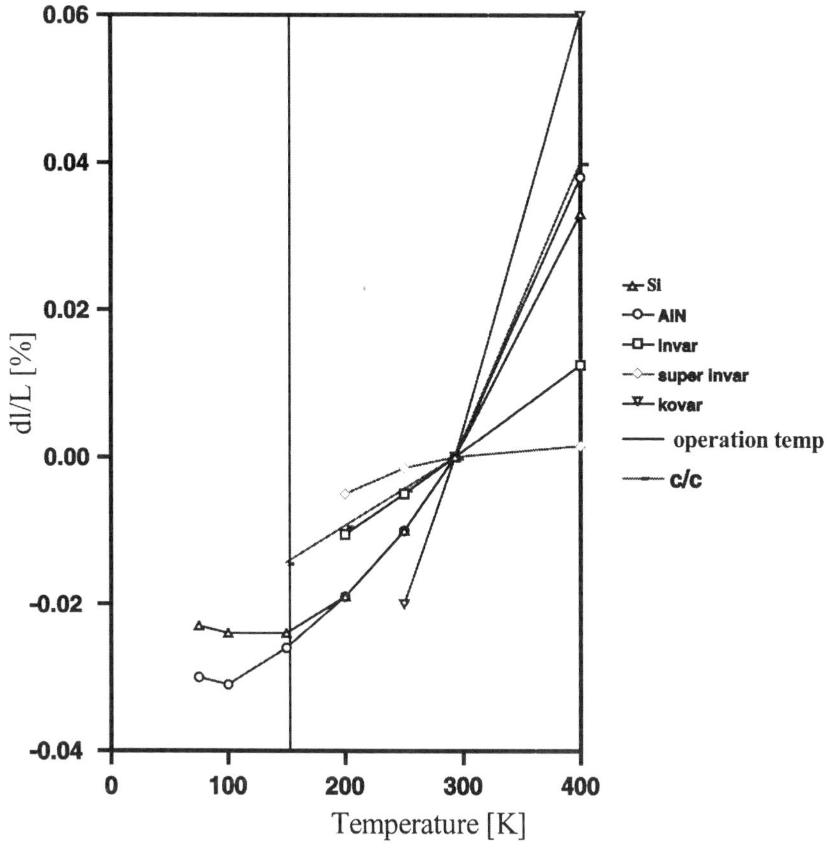

Figure 1:*expansion curves.*

4. FLEXURE CONTROL

The one axis flexure control system proposed is also driven externally by a feedthrough drive stage. It has a stroke of ± 60 μm, a repeatability of 1 μm and a resolution of 0.5 μm. This design was also adapted from work done by Peter Hastings of ROE.

The flexure is expected to support approximately 10 pounds (4.5 kg). With this load it has an expected gravity deflection of approximately 1 μm. The maximum predicted stress in the flexures is expected to be approximately 25,000 pounds per square inch (psi) at full stroke and maximum loading. Further design work will be done to try to reduce this stress by about a factor of two.

The plan will be to prototype this flexure later this year.

5. Dewar

The LN2 dewar is planned to be similar to the ones previously designed and fabricated at UCO/Lick. The dewar will have a hold time of 32 hours and will have an ion-pump, as do all our existing dewars. The LN2 will be remote from the detector head and attached by a flexible coupling to avoid changes in gravity induced flexure caused by the dewar changing weight as the LN2 boils off.

6. Mounting in DEIMOS Structure:

The planned mounting scheme is to attach the detector head directly to the camera body and attach the LN2 dewar body to the structure of DEIMOS. These details have not been designed. Most likely, the field flattener will be attached directly to the detector body, and the principle connection to the camera will be around the filter wheel.

HIGH PRECISION PHOTOMETRY FROM SPACE WITH COROT: EFFECT OF THE SPATIAL IMAGE JITTER ON THE PHOTOMETRIC PERFORMANCES OF CCDS

RACHEL GRUNEISEN[1]

ABSTRACT

COROT is an asteroseismology mission, dedicated to the study of the stars' internal structure by means of oscillation modes detection.

1. SCIENTIFIC GOALS

The goal of this mission proposed to the French space agency (CNES) is to measure oscillation mode frequencies of 5 or 6 stars as main targets and 20 to 30 secondary targets with a very high precision to give new strong constraints on the existing models of stellar structure. This measurement demands a photometric precision of one part per million reachable only **from space**.

2. PERFORMANCE REQUIREMENTS

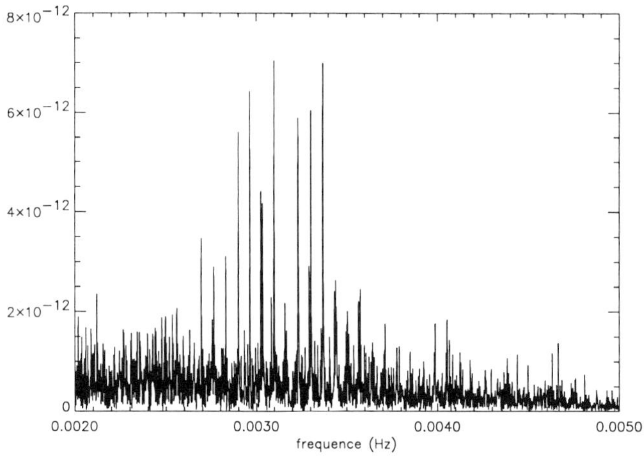

Figure 1: *Power Spectrum. The unit on the vertical axis is power [ppm²/mHz]*

To detect these frequencies a Fourier Transform is performed (see figure 1) on the total signal time series of the stars.

The *amplitude resolution* in the power spectrum is of $10^{-6}/\sqrt{\mu Hz}$, requiring a relative *random noise* level over a 5 day acquisition time of:

$$\frac{\sigma_{random}}{S} = 3x10^{-7} < \text{photon noise} \qquad (1)$$

[1] Observatoire de Paris-Meudon, France. Email:**Rachel.Gruneisen@obspm.fr**

J. W. Beletic and P. Amico (eds.), Optical Detectors for Astronomy, 217–222.

corresponding to a relative noise level of 10^{-4} for a 1second integration time acquisition. It requires also a relative *periodic perturbation* level of less than $\mathbf{6x10^{-7}}/\sqrt{\mu Hz}$. All the noises have to be kept below the photon noise. A *frequency resolution of 0.1 μHz* requires **150 days** of *continuous observation* at least.

3. STAR IMAGE AND FOCAL PLANE

A CCD used in MPP mode has a full well capacity of the order of 10^5 photo-electrons. Therefore the star image needs to be defocused over at least 100 pixels on the CCD in order to collect some $3x10^6$ photo-electrons per second without reaching saturation.

The defocus limit is imposed by confusion with the other stars in the field and by the necessity to keep the readout noise below photon noise.

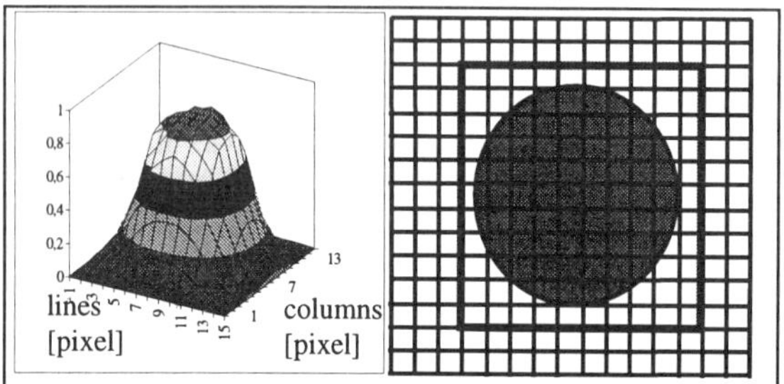

Figure 2: a)*Defocused Point Spread Function (left) and b) aperture photometry window (right)*

The defocused point spread function will have a diameter of about 14 to 20 pixels (see figure 2a).

Aperture photometry and PSF fitting will be performed to get the total signal; the signal of the stars is summed over all the pixel of a defined area (see figure 2b).

4. EFFECT OF IMAGE JITTER ON THE PHOTOMETRIC PERFORMANCES OF CCDs

Satellite vibrations cause the star image to move on the CCD from one exposure to another. The flat field uncertainties of the CCD coupled with this image jitter, gives rise to an additional noise called hereafter: **jitter noise** (see figure 3).

Even with a spatial jitter amplitude of much less than one pixel, without special care, jitter noise can dominate over photon noise.

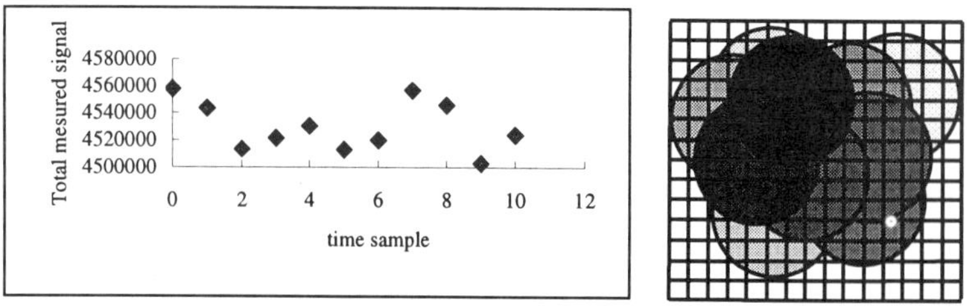

Figure 3: *PSF motion over a CCD and corresponding flux* fluctuations

5. SYSTEM PARAMETERS INVOLVED IN THE JITTER NOISE

The jitter noise level depends on several parameters which are the *dimension and profile* of the defocused star image, the *amplitude of the image jitter* on the CCD and the pixel to pixel and intra-pixel *quantum efficiency non-uniformity*: A helpful result, concerning the combination of jitter noise and the pixel to pixel response uncertainties is given (for $\delta x_{jitter} < 1$) by:

$$\frac{\sigma_{jitter}}{N_{photons}} = \sqrt{2}\,\frac{\sigma_{\eta}\,\delta x_{jitter}}{\left(d_{image}\right)^{3/2}} \qquad (2)$$

where σ_{jitter} is the jitter noise in photo-electrons, $N_{photons}$ is the number of photo-electrons, δx_{jitter} is the jitter amplitude, σ_{η} is the rms Pixel Response Non-Uniformity or, if not known, the rms flatfield uncertainties and d_{image} is the image diameter.

The following results concerning each of these parameters are all given by a computed model.

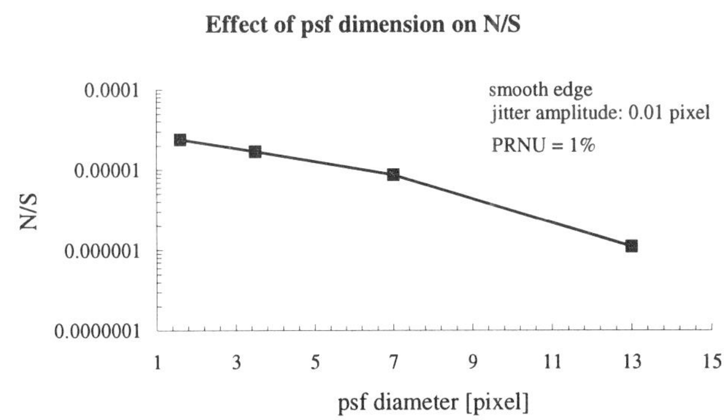

Figure. 4: *Effect of PSF dimension, simulation with pixel to pixel response non uniformity*

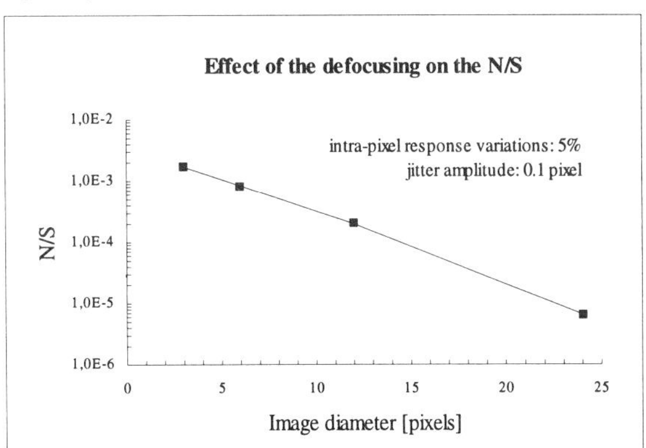

Figure. 5*. Effect of PSF dimension, simulation with* ***intra-pixel*** *response non uniformity for a backside illuminated device.*

5. JITTER NOISE AS A FUNCTION OF STAR IMAGE DIMENSION

In top of avoiding saturation a larger PSF lowers the jitter noise due to pixel to pixel non-uniformity by averaging the measured signal over more pixels (see figure 4).

By increasing the defocus one can significantly reduce the jitter noise due to intra-pixel non-uniformity, since only the peripherical pixels of the psf contribute to this noise (see figure 6). Defocusing lowers their weight with respect to the central pixels of the PSF.

6. JITTER NOISE AS A FUNCTION OF QUANTUM EFFICIENCY NON-UNIFORMITY

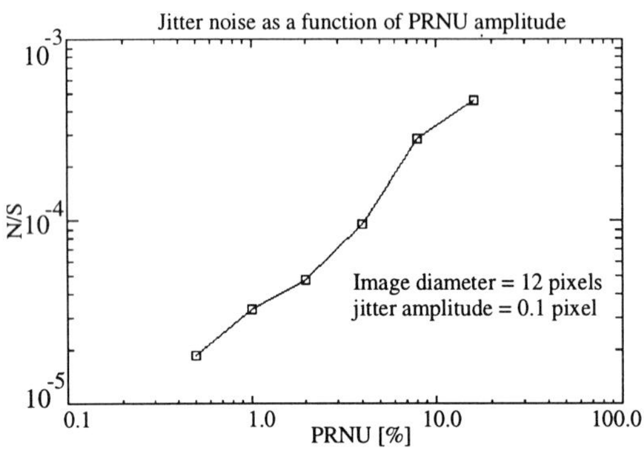

Figure. 6. *Effect of PRNU amplitude, simulation with* ***pixel to pixel*** *response non uniformity*

A Pixel Response Non Uniformity (PRNU), or a flatfield uncertainty of less then 1 or 2 % is expected to keep the jitter noise level sufficiently below that of photon noise.

For our application thinned back-illuminated CCD are required to get some signal below 400 nm. Although these CCD are known to have a PRNU of rather 5% (measured over the whole CCD) one must keep in mind that in our case the PRNU needs only to be evaluated over windows of 20 × 20 pixels (the size of a defocused PSF) reducing the measured PRNU. An order of magnitude can be gained by choosing a thinned, back-illuminated CCD, for which the intra-pixel design is much smoother than for front illuminated CCD (a factor of 5 to 10 can be expected).

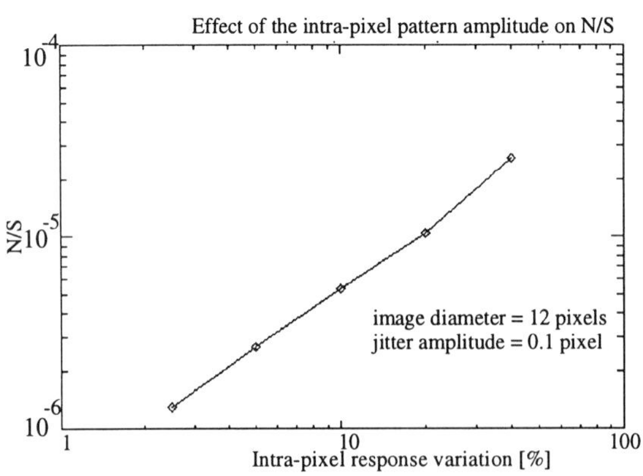

Figure 7*: Effect of intra-pixel design amplitude, simulation with intra-pixel response non uniformity*

Measurements of the intra pixel quantum efficiency is usually done using mono-chromatic light. Although asteroseismology is done in white light, we suppose that white light should rather flatten the intra-pixel design. The results shown here (see figure 7), simulated using a monochromatic profile can therefore be considered as an upper limit for jitter noise. The PRNU specification for COROT is 1.5%

7. Jitter noise as a function of jitter amplitude

The jitter amplitude is random. The value given here is the rms value (see figure 8), not the peak to peak value. This result shows the jitter amplitude that can be tolerated considering a given pixel to pixel non-uniformity and a given image size. Different positions on the CCD give raise to slightly different noise level values.

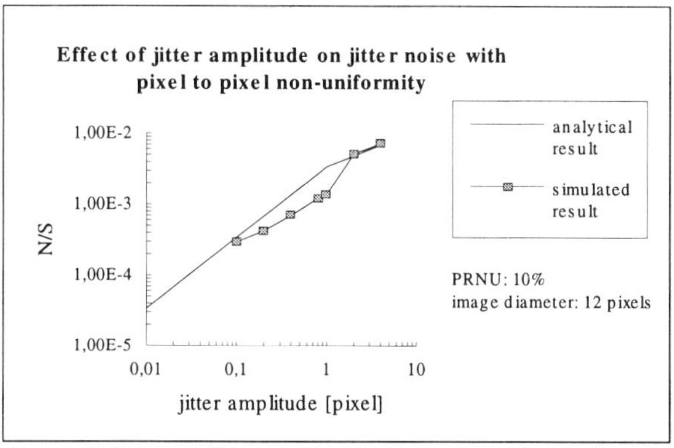

Figure. 8 Effect of jitter amplitude, simulation with pixel to pixel response non uniformity

Jitter amplitude has also a direct impact as far as intra-pixel pattern is concerned (see figure 9).

With a 20 pixel image diameter and a 1.5% PRNU, expecting a relative noise jitter of $10^{-3.5}$ implies: $\delta x \leq 0.2$ pixel, corresponding to a 1 arcsec jitter.

8. Jitter noise as a function of PSF shape

The shape of the psf has a direct impact on the jitter noise as far as intra-pixel response pattern is considered. This result (see figure 10) shows that one can gain a factor of ten by avoiding sharp edges. Using this effect we can have the effect of intra-pixel sensitivity lowered beneath the pixel to pixel non-uniformity.

A value of R-r = 3 (R = radius of the base of the PSF, r= radius of the top of the PSF; see figure 10) seems to be suitable for our purpose.

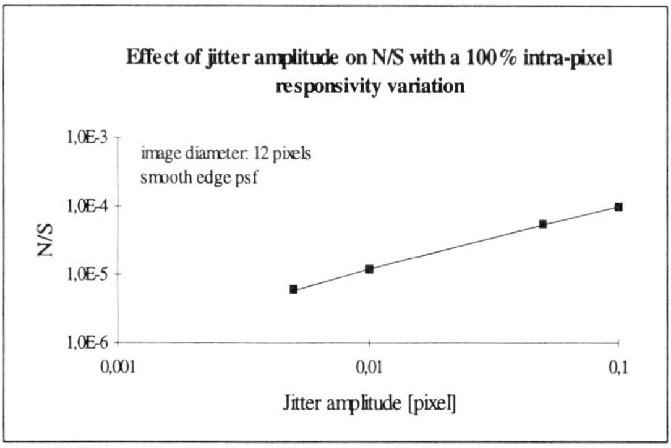

Fig. 9 Effect of jitter amplitude, simulation with intra-pixel response non uniformity

9. A simple way of building a map of the CCD jitter noise

Having determined the parameters involved in jitter noise, we want to build for a given CCD, star image and an approximate value of jitter amplitude Having determined the parameters involved in jitter noise, we want to build for a given CCD, star image and an approximate value of jitter amplitude, a map of the jitter noise level over the CCD. The first step consists in performing the convolution of the CCD flatfield with this star image. The resulting matrix corresponds to a map of the measured star flux as a function of the star's position on the CCD. Figure 11-top shows part of a typical line (24) of this matrix. It is reasonable to think that the smaller the gradient of this map, the smaller the induced flux variation for a given position variation.

The second step consists in performing a 2D gradient of this matrix (see figure 11-middle). Figure 11-bottom shows the jitter noise simulated over the same pixels of the line 24 of the CCD. The results show a good correlation between the processed flatfields and the jitter simulation.

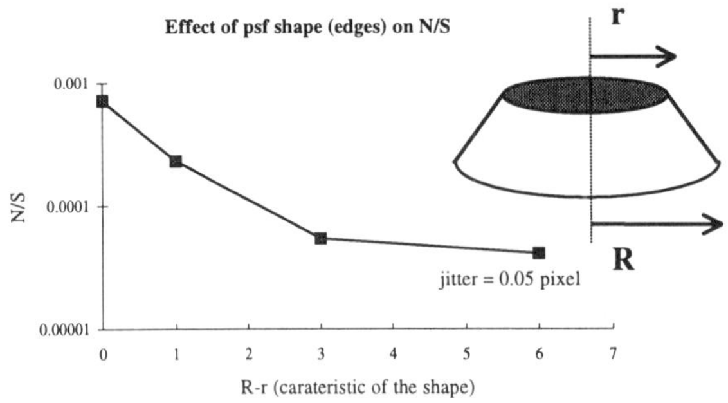

Figure 10 *Effect of PSF shape, simulation with intra- pixel response non-uniformity*

10. Conclusions

The simulation results helped defining several specifications for COROT as the dimension and shape of the Point Spread Function, the maximum jitter amplitude and the spatial non-uniformity of the CCD response. With the tool shown in the last part of this contribution, it seems possible to get, for a given CCD and a given PSF, a map of the jitter noise due to pixel to the pixel response non-uniformity. The next step will be to incorporate to this process the intra-pixel variation.

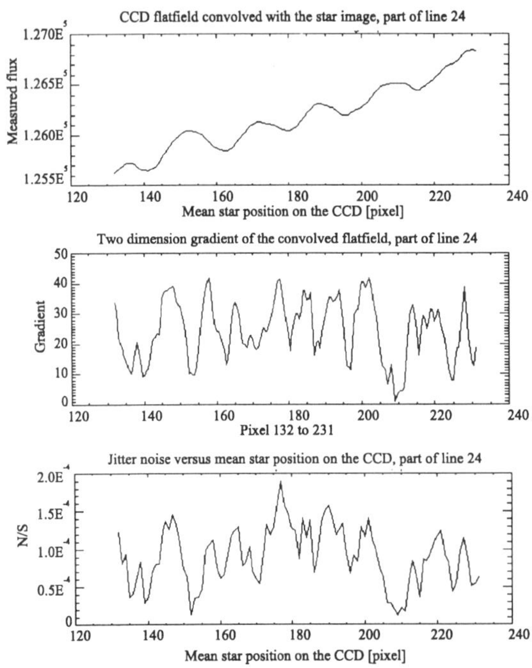

***Figure 11**: The matrix of the convolution of the CCD flatfield with the star image corresponds to a map of the measured star flux as a function of the star's position on the CCD.*
(top) part of line 24 of this matrix.
(middle) a 2D gradient of this matrix.
(bottom) the jitter noise simulated over the same pixels of the line 24 of the CCD.

CHARGE TRANSFER EFFICIENCY IN PROTON DAMAGED CCDs

T. HARDY[1], *M. J. DEEN*[1], *R. MUROWINSKI*[2]

ABSTRACT

We have performed detailed measurements of the charge transfer efficiency (CTE) in a thinned, backside-illuminated imaging CCD. The device had been damaged in three separate sections by proton radiation typical of that which a CCD would receive in space-borne experiments, nuclear imaging or particle detection. We examined CTE as a function of temperature, clock rate and radiation dose.

The dominant factor affecting the CTE in radiation-damaged CCDs is seen to be trapping by bulk states. We present a simple physical model for trapping as a function of transfer rate, trap concentration and temperature. We have made calculations using this model and arrived at predictions that closely match the measured results.

1. INTRODUCTION

In many scientific imaging applications, it is necessary to subject the detector to harmful radiative environments. Such applications include almost any space mission, x-ray crystallography, some forms of medical imaging and energetic particle detection. Radiation has several effects on the performance of CCDs, and one of the most important is a degradation of the Charge Transfer Efficiency (CTE): the amount of charge in a given packet which is successfully clocked past the series of gates from one pixel to the next. This factor is usually represented as a single fraction, typically 0.999990 for a modern CCD. It is equivalently expressed by the Charge transfer inefficiency (CTI = 1-CTE). However, the total signal loss is not simply proportional to the amount of charge in a packet and the number of transfers it undergoes. The experiments reported here, as well as our previous work [1,2] and that of others [3-5] show that the CTI is dependent on additional factors, including the history of charge packets transferred through a given pixel. A detailed knowledge of the factors affecting CTI will enable the user to better understand the data collected by CCDs and will enable designers to improve the performance of new devices.

2. RADIATION DAMAGE

In general, radiation causes two forms of damage to a buried channel CCD. The first is damage to the oxide and the bulk-oxide interface through ionization. This form of damage results in increased surface dark current, shifts in the flat band voltage, increased amplifier noise and changes in linearity. The effect of increased surface dark current can be significantly alleviated by running an appropriately designed CCD in

[1] School of Engineering Science, Simon Fraser University Burnaby, B.C., Canada. Email:**hardy@dao.nrc.ca**
[2] Dominion Astrophysical Observatory, National Research Council of Canada, Canada

J. W. Beletic and P. Amico (eds.), Optical Detectors for Astronomy, 223–230.

inverted or multiple-pinned phase (MPP) mode. Changes to the flat-band voltage can be compensated for by adjusting the operating voltages.

The second form of damage is caused by energetic particles such as protons or neutrons. These particles can collide with silicon atoms, displacing them from their positions in the lattice and creating vacancies within the bulk silicon. Most of the interstitial atom - vacancy pairs thus created recombine and result in no permanent damage. Typically, 2% of the initially generated pairs remain [6]. The vacancies that do not recombine are unstable and will migrate to more favorable positions in the lattice, often becoming trapped near impurities because of the stress these atoms cause to the lattice. The most important of the resulting defects in the CCDs we are studying appears to be the phosphorus-vacancy complex (or E-centre), which introduces a trapping state with an activation energy of about 0.4 eV below the conduction band [7-10]. However, radiation-induced traps with energies of 0.14 eV and 0.23 eV [11] as well as 0.12 eV and 0.30 eV [9] have also been reported. We found evidence for a trap at around 0.2 eV in our own measurements, which suggests the presence of an oxygen-vacancy complex (A-centre) [12]. The trapping states created by displacement damage result in an increase in CTI and an increase in the bulk dark current.

Since the form of the permanent displacement damage is not dependent on the energy of the incident particles, for experimental purposes we can simulate the effects of any desired energy spectrum using a monoenergetic beam by adjusting the fluence to create an equivalent number of displacements [2].

3. A SIMPLE PHYSICAL MODEL

The charge contained in the potential well beneath a CCD gate is transferred to the next and following gates by three basic mechanisms: thermal diffusion, drift due to the fringe field at the gate edges, and self-induced drift due to the mutual electrostatic repulsion between charges [13]. The efficiency of the self-induced drift will be affected by the charge packet size, and all three will be sensitive to temperature. However, the time constants of these processes are on the order of a few nanoseconds, and at the clock rates that we used for our experiments (~1 μs per serial transfer) their effects on the CTI are negligible.

In such slow-scan applications, the effect of trapping by bulk states is more important. Trapping is certainly the most important for any CCD that has suffered bulk damage from energetic particles. The following analysis, after Kim [14], can serve as the basis of a charge deferral model resulting from bulk states. A similar analysis is given in [10].

The capture and emission time constants τ_c and τ_e for the bulk states are

$$\tau_C = \frac{1}{\sigma_n v_{th} n_e} \tag{1}$$

$$\tau_e = \frac{e^{\frac{E_t}{kT}}}{\sigma_n v_{th} N_c} \tag{2}$$

where E_t is the trapping state energy level below the conduction band, σ_n is the trapping cross section, v_{th} is the thermal velocity of carriers, n_e is the density of electrons in the conduction band and N_c is the effective density of states in the conduction band. [15]'

$$v_{th} = \frac{A^* T^2}{qN_C} \tag{3}$$

where A^* is the effective Richardson constant, T is the absolute temperature and q is the electronic charge. We can therefore express the emission time constant as a function of temperature:

$$\tau_e = \frac{e^{\frac{E_t}{kT}}}{\sigma_n \frac{A^*}{q} T^2} \tag{4}$$

When a charge packet is present, the number of charges n_t held in bulk states within the packet volume V_s containing N_t traps per unit volume is (assuming all traps are full)

$$q_t \approx qN_t V_s \tag{5}$$

Consider one gate of the typical clocking scheme of a three phase CCD (not operated in MPP mode) such as p1 of Figure 1.

We assume that the charge packet has substantially arrived under this gate at time T_1. Traps in the volume occupied by the charge packet will be filled with time constant τ_c, and because this time constant is relatively short, we will assume the states are filled at T_1. At time T_2 the charge packet becomes shared with *p2*, the volume of the charge packet under *p1* diminishes and the traps under *p1* begin to emit. By time T_3 the charge packet has moved on completely to *p2*. The traps under *p1* will continue to emit until T_6, or, in the case of a sparsely illuminated CCD, until the next non-empty charge packet arrives. At that time, the states which have emitted their charges during the period since T_2 will be filled from the new packet, resulting in a charge loss from the new packet of

$$Q_{trapped} = qN_t V_s \cdot \left(1 - e^{-\frac{T_{emit}}{\tau_e}}\right) \tag{6}$$

where T_{emit} is the total emission time from T_2 to the arrival of the new charge packet.

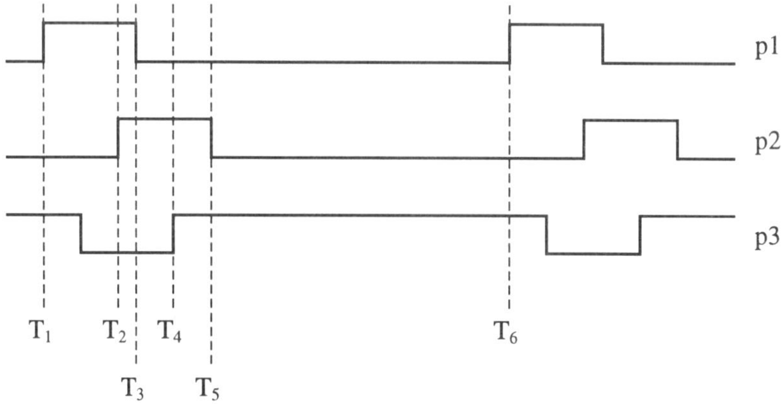

***Figure 1**: Typical clocking in a three phases CCD.*

A portion of this charge, however, is regained. Any charges emitted between T_2 and T_4 will rejoin their parent packet due to the fringe field. In fact, some of the charges emitted between T_4 and T_5 will also join the parent packet, but because of uncertainty of the partition function, we will assume that the time period during which the charges can join the parent packet is $T_{join} = T_4 - T_2$. The amount of charge that is reintroduced to the packet by these emissions is

$$Q_{join} = qN_tV_s \cdot \left(1 - e^{-\frac{T_{join}}{\tau_e}}\right) \tag{7}$$

and therefore the net loss from a charge packet for one transfer through a *p1* gate is

$$Q_{loss} = Q_{trapped} - Q_{join} = qN_tV_s \cdot \left(e^{-\frac{T_{emit}}{\tau_e}} - e^{-\frac{T_{join}}{\tau_e}}\right) \tag{8}$$

where T_{emit} is the emission time from the previous packet. A similar evaluation using appropriate values for T_{emit} and T_{join} can be performed for the other two phases or for the case of a CCD operated in MPP mode. If more than one trap level exists, additional terms of the form of the right side of (9) can be added, using the corresponding values of τ_e.

4. MEASUREMENTS

A useful and simple method often used to measure the CTI of a CCD is the Extended Pixel Edge Response (EPER) technique [7]. In this technique, the CCD is illuminated uniformly and read out in such a way as to obtain an image which extends over one or both of the edges opposite the output amplifier. The signal measured in the extended regions gives an indication of the amount of charge that has been deferred from the pixels in the imaging area of the CCD through the clocking process.

EPER is perhaps more properly considered a technique for measuring charge *deferral*, since it only gives an indirect measure of the CTI. It cannot detect losses due to charge, which is completely absorbed or deferred for times much longer than the clocking periods. However, if the CTI is dominated by trapping at a single energy level, which it appears to be for radiation damaged CCDs, EPER gives an accurate indication of the CTI. The EPER technique is particularly useful in probing how the CTI changes with charge packet size, since the illumination level can be easily varied. A further advantage is that it requires no specialized equipment.

If we assume that charges are not lost during transfer, but are simply deferred by the trapping mechanism described earlier, we can see that a flat-field image such as that used for EPER will show no discernible CTI degradation. Since each charge packet is roughly the same size, the packets will interact with the same traps within the pixels, and the number of charges trapped will be the same. During each transfer, the charges emitted by the traps which do not rejoin the original packet will join the following packet and be used to refill the traps so that there is no net loss from that packet. However, the charges captured from the last packet in a row or column will be emitted into the empty packets of the extended region. These charges will not be able to interact with the traps available to the larger field packets because they interact with a much smaller volume of silicon. The number of charges emitted into the extended pixels indicates the number of empty traps that would be available to capture charge from a subsequent signal packet, and therefore this number can be used to estimate the CTI.

For our results, we calculated the CTI by dividing the total amount of charge in the extended pixels by the amount of charge in the edge pixel and then dividing by the total number of single-phase transfers involved in the read out. This gives a "worst case" CTI figure equivalent to that experienced by an isolated feature in a completely dark field.

The CCD we measured is a TK512, made by Tektronix, with 512 × 512 square pixels, each of 27 μm size. This device is a thinned, backside-illuminated buried-channel device which was subjected to bulk damage by means of a 3 MeV proton beam in a manner similar to previous experiments [2,16]. The imaging area and serial registers were divided into three zones which were subjected to 6.0×10^9 protons/cm^2, 1.5×10^9 protons/cm^2 and no radiation respectively, resulting in two damaged regions containing about 9.0×10^8 and 2.3×10^8 displacements/cm^2 and an undamaged control region.

5. Results and Discussion

Figure 2 shows the parallel EPER CTI of the radiation damaged TK512 device as a function of temperature. The three sets of data plotted indicate measurements of the three CCD areas, the first of which received a high amount of radiation, the next received one quarter of this amount and the last was undamaged. We note not only a general increase in CTI with increasing amount of damage, but also that the CTI is more strongly dependent on temperature in the damaged sections. Unfortunately, we could not get data at higher temperatures because the dark current was so high, especially in the high radiation zone, that it was distorting the results.

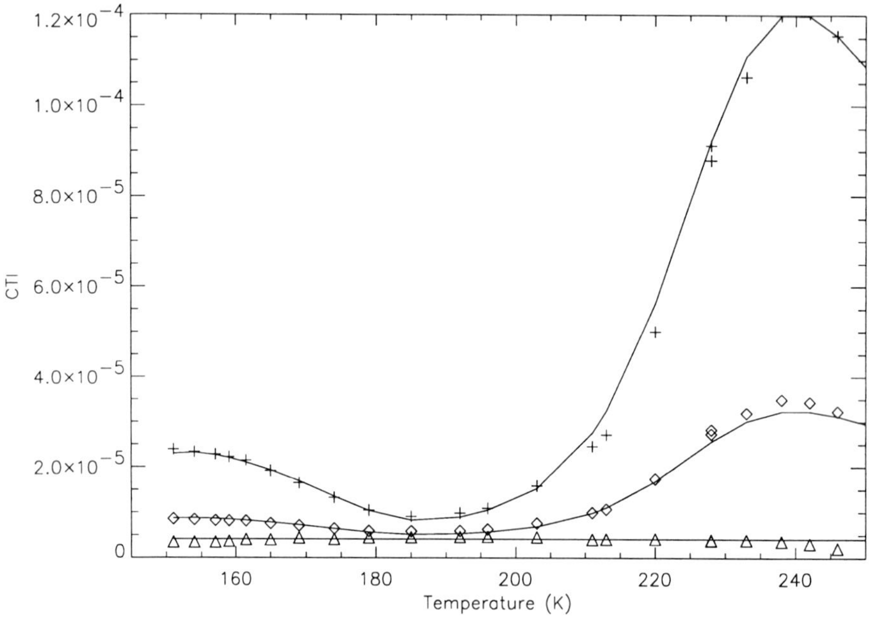

***Figure 2**: CTI as a function of temperature. The upper, middle and lower symbols are experimental data from the high (6×10^9 protons/cm^2), low (1.5×10^9 protons/cm^2), and no radiation sections of the device respectively. The lines indicate model results evaluated for two traps ($E_{tA} = 0.223$ eV, $\sigma_{nA} = 4{\times}10^{-16}$ cm^{-2}, $N_{tA}V_s = 3.90$ (high), 0.975 (low) and $E_{tB} = 0.42$ eV, $\sigma_{nB} = 5{\times}10^{-15}$ cm^{-2}, $N_{tB}V_s = 23.8$ (high), 5.95 (low)).*

Since it could be assumed that the additional CTI experienced by the TK512 from radiation damage is the result of bulk trapping, we decided to see how well the trapping model described above would fit the data. First, the CTI in the undamaged part was fit to a constant ($5.16{\times}10^{-7}$), which is shown as the lowest line in Figure 2. Then the excess CTI in the high radiation data was fit to the model described above. The uppermost line drawn on the graph indicates the model evaluated for two traps: one at E_t = 0.223 eV with $\sigma_n = 4{\times}10^{-16}$ cm^{-2} (Trap A) and the other at E_t = 0.42 eV, with $\sigma_n = 5{\times}10^{-15}$ cm^{-2} (Trap B), and making use of clock timings measured from our CCD electronics. In this experiment we used the non-MPP clocking scheme shown in Figure 1 with a clock pulse width of 96 μs. The period of the three-phase clocking cycle was 200 μs for the first 500 cycles, during which the serial register was simply flushed, and 22.2 ms for the final 13 cycles when the serial register was being read out between parallel transfers. Trap B was modeled as having six times the trapping state density N_t, resulting in N_tV_s values of 3.90 for Trap A and 23.8 for Trap B. If we estimate the effective volume of the charge packet to be 27 μm × 9 μm (gate area) × 0.5 μm (buried channel depth), the corresponding trap density N_t is about $2{\times}10^{11}$ cm^{-3} for Trap B and $3{\times}10^{10}$ cm^3 for Trap A. Finally, assuming the density of trapping states is proportional to the proton dose, the model was evaluated for one quarter of the N_t values used in the previous case, and that model was plotted as the middle line for comparison with the low radiation data.

Trap A is in line with previously published figures for the oxygen-vacancy complex [12], and Trap B matches the published figures for the phosphorus-vacancy complex, which is usually given as the dominant trap level in irradiated CCDs. Energy levels for this trap have been reported in the range 0.4 - 0.44 eV [7-10]. Using our model to match the CTI vs. temperature data, it is difficult to decouple the effects of the trap energy and the trapping cross-section, so we could not unambiguously determine these two parameters.

The model can now be used to choose combinations of temperature and clocking speed to minimize the CTI for a radiation damaged CCD. For example, the minimum at around 190 K in our data could be moved to 200 K by doubling the clocking speed.

6. Conclusion

We have described the temperature and signal dependence of the CTI in proton-damaged CCDs and have presented a simple physical model to predict the CTI as a function of temperature for degradation due to trapping states. The model predictions are in good agreement with the experimental results. The data clearly indicates the introduction of two dominant bulk trapping levels as a result of proton radiation. We have estimated the activation energies to be about 0.42 and 0.223 eV respectively, with trapping cross sections of 5×10^{-15} and 4×10^{-16} cm^{-2}. The number of traps introduced appears to scale linearly with proton fluence and the ratio of concentrations of traps at the two levels is approximately 6 to 1. After irradiation by a 3 MeV proton beam with a fluence of 6.0×10^{9} protons/cm^{2}, the trap concentrations were estimated to be 2×10^{11} cm^{-3} for the 0.42 eV trap and 3×10^{10} cm^{-3} for the 0.223 eV trap.

Acknowledgments: We would like to thank both the Canadian Space Agency and Dr. John Hutchings at the Dominion Astrophysical Observatory for their support of this work. We are also grateful to the Natural Sciences and Engineering Research Council of Canada for partial financial support.

8. References

[1] R. Murowinski, M.J. Deen, T. Hardy, "Charge transfer efficiency in low temperature charge coupled devices", *Proceedings of the Symposium on Low Temperature Electronics and High Temperature Superconductivity,* Proceedings Vol. 95-9, C. Claeys, S. Raider, R. Kirschman and W. Brown, Editors, The Electrochemical Society Press, New Jersey, pp. 369-382 (1995).

[2] R. Murowinski, G. Linzhuang, M.J. Deen, "Effects of space radiation damage and temperature on CCD noise for the Lyman FUSE mission", in *Photonics for Space Environments,* E. Taylor, Editor, Proc. SPIE 1953, pp. 71-81 (1993).

[3] A. Mohsen and M. Tompsett, "The effects of bulk traps on the performance of bulk channel charge-coupled devices", *IEEE Trans. on Electron Devices* Vol. ED-21 (11), pp. 701-712 (1974).

[4] I. Zayer, I. Chapman, D. Duncan, G. Kelly, K. Mitchell, "Results from proton damage tests on the Michelson Doppler Imager CCD for SOHO", *Proceedings of the SPIE* Vol. 1900, pp. 97-107 (1993).

[5] I. H. Hopkins, G. Hopkinson and B. Johlander, "Proton-induced charge transfer degradation in CCDs for near-room temperature applications", *IEEE Trans. on Nuclear Science* Vol. NS-41 (6), pp. 1984-1991 (1994).

[6] V. A. J. Van Lint, "The physics of radiation damage in particle detectors", *Nuclear Instr. and Meth.* Vol. A253, pp. 453-458 (1987).

[7] J. Janesick, G. Soli, T. Elliot and S. Collins, "The effects of proton damage on charge-coupled devices", *Proceedings of the SPIE* Vol. 1447, pp. 87-108 (1991).

[8] M. S. Robbins, T. Roy and S. Watts, "Degradation of the charge transfer efficiency of a buried channel charge coupled device due to radiation damage by a beta source", RADECS 91, *IEEE Proceedings*, pp. 327-332 (1992).

[9] A. D. Holland, "The effect of bulk traps in proton irradiated EEV CCDs", *Nuclear Instr. and Meth.* Vol. A326, pp. 335-343 (1993).

[10] C. Dale, P. Marshall, B. Cummings, L. Shamey and A. Holland, "Displacement damage effects in mixed particle environments for shielded spacecraft CCDs", *IEEE Trans. on Nuclear Science* Vol. NS-40 (6), pp. 1628-1637 (1993).

[11] N. S. Saks, "Investigation of bulk electron traps created by fast neutron irradiation in a buried n-channel CCD", *IEEE Trans. on Nuclear Science* Vol. NS-24 (6), pp. 2153-2157 (1977).

[12] Z. Li, H. Kraner, E. Verbitskaya, V. Eremin, A. Ivanov, M. Rattaggi, P. Rancoita, F. Rubinelli, S. Fonash, C. Dale and P. Marshall, "Investigation of the oxygen-vacancy (A-center) defect complex profile in neutron irradiated high resistivity silicon junction particle detectors", *IEEE Trans. on Nuclear Science* Vol. NS-39 (6), pp. 1730-1738 (1992).

[13] E. K. Banghart, J. Levine, E. Trabka, E. Nelson and B. Burkey, "A model for charge transfer in buried channel charge-coupled devices at low temperature", *IEEE Trans. on Electron Devices* Vol. ED-38 (3), pp. 1162-1174 (1991).

[14] Ch.-K. Kim, in *Charge-coupled Devices and Systems*, M.J. Howes and D.V. Morgan, Editors, Wiley-Interscience, New York, p 57 (1979).

[15] S.M. Sze, *Physics of Semiconductor Devices*, Wiley-Intersciences, New York, p 261 (1981).

[16] R. Murowinski, G. Linzhuang, and M.J. Deen, "Effects of space radiation damage and temperature on the noise in CCDs and LDD MOS transistors", *IEEE Trans. on Nuclear Science* Vol. NS-40 (3), pp. 288-294 (1993).

Low Smear Drift Scanning CCDs

D. Ives[1] *& A. Bennett*[1]

Abstract

A new CCD drift mode has been developed called Low Smear Drift Mode. It allows a reasonably sized windowed (1124x10) frame to be readout continuously at approximately one windowed frame per second or slower. A novel clocking scheme has been implemented in conjunction with an aperture mask to ensure low level smearing of the image while it is being clocked with the shutter open (<0.1% for 1s integration). Absolute time stamping of each image has also been implemented.

1. Drift Scanning CCDs

- This involves repeatedly exposing the CCD through a mask. Between each of the exposures n rows of the CCD are read out (were n is equal to the size of the mask window). By doing this repeatedly each exposed region will move down the chip and be finally readout and processed.
- This method can be used to reduce the dead time between multiple exposures as the CCD is not cleared between each successive exposure and only the required data is read out.
- Standard drift scanning has several problems.
- As the CCD is exposed for the whole duration of the integration and readout period, this results in the current exposed region and the next being smeared during the readout process. The amount of smearing is dependent on the x size of the window. This is more pronounced at shorter exposure times.
- Unless the mask window is positioned exactly a whole number of window sizes up the chip, each exposure will be spread across two readouts
- The cosmic ray events for an exposure are related to:

$$\text{exposure time} \times \left(\frac{chip\_y\_size}{window\_y\_size} \right) \quad (1)$$

which is much greater than the same exposure time for single readout of the same window.

[1] Royal Greenwich Observatory, Madingley Road, Cambridge CB3 0HJ, UK
Telephone (01223) 374000; Fax (01223) 374700 Email: **ives@ast.cam.ac.uk**

J. W. Beletic and P. Amico (eds.), Optical Detectors for Astronomy, 231–235.

2. LOW SMEAR DRIFT

To limit the amount of smearing the data is shifted down by the size of the mask aperture before and after each readout sequence. For this to operate correctly the start position of the image area MUST be:

$$n((2 \times mask\_aperture + window\_size) + mask\_aperture) \qquad (2)$$

where n is a whole number. To use this mode a user defines a window on the chip and this is used as the definition for the mask aperture size and its position on the chip. The software then finds the best fit for the number of rows to shift etc. It is allowed to increase the shift amount from the size of the window and move the start of the window down effectively growing the size of the window (but these extra rows are skipped during the readout, so that the correct window is read out).

- Each image row is only smeared by an amount equal to the vertical clock time, which is approximately the same order of magnitude as the pixel routine, so the amount of smearing is reduced by the number of x pixels in the window (i.e. approximately 1000 times smaller for the 1124x1124 chip used in the experiments).
- Less integrations before the data reaches the output - more useful data is returned (27 instead of 10 in the example).
- Effective exposure to cosmic rays is lower (3 seconds instead of 18 in the example).

This new mode has been successfully implemented and used for observations during May 1996. It can be used in conjunction with on chip binning; multiple windows can also be used. The maximum separation between windows depending on the number of CCD rows under the bottom mask.

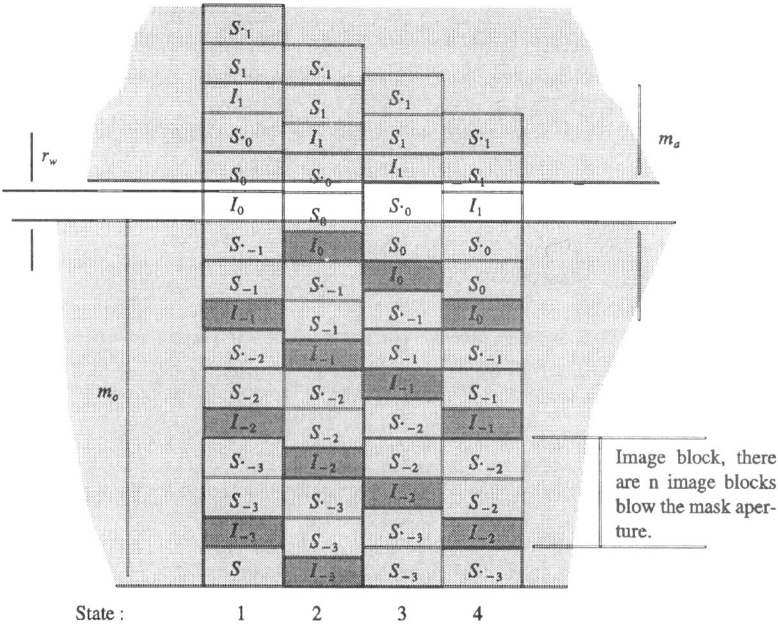

Figure 1

Operation
State 1: Integrate onto section I.
State 1 to 2: Fast shift down S rows.
State 2 to 3: Read out data.
State 3 to 4: Fast shift down S rows.
State 4: Start next integration.

The software is presented with the values for and calculates the values of S from the relationship below:

S = Size of shift region

r_{aw} = Size of readout window (either size of single window or bounding size of multiple windows, in y) prior to any binning is applied.

n = Number of image blocks below the mask aperture.

m_a = Mask aperture.

m_o = Mask aperture offset from start of chip (bottom window offset).

$$S = r_w + r_f$$

$$r_f = \frac{(m_o - r_w(3n+1))}{2n+1} \quad (3)$$

For correct operation $S \geq m_a$

3. Comparison Between Normal and Low Smear Drift

The exposures were identical for each mode; 30 exposures, each 500 ms in length of a 1124x35 window on a 1124x1124 TEK CCD, of a spot projected at (428,690), with a diameter of 6 pixels, vertical clock of 160 μs and pixel routine length 21.5 μs.

Figure 2 shows a cut through the center of the spot for all of the exposures for each drift mode. Note that the normal smear mode produces 10 images whereas the LSD mode produces 27 images.

The table below compares the actual smearing measured for the two drift modes. Note as the object is 6 pixels in diameter the smearing is approximately 6 times more than if the object was only one pixel.

Type of drift scanning used	**Number of counts in Object**	**Number of counts in Smear**	**Amount of smear %**	**Smear adjusted for object size**
Normal	16300	14300	88	15
Low smear	16400	215	1.3	0.22

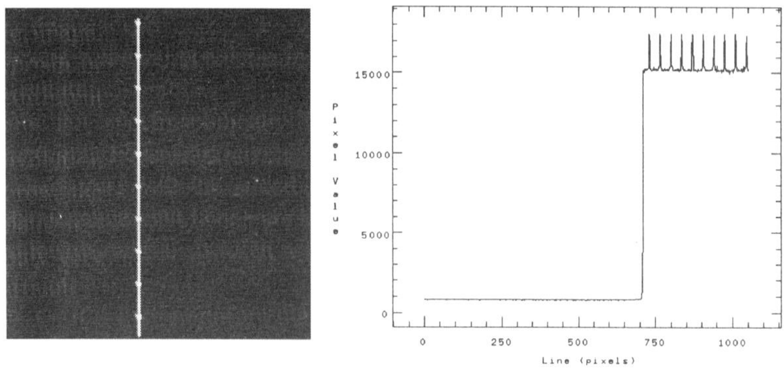

Normal Drift mode

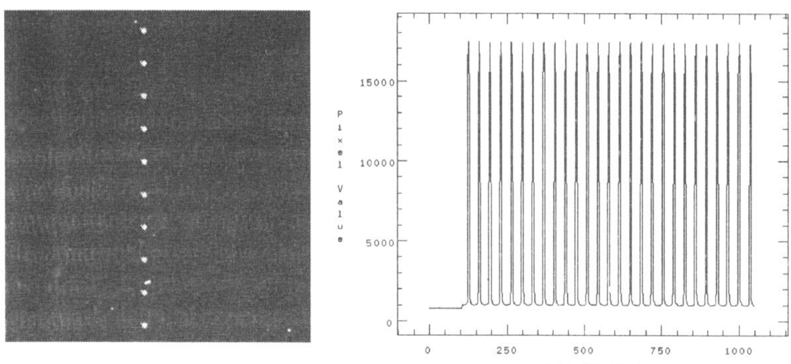

Low Smear Drift (LSD)

Figure 2

4. EXAMPLE OF DATA OBTAINED USING THE LOW SMEAR DRIFT MODE

The spectrum shown in Figure 3 is an extract from a 20 minute observation of Sco X-1 (May 1996). It comprises 60 individual exposures of approximately 1 second (with approximately .5 second between them. This observation was carried out using the LSD drift mode in combination with a GPS receiver to time stamp each exposure so that the data could be compared with data from an X-ray satellite and other simultaneous earth bound observations.

It is hoped that the fluctuations in the optical intensity can be related to the fluctuations in the X-ray intensity and thus provide information about the physical structure of Sco X-1.

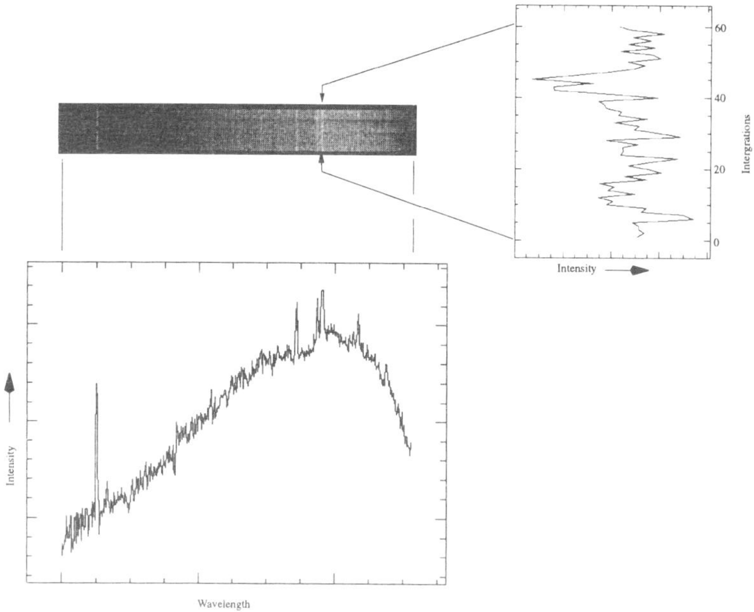

Figure 3

PARALLEL PHASE TEST

RALF KOHLEY[1], KLAUS REIF[1], PHILIPP MUELLER[1]

ABSTRACT

Contribution to the Roundtable: CCD Characterization and Optimization

There's an easy way to determine the intrinsic properties of the parallel phases, if your controller allows voltage settings between shutter closure and readout.

In our case with a 3-phase Loral CCD we had to find out which of the phases is additionally doped to allow MPP mode, at which voltage the phases get out of inversion and how much the voltage offset would be between the normally and additionally doped phases.

The test is to illuminate the CCD in MPP mode with a point-like source up to full well, closing the shutter and applying a predefined voltage to one of the phases for a short time (0.1 s) and then switching back to MPP voltages. After normally reading out the CCD the images can be analysed in the following way:

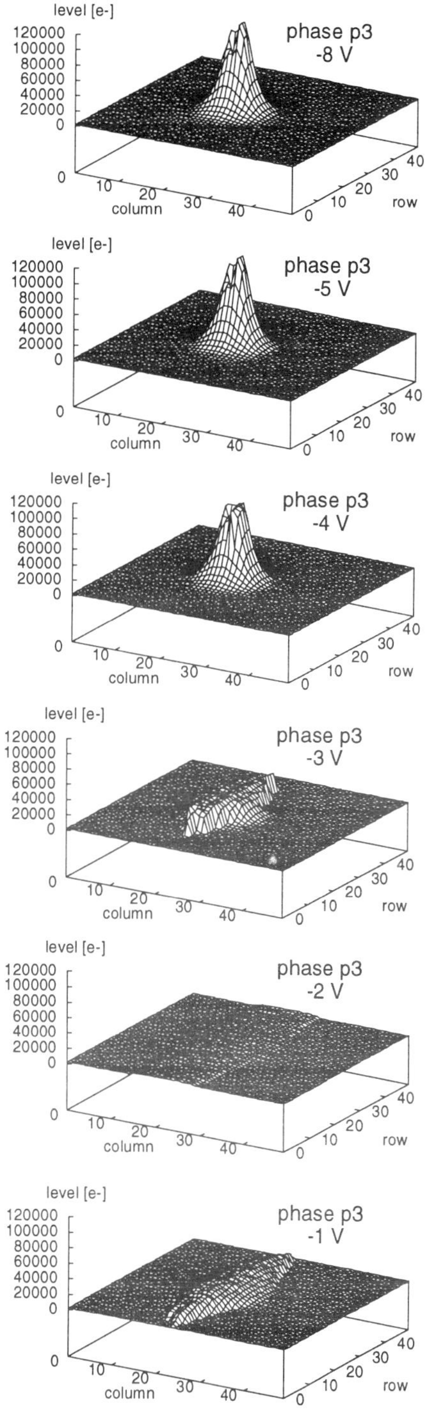

Figure 1:*From top to bottom: 1) max. level (~110000 e⁻) close to MPP full well capacity, no blooming; 2) phase p3 3 volts higher, no change, p3 still in inversion and pinned; 3) full well decreased, blooming starts, p3 comes out of inversion; 4)strong blooming, p3 is totally out of inversion, but still barrier phase;5) total blooming, p3 potential is equal to p1 and p2 potentials; 6)p3 becomes collecting phase, blooming occurs while switching from barrier to collecting state*

[1] Sternwarte der Universitaet Bonn, Radio astronomisches Institut der Universitaet Bonn, Auf dem Huegel 71, D-53121 Bonn, Germany.
Email: rkohley@astro.uni-bonn.de

J. W. Beletic and P. Amico (eds.), Optical Detectors for Astronomy, 237–238.

The illustration (Figure 1) shows the case for p3, the additional doped phase.
Between -8 V (MPP voltage for all phases) and -5 V no change can be seen. Since the maximum of the distribution reaches nearly full well capacity, a slight change in the barrier phase would cause blooming. No change means, that the barrier phase (p3) is still in inversion and pinned.
Blooming occurs at -4 V with p3 coming out of inversion. Between -4 V and -2 V, blooming becomes successively stronger, resulting in total blooming at -2 V. Total blooming occurs when all potentials have the same depth and therefore no barrier exists.
Applying these voltage changes to the phases p1 and p2 doesn't show any change in the images, because the integration potential of these phases is increased and the barrier stays the same.
To find out which voltage to apply to p1 and p2 to drive these phases out of Inversion, -2 V has to be applied to p3 to erase the barrier during integration time. At the p1 or p2 voltage, when charge collection becomes visible, the corresponding phase is out of inversion.

CONVERTING A SCHMIDT TELESCOPE TO CCDS: THE OCA/DLR ASTEROID SURVEY

ALAIN MAURY[1], JEAN PIERRE RAGEY[1], HERVÉ VIOT[1]

HARALD MICHAELIS[2], THOMAS BEHNKE[2]

ABSTRACT

The OCA Schmidt telescope is a classical Schmidt telescope, 90cm aperture, 1.50 meter mirror and 3.14 meters focal length. It was put into operation in 1978, at a time when astronomical CCDs were in their infancy, and has been designed as a photographic telescope. It has been used for a variety of projects, involving photographic plates, and later photographic films. Intended mostly as a technological development, a small CCD camera was installed inside one of our plate holders has been done in 1992. Thereafter, the development of larger CCD cameras for our telescope has been motivated by the need to survey large area of sky in order to discover Near Earth Asteroids. Since 1995, this effort is a collaborative effort between our team and DLR Berlin.

1. ASTEROID SEARCHES

All things being equal, the efficiency of a survey detector equals the product of its DQE over the same passband by its area.

While this simple figure still gives a large advantage to photographic film over existing single CCDs, the nonlinear behaviour of the photographic emulsion DQE gives a large advantage to CCDs when it comes to detecting moving objects like asteroids. This was proven in 1989 by the Spacewatch group, when using one of the first available thick 2K Tektronix chip. This group discovered much more, and fainter asteroids than had ever been done using photography [3]. This message came loud and clear to the community of asteroids observers, which, while still looking at their photographic plates visually for asteroids, started to look into CCD technology for the future search programs.

After 1988, the planetary community realised than there ought to be about 2000 yet undiscovered Earth Crossing Asteroids larger than 1km, and that we needed to start a survey aimed at discovering all these objects in order to become certain that none was going to hit us anytime soon [6]. This informal effort has now been started by several groups worldwide and is dubbed the Spaceguard Survey. The crash of comet Shoemaker Levy 9 in Jupiter in 1994 helped to visualise what a cosmic impact was, and made us clearly understand why we needed more than one CCD in the world looking at the sky for these objects.

---

[1] Observatoire de la Côte d'Azur - Caussols - 06460 Saint Vallier de Thiey - France.
Email: **maury@obs-azur.fr**

[2] Institüt für Planet Erkundung Forschungzentrum - Rudauer Chaussee 5 - D12489 DLR Berlin

J. W. Beletic and P. Amico (eds.), Optical Detectors for Astronomy, 239–244.

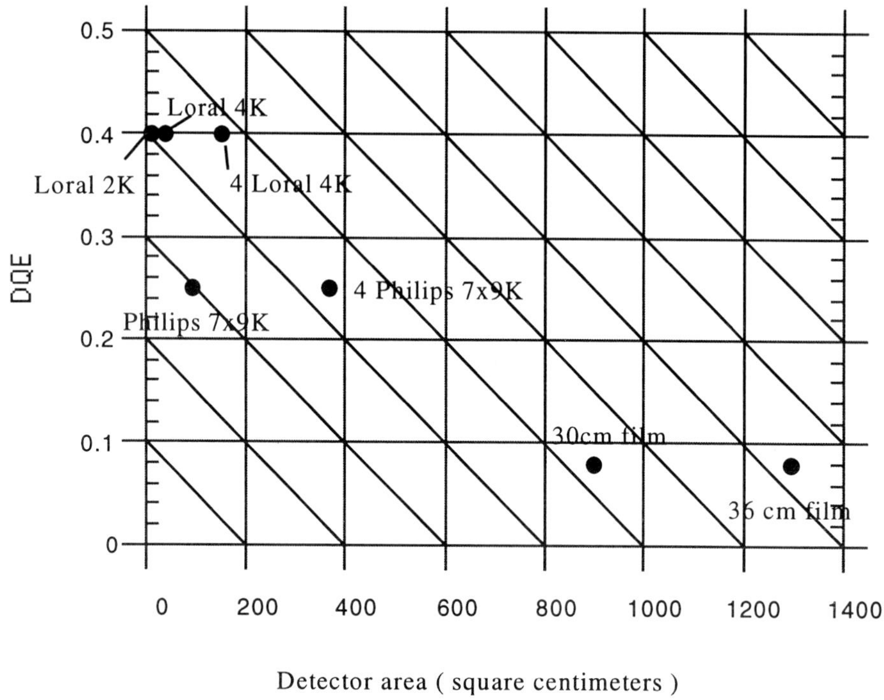

Figure 1: plot of the estimated DQE over the sensitivity range of CCDs (i.e. 400nm to 900 nm) and films (i.e. 300nm to 700 nm) versus their area in square centimeters.

2. CCD VERSUS PHOTOGRAPHY AS A WIDE FIELD DETECTOR.

Most of the Schmidt telescopes who have still not converted to CCDs have converted to the use of photographic film. Kodak 4415 Technical Pan film was introduced in 1977, but did gain general use in large Schmidt telescopes only in the late 1980s. Films are much less expensive than glass plates, and for small facilities it means a non-negligible economy of several thousands of dollars per year. Film base is not as dimensionally stable as glass, but is equivalent to a sheet of steel of the same thickness. The main feature of hypersensitised 4415 film is its ability to reach a signal to noise ratio between 120 and 150 on sky limited exposures, compared to about 50 for the former generation of glass plates. Unfortunately, these values are reached after long exposures, and do not translate directly into the short exposure times such as required for the detection of fast moving objects. In practice, a long exposure on the ecliptic will reveal about 200 asteroids in a 45 minute exposure, whereas a single exposure on a thick 2K detector will typically detect 10 asteroids in a single minute of exposure time in the same zone. While film still remains the detector of choice for Schmidt telescopes, having better efficiency and uniformity, CCDs are progressing fast, and while more expensive to build, are already enabling much better performance in certain areas of choice, such as the detection of moving targets and massive stellar photometry.

Examining the numbers for a few currently available large CCD combinations and existing films gives the following table.

	DQE	Area (sqr cm)	Merit Factor	Current detector price [K$]
Loral 2K	0.4	9.4	3.7	3
Loral 4K	0.4	37.8	15	15
4 Loral 4K	0.4	151	60	60
Philips 7x9K	0.2	92.4	18.5	150
4 Philips 7x9K	0.2	369.6	74	600
30cm 4415 film	0.08	900	72	0.01
36 cm 4415 film	0.08	1296	103.7	0.02

3. Converting to CCDs.

While cost is not a real limiting factor for most large telescopes, it is for a 1 meter class, which moreover is considered an old facility. Another factor is that only 4 persons operate, maintain and improve the telescope, detector and software environment.

Whereas it has been possible to fit a CCD camera to smaller Schmidt telescopes using a flat 45° mirror ("à la Newton"), this simple solution is impractical for larger Schmidt telescopes, and special efforts have to be expended in order to fit a CCD camera into an existing plate holder assembly.

In 1992, we fitted a small Thomson 7882 CCD (384 × 576 pixels) into one of our plate holders, with the two goals of learning this new (to us) technology, then to see whether it would be practical to perform photometric calibration of sub areas of photographic plates using the same instrument for plates and calibration. The controller used at the time was a slow but low noise controller of the Palomar design [4].

This detector was replaced at the end of 1993 with a thick 2K Loral CCD 442A, with an overall improvement of the hardware and software around the CCD camera, including a filter wheel fitted with the "standard" Bessel combination made of Schott filters [2]. We soon learned that using filters shorter than V with our thick CCD was hopeless.

At this time, the efforts around the telescope started to concentrate more on CCD development rather than on photographic work. More recently, the CCD camera became the only detector allowed on the telescope. While less efficient than photographic plates for large field work, the CCD gives several advantages, among others: lower operating cost, better photometric quality, and better usage of average nights. (In photography, long exposures require that the sky quality stays constant, whereas with a CCD short exposures can be taken almost though holes in the clouds. It is possible to use a partially cloudy night with a CCD, whereas this is not possible with photography).

4. The current CCD camera at OCA.

"Regular" CCD controllers are not well matched to the use of Schmidt telescopes.

Because the telescope is a relatively fast optical system (F/3.5), all but the shortest of exposures are photon noise limited. We developed a very simple model showing that spending the night gathering light was much more efficient than waiting for the CCD to readout very slowly in order to save a few electrons of noise. Our current CCD is in fact not a "scientific" CCD with a very low noise output amplifier, but rather a "commercial" type device. Our current performances are 6 microseconds per pixel, providing a 25 seconds readout time using a single CCD amplifier, and 25 e^- readout noise.

Another problem we have had to address was that the camera, and a good part of its controller (clock drivers and acquisition chain) had to fit into an old plate holder so as to be able to be installed inside the telescope without having to go through a major redesign of the focus assembly. This required special care in order to have compact electronic boards. The data acquisition board is in fact a 6 layers printed circuit board design.

Our current controller is so far not the final evolution of a series of prototypes. It is composed of 2 Euroboards (100 × 160 mm): one for a 87C750 based CCD camera sequencer, and another for a clock voltage generation board. These boards are in a chassis located outside the telescope tube. two smaller boards (60 × 120 mm) are placed near the CCD cold box. The first one contains Maxim Max333A voltage translators for the CCD, and the other contains all of the acquisition chain, including PT100 temperature measurement. This last board is based on OPA627 op amps, and DSP102 dual channel A/D converter. While this converter is 18 bits in theory, we trim the data to 15 bits in the PC. Communication to and data reception from the CCD camera is done through 5 optical fibers and data acquisition is done using a PC, and a memory buffer board and serial to parallel converter supplied to us by the DLR Berlin electronic group.

One problem of our installation is that most of the data coming out of the telescope has to be reduced in almost real time, since it is not acquired for a particular astronomer in mission, but for one of the programs currently scheduled at the telescope. While we looked at Unix workstations, DSPs, and other more exotic technical solutions, we recently decided that it was more cost efficient to use high end PCs. While Unix workstations are very difficult to maintain without a local Unix engineer, we found out that a technician is able to take care of system maintenance for our network of Windows 95 and NT based PCs, and that the price per megaflop was much lower using PCs. Apart from an old HP9000/710 HP workstation, all the computers used at and around the telescope are PCs.

For data reduction, we are using commercially available PC packages (QMIPS32 and PRISM/Atlas des Pises), as well as PC ports of the Sextractor detection package [1],and an asteroid detection package written by Magnus Lundstroem and Stefano Mottola at DLR.

In order to archive data, we started recording our frames on Exabyte tapes. After three years of operations, we have found out that we almost never looked back at old data. We did so in order to find out that some tapes which had not been stored properly already showed signs of data loss (some pixels in an old image reading as 32767). We also found out that Murphy always places interesting frames at the end of the tape, and that access time was usually very long. More recently we have decided to compress our data using the STScI H compress algorithm. While this process is a lossy process, blinking an original image and a compressed/decompressed frame showed that very little is indeed lost on regular frames. This will allow us to store about 6 nights of observing on a single CD ROM, with a better probability of being able to recover the interesting information in these frames 20 years from now.

Since the focal plane of a Schmidt telescope is curved, we are using a field flattening lens, which is a simple plano convex lens instead of a plane parallel window [5]. This introduces very little aberrations to the optical system.

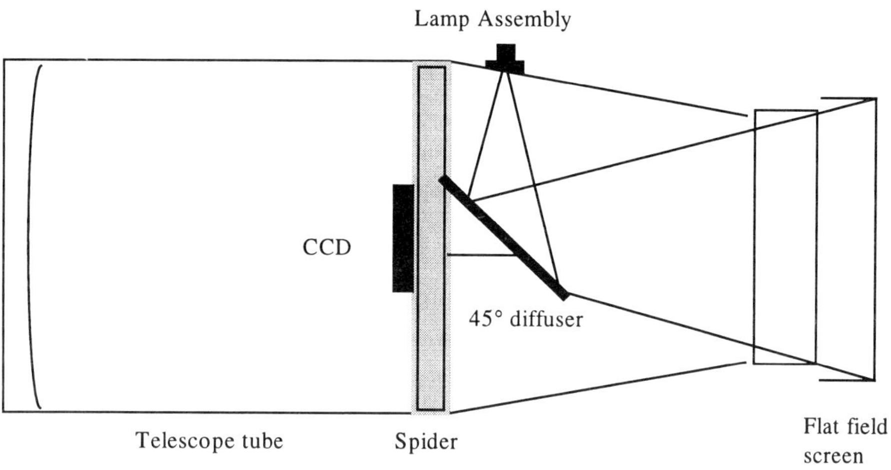

***Figure 2**: Flat field illumination configuration*

When we were still using the slow Palomar controller on our 2K chip, we found out that it was almost impossible to take twilight flat fields in all colors used during the night during the short interval allowed by sky brightness. We developed a dome flat field system described in the following diagram. While we started to use a back surface mirror as a reflector, we found out that any dust on it was showing very clearly on the flat field screen. We replaced it by a piece of white paper, and the diffuseness of this system has worked very well, with flat field screen non uniformity limited to less than a few percent. Color calibration of the light source to standard illuminant D55 has been performed using a thermocolorimeter.

5. CURRENT OBSERVING PROGRAMS

While the OCA/DLR collaboration takes around 50% of the observing time, there is another 30% taken by a geostationary satellites detection program. The remaining 20% are shared by the OCA astronomers. Different programs related to asteroid photometry, transneptunian asteroids astrometric follow up, variable star detection, high precision astrometry and galaxy cluster photometry are currently on the schedule.

6. THE SECOND GENERATION CAMERA

While we studied the feasibility of paving the focal plane with 2K CCD camera modules, we soon found out that this procedure would be nightmarish, and that the use of relatively large CCDs with correspondingly less CCD controllers and wiring would be a better solution. Taken into account the very limited space in the focal plane, it seems that a very interesting first step is to replace the current 2K detector by a 4K made by the same dealer. This project is a collaboration with DLR, with OCA making the CCD cold box and cooling system, and DLR providing a 12 bits, 2.5 MHz controller and installing the CCD in its cold box.

This should become operational in June 1997, and while there does not seem to be a lot of problems to be expected as far as running the camera, dealing with 4 times more data is expected to be the main problem. While we should be able to discover 4 times more asteroids, we should expect around 6 gigabytes of data per night to be reduced in real time.

Scan will not be used since differential trailing over the CCD limit the area of sky to +/- 1.5 degrees across the equator.

Cooling will still be performed by Peltier cooling since liquid nitrogen cooling inside the telescope focal plane assembly might lead to a very complex and costly design, and because it is possible to cool efficiently such a large CCD using currently available 2 stage coolers (the model envisaged for this application is a dual stage 66 mm at the base, 60W module).

The field flattening lens will be a 160mm silica lens, still able to correct the optics adequately taken into account the pixel scale.

7. REFERENCES

[1] Bertin E., Arnouts S., *SExtractor: software for source extraction*, 1996 *A&AS* **Vol.117** 393-404.

[2] Bessel M.S. 1990, *P.A.S.P.***102**, 1181

[3] Gehrels, T. 1991. *Scanning with CCDs. Space Sci. Rev.* **Vol. 58**, 347-375

[4] Gunn et al. June 1987, *P.A.S.P.*

[5] Linfoot E.H., Wayman *P.A.* - *AJ* **Vol.109** 535-556

[6] Morrison D., Ed. (1992), *The Spaceguard Survey : Report of the NASA international Near Earth Object detection workshop, NASA, Washington D.C.*

8. WEB SITES OF INTEREST

1. OCA Schmidt telescope : http://wwwrc.obs-azur.fr/schmidt/Welcome.html
2. EARN (European Asteroid research Network) : http://129.247.214.46/
3. Asteroid and Comet Impact Hazards Page: http://ccf.arc.nasa.gov/sst/
4. Hcompress Image compression software : http://www.stsci.edu/software/hcompress.html

ON THE OPTIMIZATION OF CCD READOUT NOISE.

RICARDO E. SCHMIDT[1]

ABSTRACT

The noise performance for a SITe 2K × 2K CCD was optimized focusing on the systematic variation of a single parameter, the Reset Drain Voltage (VOD). A graph containing a set of curves for the FET output amplifier was used to illustrate the region in which the CCD output stage operates as VOD is varied. Additionally the effect of different CCD grounding schemes on readout noise (RON) was investigated. Finally, load resistor RL and CCD operating temperature were also varied to explore their influence on RON.

1. INTRODUCTION

CTIO recently purchased a back-illuminated SI-424AB CCD from SITe. This device is one of SITe's new 2K × 2K CCDs with a Boron implant under the Transfer Gate. At CTIO we tagged it T2K5. T2K5 was bought as a chip with two out of four working amplifiers. This new CCD is intended for spectroscopy, and hence low readout noise is especially important. We already had two low noise Tektronix 2048EB, back-illuminated 2K × 2K CCDs in regular operation. These older devices we identify as T2K3 and T2K4 and each has four working amplifiers.

The manufacturer's test data sheet for T2K5 CCD quotes a RON = 9e$^-$ RMS, measured at a temperature of -90 Degrees C and at a data rate of 50 Kpixels / sec. This figure turns out to be higher than the RON of 5e$^-$ RMS typically quoted for similar test conditions in the data sheets for T2K3 and T2K4. It was hence unclear if T2K5 was suitable for spectroscopic work. The optimization work reported below was carried out to answer that question.

A set of operating voltages was chosen after iterating on bias and clock settings. At this point T2K5 had demonstrated satisfactory performance, including good charge transfer and cosmetics, linearity to better than 0.5%, a Full Well of > 200Ke$^-$, etc., but showed a readout noise exceeding that of T2K3 and T2K4, as is illustrated in Figure1.

Shown in Figure 1 is the readout noise of each CCD, i.e. averaged over the working amplifiers. Readout noise figures for working amplifiers on a given CCD, are typically found to be within 15% from each other.

---

[1] Cerro Tololo Inter-American Observatory National Optical Astronomy Observatories * 950 N. Cherry Ave., Tucson, AZ 85719. Email: **rschmidt@noao.edu**

The Cerro Tololo Inter-American Observatory is part of the National Optical Astronomy Observatories which are operated by the Association of Universities for Research in Astronomy (AURA Inc.), under cooperative agreement with the National Science Foundation (NSF).

J. W. Beletic and P. Amico (eds.), Optical Detectors for Astronomy, 245–250.

2. READOUT NOISE RESULTS AFTER THE FIRST OPTIMIZATION ATTEMPT

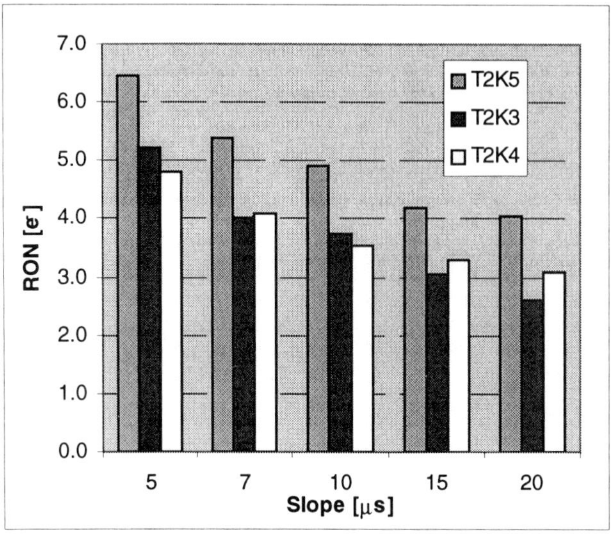

Figure 1:Readout Noise vs. Slope Time

For all these CCDs, the amplifier gains were adjusted to produce a uniform 1 e^-/ADU conversion factor for a Slope time of 20 μs. Hence a Slope of 5 μs corresponds to 4 e^-/ADU. MPP operation was chosen after measuring dark current to be 6 e^-/pixel/hour (MPP) vs. 57 e^-/pixel/hour (non-MPP). No correlation with readout noise was found for variations of Serial Clock and Summing Well High and Low rails, Output Gate, and Reset Clock High and Low rail. The settings for each of these parameters was tested by way of voltage excursions around the chosen operating point. Also, lack of abrupt changes in readout noise when varying the Output Drain, with VDD set to 24V, was interpreted to mean that VDD as not being near or in an avalanche condition.

3. OPTIMIZATION: PHASE TWO

The data in Figure 1 suggest that T2K5 is a poor choice for spectroscopy, when compared to T2K3 or T2K4. However one wants to rule out that the higher noise level is intrinsic to T2K5. better characterization of the output FET is necessary and we decided to use a set of FET output curves to help interpret its results. The manufacturer confirmed the SI-424AB output stage design to be the same that was used in model 2048EB, but had no curves available. We generated pseudo curves using a set of voltage measurements for gate (VOD), drain (VDD) and source (V_OUT), to adjust some curves we had for an older CCD. An external unity gain buffer was used to protect the CCD output while measuring (V_OUT).

The curves are shown in Figure 2. In order to maximize the output FET gain, the Output Drain voltage VDD was set to 25V. The highest VDD generated by our controller is 25.5 V, value which was first used to confirm absence of any avalanche problems, and then was lowered for safety by 0.5 volt.

VDD at 25V and a RL of 20 KΩ, fix the load line (middle one in Figure 2). RON will be measured as a function of the Reset Drain voltage, VOD. VOD, which represents the output FET gate voltage referred to ground, does not appear explicitly on Figure 2. VDG curves, which represent the output gate voltage referred to drain voltage, appear instead. FET voltage noise is inversely proportional to FET transconductance g_m, and g_m increases with drain current ID, so suitable CCD operating points should be found towards the left on the load line, as long as they still are on the saturation region of the ID curves, and as long as the channel is prevented from reaching the surface (possibly near VGS = -3V).

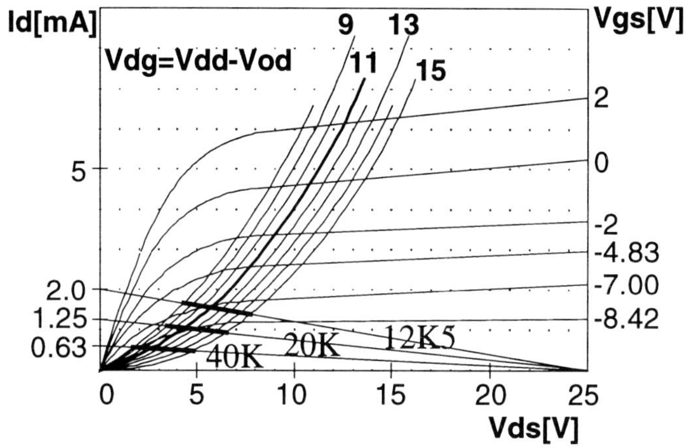

Figure 2: Output curves

Given an operating point on the load line, any pixel charge represents an excursion towards the right since charge causes a gate voltage decrease and hence a source voltage decrease (the FET is operated as a voltage follower) which gets reflected as a VDS increase in Figure 2 given that VDD is fixed (at 25V).

A third T2K5 amplifier turned out to work as well, and so the readout noise, RON [e^-], was measured as a function of VDG for three T2K5 amplifiers: UL, UR and LR. The results can be seen in Figure 3. For each VDG setting the values for gain [e^-/ADU] and RON [e^-] were obtained. The RON values given in Figure 3 are higher than those in Figure 1, due to excess noise caused by the Extender Board needed during these measurements for the Video Board. Also, Linearity and Full Well were measured for each VDG setting, to avoid low RON values that might compromise these quantities. For the VDG values shown in Figure 3, typically linearity is less than 0.5% and Full Well (MPP) exceeds 200 KΩ. Figure 3 indicates that RON is lowest at VDG = 10V for UL and UR, and RON is lowest at VDG = 11V for output UR.

These measurements were repeated for load resistors of 12.5 KΩ and 40 KΩ, for UL and LR respectively. Results are shown in Figure 4. Again RON is lowest for VDG = 10 V in each case. However in this instance there is no further decrease in the actual RON value over the RL = 20 KΩ case. This agrees with the fact the FET gain (defined as ΔV_OUT/Δ[Vague = VOD]) was measured to largely be independent of the RL value. For example for a VDG = 10V, FET gain turned out to be 0.62 (12.5 KΩ), 0.63 (20 KΩ), and 0.63 (40 KΩ) respectively. Thus RL = 20 KΩ was used for all outputs.

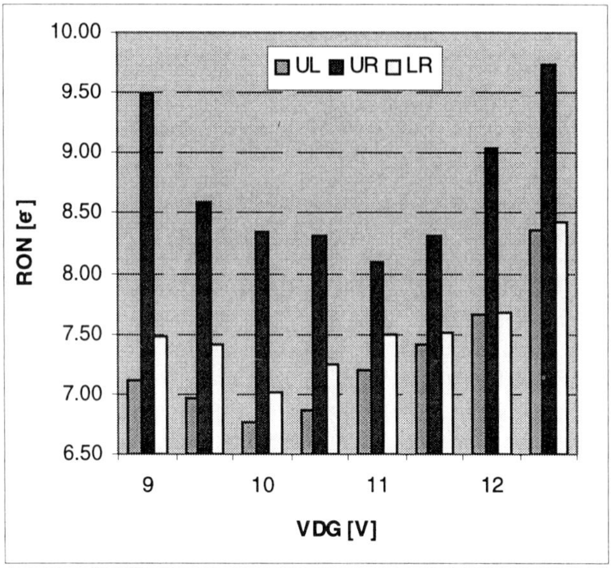

Figure 3: *RON [e⁻] vs. VDG [V]*

Another parameter tested for its influence on RON was the CCD operating temperature.

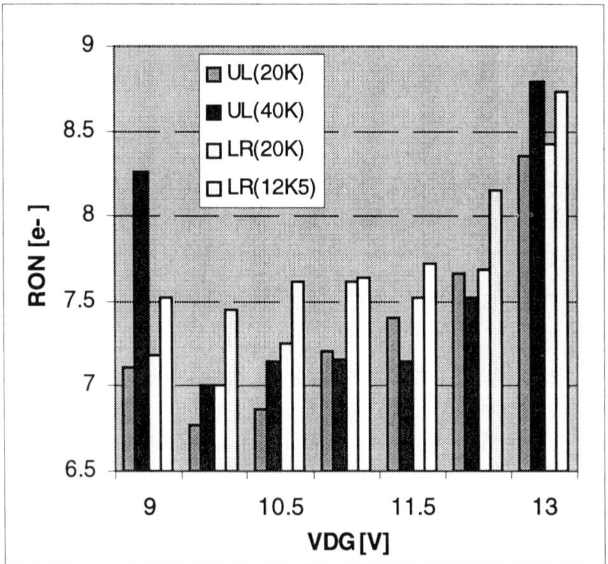

Figure 4: *RON vs. VDG [V]*

Figure 5 displays the results: RON decreases as temperature increases. However dark current increasing with temperature -a value of 90 e/hour/pixel was measured at

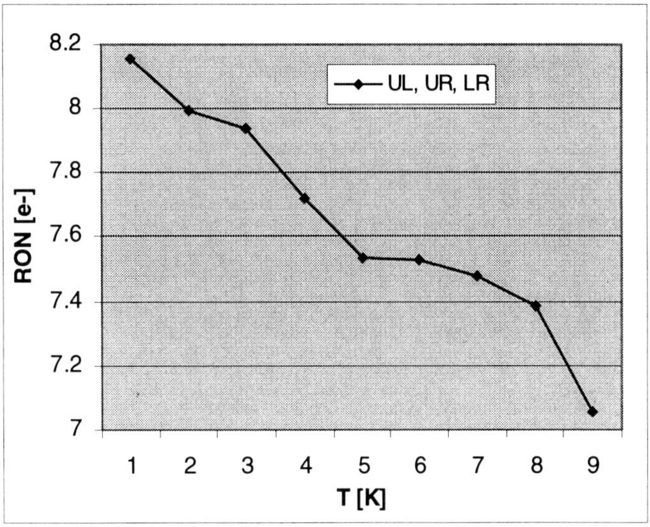

Figure 5: *RON [e$^-$] vs. Temperature [^{o}K]*

200 °K, negates this lower RON. The chosen CCD operating temperature was the initial 160 °K.

In Figure 6, RON [e$^-$] versus Slope [us] is summarized for T2K5, with the black bars representing the initial situation. The white bars in Figure 6 show the present status after optimizing of VDG to: 10V (VOD = 15V) for UL and LR, and 11V (VOD = 14V)

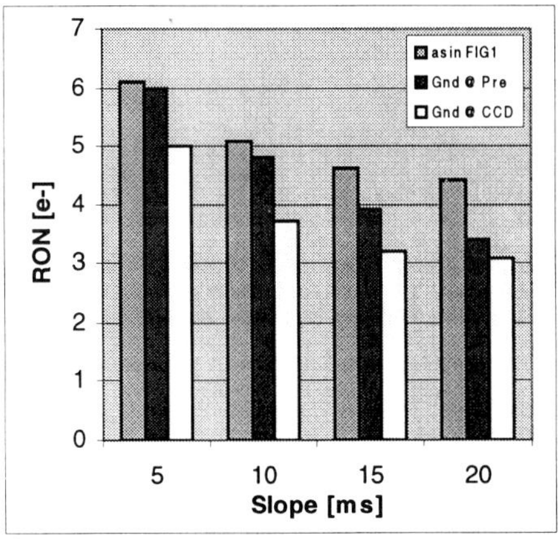

Figure 6: *RON [e$^-$] vs. Slope [ms] for t2K5*

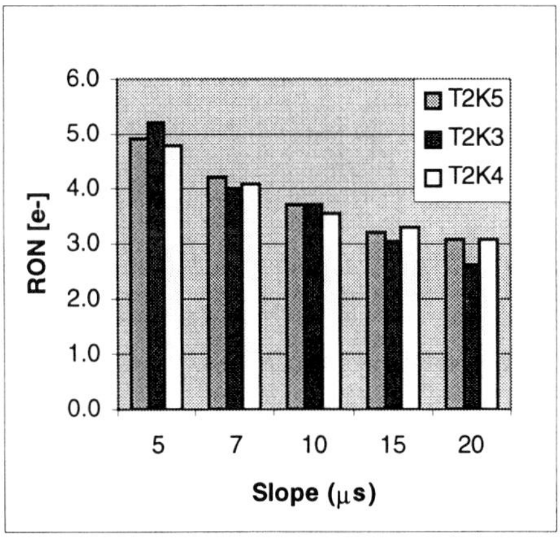

Figure 7: *Final RON vs. Slope Time*

for UR. Note that RON has significantly improved for the longer slope times such as 20 us slope which represents a gain of 1 e$^-$/ ADU, but that there has been only a modest improvement for the short slope times such as 5us which represents a gain of 4 e$^-$/ ADU. The "excess noise" which still exists with respect to T2K3 and T2K4 is caused by the

CCD Output Ground reference pins being grounded at the preamplifier end. Tying these pins directly to ground at the CCD, the final result shown by the cross-hatched bars in

Figure 6 is achieved. The relative performance of T2K3,T2K4 and T2K5 is given inFigure 7. Compare these results with those of Figure 1.

4. CONCLUSIONS

After optimization, T2K5 is suitable for spectroscopic use. To optimize the output stage for RON, VDD was set to the highest value that still avoided avalanche problems and VDG = VDD-VOD was adjusted for lowest RON. The main effect of this optimization is seen with long slope times. To get lowest RON for all slope times, the CCD Output Ground reference pins also have to be grounded directly at the CCD.

How Linear Are Typical CCDs ?

Roger M. Smith[1]

Abstract

The typical linearity achievable with CCDs was questioned at the previous ESO Optical Detector Workshop in 1993. This paper describes the efforts at CTIO since then to develop a technique for measuring linearity accurately, and rapidly enough to make the exploration of CCD operating parameters practical. A simple circuit has been adopted which stabilizes the intensity of a pulsed LED. It is compact enough to be installed behind the shutter so that linearity can be monitored at the telescope while the dome lights are on. Methods are described for verifying the linearity and stability of the measurement system itself. These requirements are very stringent, since CCD gain variations have typically been measured to be 0.2% or less over most of the operating range. Methods are proposed for reducing non-linearity further, and for generating each new linearity curve from a single exposure, after having taken several calibration exposures. Several discrepancies have been seen between linearity of point sources and linearity measured with flat illumination: the amplitudes were similar but the gain variations had opposite slopes over the central part of the curve, and a more pronounced signal deficit has been noted at low light levels for point sources, which is equivalent to the subtraction of few electrons from every pixel.

1. Historical Context

A discussion session at the previous ESO Optical Detector Workshop in 1993 revealed that no-one present could clearly state what is the best linearity achievable with typical astronomical CCDs, or whether the linearity problems they had experienced were due to the controllers or the CCDs themselves. It was clear that systematic errors could easily overwhelm the non-linearities being measured. This was especially true for tests made on the telescope where illumination varies as flat field lamps warm up and shutter timing corrections are necessary. We exchanged stories of gross non-linearities (3-10%) having escaped notice for alarmingly long periods of observing.

By 1993, the deployment of multi-amplifier CCDs allowed an increasing number of users to compare the transfer functions of the amplifiers. Differences produce a discontinuity at quadrant boundaries when images containing appreciable sky background are divided by flat-fields taken at a radically different brightness. Differences as small as 0.2% are visible to the eye in scenes with low intrinsic contrast and can easily be extracted from the noise by plotting the average of many lines or

---

[1] Cerro Tololo Inter-American Observatory, Casilla 603, La Serena, Chile. Email: **rsmith@noao.edu**
CTIO is a division of the National Optical Astronomy Observatories, operated by the Association of Universities for Research in Astronomy, Inc., under cooperative agreement with the National Science Foundation of the USA.

J. W. Beletic and P. Amico (eds.), Optical Detectors for Astronomy, 251–264.

columns. There is no uncertainty introduced by temporal variations in illumination, or spatial variations, which are generally slow or at least uncorrelated with quadrant boundaries.

Unfortunately, although having multiple amplifiers provides our customers with an easy way to detect linearity problems, it is not a sufficient tool since it requires only that the linearity characteristics be the same from amplifier to amplifier. A technique is needed which allows us to determine the absolute linearity to 0.05 % precision without removing the CCD from the telescope, and which can be automated to encourage routine checking. To be a practical tool for exploring the effects of CCD and controller parameters, each curve should take only a few minutes to obtain. Traditional methods are very time consuming.

2. MEASUREMENT METHODS

Non-linearity is a variation in gain as a function of signal level. Gain can be defined either as the first derivative of the transfer function or as the ratio of output to input. In this paper, the latter definition is used since it is not only easier to measure but also serves as a correction factor. Non-linearity ,γ, is simply defined as :

$$\gamma(\phi) = \frac{M}{\phi} - 1 \tag{1}$$

where M = measured signal and ϕ = expected signal; i.e.:

$$\begin{aligned} \phi &= \frac{M}{(\gamma(\phi)+1)} \\ \phi &\approx \frac{M}{(\gamma(M)+1)} \end{aligned} \tag{2}$$

since $M \approx \phi$, and $\gamma(\phi)$ varies slowly.

It is convenient to characterize CCD non-linearity with the single parameter, "peak-to-peak non-linearity", which is the variation of $\gamma(\phi)$ over the useful signal range (Maximum signal is where variance departs from photon statistics by 5%.). We will see that the common practice of assuming that $\gamma(\phi)$ is linear when applying a linearity correction is not always adequate, especially at low light levels where most astronomical data lie. Measurement methods fall into two broad categories:

2.1 Vary intensity holding exposure time fixed.

The true intensity is measured with another detector system, usually a photodiode, which has been independently verified to be linear. Since the "expected signal" is measured concurrently, intensity stability of the light source is not an issue. Shutter time is held constant, so one need not be concerned with timing correction although timing must be reproducible. A shortcoming of this method is that one must rely on the manufacturer's assurances that the linearity of the photodiode system meets the stringent linearity and drift requirements or devise a precise way to validate its performance. Dynamic range limitations in the photodiode/preamp/ADC may require gain or exposure time switching, then normalization of ranges to produce a composite curve.

2.2 Vary exposure time holding intensity constant.

This method relies on a stable light source and reproducible shutter timing. The large shutters required can take tens of milliseconds to move, resulting in significant exposure time differences between the center and edge of the field, which must be measured and corrected. The shutter timing correction is derived by comparing an exposure, in which the shutter has been opened and closed many times before reading out, with a conventional image of equivalent total exposure time. The difference in total dark time must also be accounted for if dark current is not negligible. An independent correction must be applied for each pixel. The shortest exposure time is limited to several thousand times the residual error after correction, and the longest exposure time depends on the dynamic range being spanned. This method is either very slow, or is inaccurate at low signal levels.

Since the test takes a long time to complete, it is more sensitive to light source intensity variations due to ambient temperature. Incandescent bulbs typically used for dome flats are nowhere near stable enough. *Linearity curves derived from sequences of dome flats are to be regarded with suspicion.* A light meter must be used to monitor and correct for lamp brightness variations with 0.05% precision, and the shutter timing correction must be carefully measured and competently applied, taking care to also monitor reproducibility.

"Beta lights" meet the compactness requirement for daytime testing on the telescope. They are generally considered to be very stable, which is true of their radioactive decay rate that provides predictable aging characteristics, however their phosphors are quite temperature sensitive (-0.3%/C). They require a dark slide or light-tight filter wheel to block dome lights while the shutter is opened, and still require shutter timing correction. They also represent a serious radiation hazard if the glass container holding the Tritium is broken.

3. STABILIZED PULSED LED

LEDs exhibit poorer thermal stability (-1%/C) than Beta Sources, even when excited by a constant current [2] however their intensity can easily be stabilized by the servo circuit shown in figures 1, 2 & 4. A Burr Brown OPT211 integrated photodiode/preamplifier measures LED brightness. An operational amplifier integrates the error between the photodiode output and a reference voltage to adjust the LED drive current. Burr Brown guarantee the gain stability (quantum efficiency and preamp gain) to be better than 200 ppm/C, so that ambient temperature could vary as much as 2.5 degrees Celsius before consuming the 0.05% error budget, making this circuit 50 times less sensitive to temperature variations than an unstabilized LED.

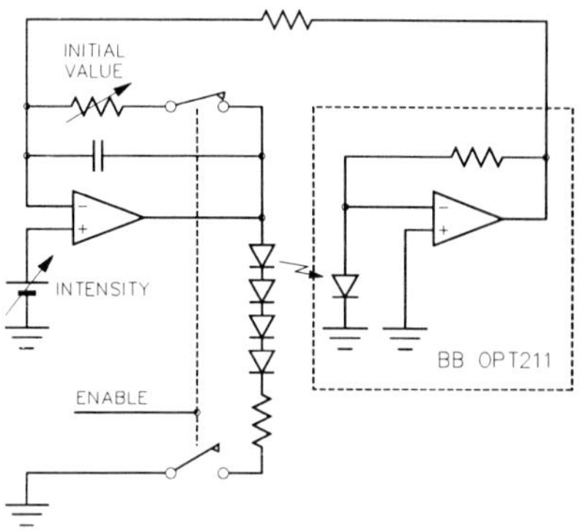

Figure 1: Simplified circuit of stabilized pulsed LED

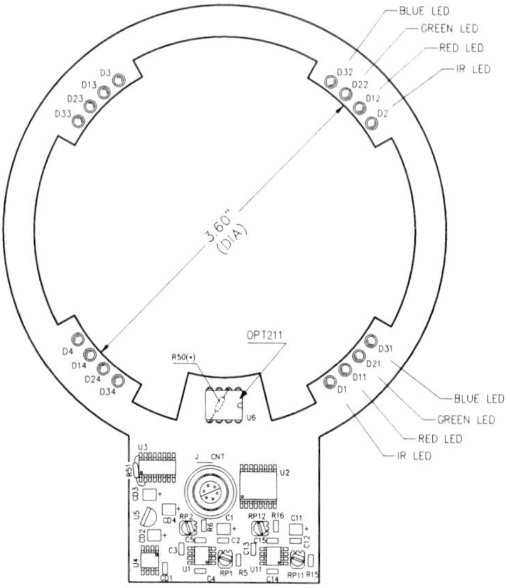

Figure 2: LED Stabilizer PCB outline and component placement

LEDs also have the advantage that they can be turned on and off about 10,000 times faster than a mechanical shutter can be opened or closed. Since the exposure times must be long enough to eliminate errors due to exposure timing, LEDs greatly reduce the time required to measure a full linearity curve. The speed advantage is not as large as this implies since the total light output can deviate from proportionality to pulse time as the LED loses efficiency due to self heating of its diode junction. This effect also creates a dependence on duty cycle. Fortunately the closed loop bandwidth of the OPT211 (150 kHz) is adequate to correct for this too.

An annular dual sided printed circuit board (figure 2) has been made which can be mounted between the shutter and the dewar window to hold the photodiode, servo circuit and LEDs. The ability to test the CCD with the shutter closed means that testing can proceed during the daytime when other activities are taking place in the dome. LEDs are now available with emission wavelengths spanning most of the CCD's response range. Four sets are mounted on the PCB to allow CCD QE stability to be checked well enough to diagnose UV flooding problems, etc.

Even a stability measurement of CCD system throughput at a single wavelength is useful. To date we have only used the red set.

Turn-off is effectively instantaneous taking 1 μs (or less, depending on LED colour) after opening the series analog switch. To obtain the fastest possible turn-on, the integrator capacitor is precharged (while the LEDs are off) to the approximate initial voltage required across the LEDs. This is achieved simply with an analog switch and potentiometer in parallel with the integrator capacitor. The proportionality of flux to pulse time is verified by taking a series of flat-field exposures during which the LEDs are pulsed N times for T/N seconds. N is varied so that individual pulses span the range to be used to measure CCD linearity. T/N_{MAX} is the minimum and T is the maximum pulse time to be used in the linearity test. If the LED is linear and the pulse times are accurate then each exposure will produce exactly the same mean value, whether the CCD and signal chain are linear or not. The delay *between* pulses can be varied to measure the effect of duty cycle. The example in figure 3 shows a slight linear dependence on the number of times the 10 second LED flash is subdivided. The 0.04% error at 1000 pulses corresponds to a turn on delay of about 4 μs. The "initial condition" potentiometer (Figure 1) is used to remove such slopes to achieve LED linearity error better than the 0.02% peak to peak.

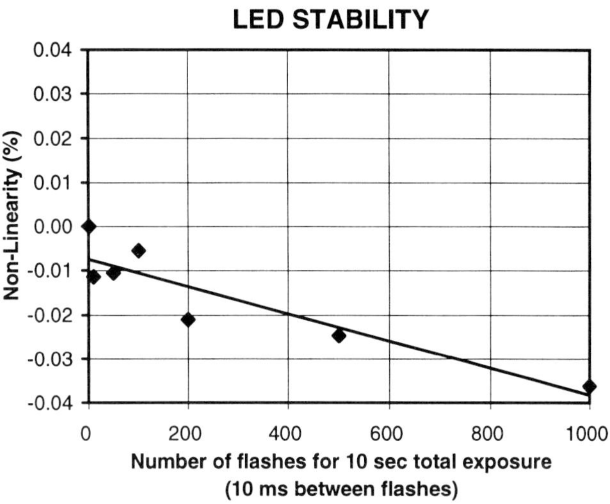

Figure 3: *LED linearity error versus inverse of pulse duration. The slope is due to turn-on being slower than turn-off and is compensated by adjusting the "initial value" plot in figure 1.*

To measure CCD non-linearity, a series of flat fields is taken over a range of LED pulse times sufficient to reach full well. This has been fully automated by an IRAF task, which both acquires the data and processes the images. To reduce readout time, only a region-of-interest is read out. An area is selected within this which is free from cosmetic defects, line/region start transients and so on. The overscan is subtracted from each frame. A similarly processed bias frame (zero length dark frame) is also subtracted to remove the effects of spurious charge injection which become significant for images with less than 5000 electrons of charge (since we are looking for only 0.1% effects). Dark current is not usually significant for low temperature operation, nonetheless bias frame subtraction also corrects it to first order since the readout time dominates the total dark time.

4. CCD LINEARITY RESULTS

When interpreting these linearity curves, it should be noted that the vertical offset is arbitrary! The independent variable in the experiment is proportional to the expected signal *(ϕ)* but the normalization is only known approximately. Signal (M) is measured versus LED pulse time (T), then linearity error is calculated as,

$$\gamma(M) = \frac{\left(M(T)/T\right)}{\left(M(T_i)/T_i\right)} - 1 \tag{3}$$

where T_i is a mid-range pulse time.

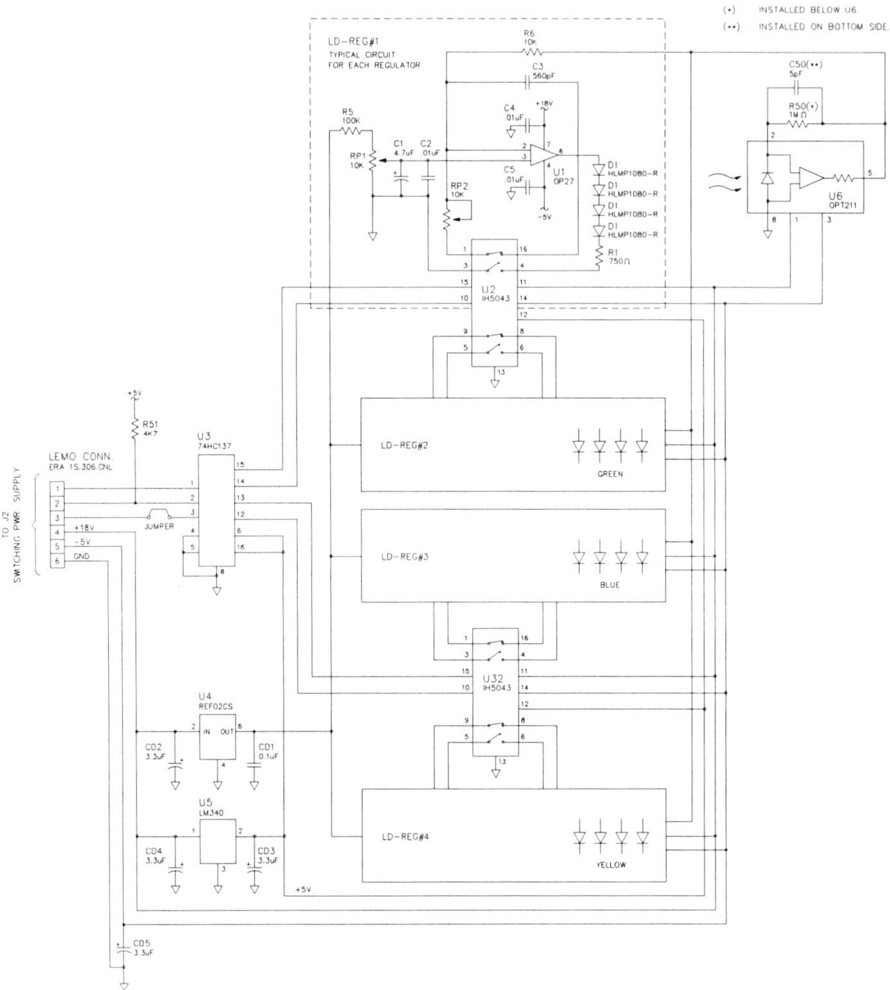

Figure 4: *Full circuit for multicolour stabilized LED*

The normalization factor *($M(T_i)/T_i$)* has been replaced by the slope of a linear fit to the central points of the curve *$M(T)$* for the data in figures 5, 6 & 7. Photometric accuracy is achieved by observing standard stars.

Figure 5 shows the non-linearity curve for a Tek 1K CCD obtained with a pulsed LED using constant current excitation instead of servo controlled intensity.

I interpret the decay in response as LED efficiency loss due to self heating, with the curve flattening as the junction reaches thermal equilibrium. This could be tested by examining the effect of varying LED brightness and thus power dissipation. The gross departure from linear at high signal can be positive or negative going if it results from blooming on the CCD rather than saturation of the output amplifier or signal chain. This is determined by whether there is a net flow into or out of the region used for measuring the flux, and is determined by the relative illumination, QE and full well capacity of the pixels surrounding the region. If the illumination is flat then it is often the case that the linearity curve remains flat *beyond* the full well capacity. Charge blooms from sensitive pixels to less sensitive neighbours without there being significant change in the total signal registered over the region. It is therefore necessary to use a variance curve or non-uniform illumination to determine full well capacity accurately.

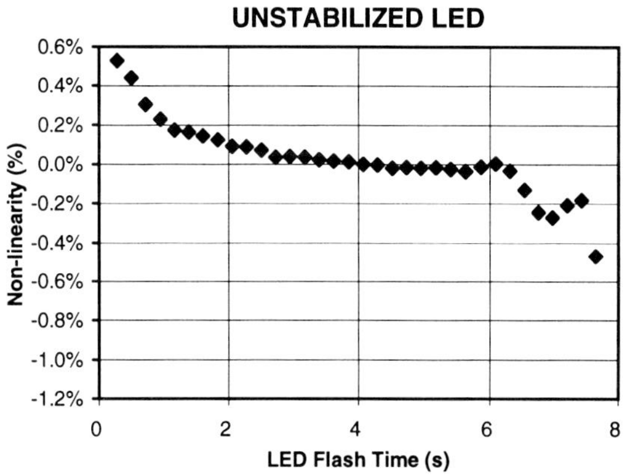

Figure 5: *CCD non-linearity measured with constant current excitation of pulsed LED. (Tek 1K)*

The increase in non-linearity towards low signal disappears or becomes a slight turn-down, when the LED intensity is stabilized and a bias frame is subtracted to remove spurious charge. We have measured the linearity of all CCDs in active use at CTIO: three TEK2048EB (SITe 2K) CCDs, a 21 μm pixel 2K CCD made by Tektronix for the STIS program, a Tektronix 1K, Thomson 1K and Loral 800x1200 and Loral 3Kx1K CCDs. These devices are operated in quad or dual readout mode except the Lorals. When correctly adjusted, all 26 amplifiers have remarkably similar linearity characteristics. Figure 6 shows the non-linearity curves obtained with the stabilized LED for the four output amplifiers of one SITe 2K (CTIO#4), which have been obtained concurrently, plus a curve for a Loral 1200x800. We found that the gain typically declines linearly by about 0.2% up to full well. The best amplifier measured at CTIO (Tek2K #4, amp D) is the heavier solid line.

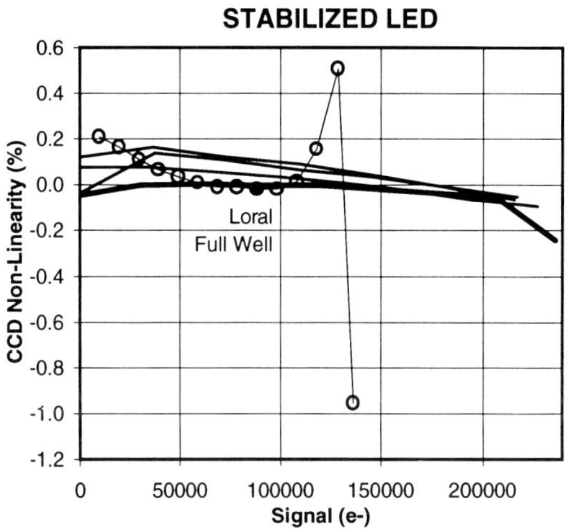

Figure 6:*Representative CCD non-linearity curves: four amplifiers of a SITe 2K (lines) and a Loral 1200x800 (circles). The heavy line is the best linearity measured at CTIO. Note that full-well occurs well before the knee*

Figure 7 shows the non-linearity curve measured by projecting the light from the pulsed LED through a pinhole mask and processing the data with the IRAF aperture photometry routines. The aim of this experiment was to show whether the morphology of the illumination affects the measured linearity. The amplitude of the non-linearity variation is encouragingly similar in the central part of the curve, however the 0.2% tilt is in the opposite sense to that measured using flat fields! The dip at low signal also appears to be more pronounced. Although this experiment needs to be repeated with many pinholes to obtain better signal to noise and to lower susceptibility to localized trapping sites, the discrepancies are still disturbing and merit further investigation.

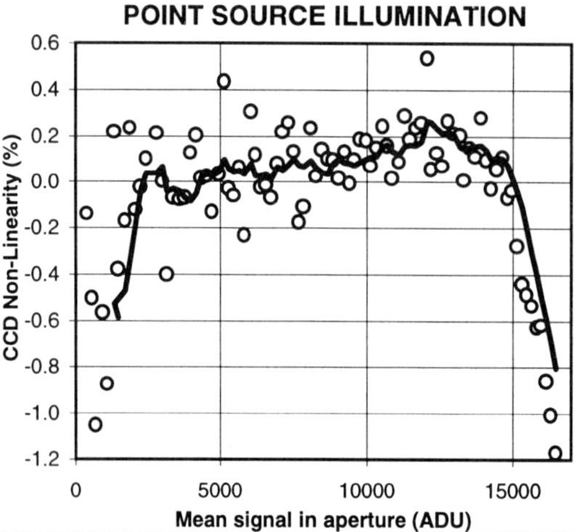

Figure 7: CCD non-linearity measured through single pinhole illuminated by stabilized LED. Loral 1200x800 CCD. The solid line is the 6 point boxcar average. Compare this to flat illumination (circles in figure 6).

5. DISCUSSION

My first interpretation of the signal deficit at low levels was some kind of systematic measurement error. For example, the IRAF aperture photometry routines are rumoured to exhibit systematic errors when the background level is very low. However, a similar deficit was reported by several other groups during informal discussions at this workshop. To correct aperture photometry with the Wide Field / Planetary Camera on the Hubble Space Telescope, STScI also recommend that observers add a few electrons of signal to every pixel lying within the aperture during reduction of their data (http://www.stsci.edu/ftp/instrument_news/WFPC2/Wfpc2_cte). Clearly, there is a prima facie case for the deficit at low light levels being real. The pinhole test suggests that the effect may be more pronounced for point like sources. Such a deficit would have serious consequences for precise photometry if the background level is low. At the very least, we hope to provide a means to measure the effect at the telescope so that a correction can be attempted during data processing.

The signal level at which the "deficit" is seen is very small compared to the bias conditions of the output FET and so the turn down at the low end of the curve is more likely to be due to a charge loss mechanism than gain modulation in the FET. Charge loss could be due to trapping in the image area or serial register. If it were due to recombination with holes before the electrons are collected then one would have to explain why the charge deficit is constant rather than a fixed fraction of the signal. If trapping is the explanation, then there should be some correlation between low level non-linearity and distance from the measurement region and the output amplifier, and a correlation with the overall CTE and trapping site distribution. It might be instructive to compare linearity curves for various combinations of serial and parallel binning factors.

Further research may show that a more physical model can be derived than the empirical "$\gamma(\phi)+1$" plots in this paper. It may be possible to separate the non-linearity correction into a small offset (ε) due to charge trapping, dependent on average sky level (B), plus a gain variation which might be approximated well by a single linear term (α), or at worst a 2nd order polynomial, so that the formula for correction becomes:

$$\phi = \left(M + \varepsilon(B)\right)\left(1 + \alpha M\right) \tag{4}$$

6. CCD ADJUSTMENTS AFFECTING LINEARITY

Normally obtaining best linearity from the CCDs at CTIO has been quite straightforward. I begin by choosing standard values, check that the shape of the preamplified video signal looks normal (i.e. no punch through - described below), then make marginal adjustments in units of 0.5V, though in rare cases V_{RD}-V_{LG} is critical at the 0.2V level. The parameters are listed here in order of importance in terms of effect on linearity:

V_{DD}: The most common problem has been that Output Drain voltage has needed to be increased, to provide 4.0 to 5.5 V from drain to source. For V_{OS} typically 20 V, this implies V_{DD} = 24.0 to 25.5 V.

V_{RD}: Reset Drain voltage is also important, and usually set between 13 and 15V for SITe and Loral CCDs. V_{DD} must be re-adjusted after changing V_{RD}. Setting V_{RD} too high will cause poor linearity (and loss of output node sensitivity), even when drain-source voltage is adequate. Setting V_{RD} too low relative to V_{LG} results in "punch through" to the SW. This appears on the video signal as a positive shift in the level after CCD-Reset and an attenuation and change in the shape of the SW (serial clock) feedthrough pulse.

V_{LG}: Last Gate (also called Output Gate or Output Transfer Gate) affects linearity primarily by affecting the range over which V_{RD} can be adjusted to optimize linearity. We usually use values in the range -3V to +1V. Setting it too low relative to the low level on Summing Well degrades serial CTE, and setting it too high induces punch through.

Clocks: Clock levels only affect linearity at the extremes of the signal range. Clock levels and wave shapes effect charge transfer which may cause low level non-linearity at low light levels. They also determine the well capacity, which generally limits the linear range.

7. SOURCE FOLLOWER LINEARITY PREDICTION

The mid-range slope in the linearity curve might be due to variations in output node capacitance, or due to the characteristic non-linearity of the FET. In principal it should is possible to measure the output FET's linearity characteristics directly by measuring the variation in V_{OS} in response to V_{RD} while keeping the reset switch closed. The results are straightforward to predict by small signal analysis of a standard FET model, since the percentage variation in source follower gain is very small. The gain of the source follower is

$$A = \frac{gR}{(1 + gR)} \approx 0.7 \tag{5}$$

where R is the load resistance, typically 20 KΩ, and g is the transconductance of the FET ($\equiv dI_D/dV_{GS}$), typically about 1/(8.6 KΩ). Differentiating equation 5:

$$\frac{dA}{dg} = \frac{R}{(1 + gR)^2} \tag{6}$$

Dividing by equation 5 we obtain the percentage change in gain expressed in terms of percentage change in transconductance:

$$\frac{dA}{A} = \frac{(dg/g)}{(1 + gR)} \tag{7}$$

This implies that transconductance must change by 0.7% to produce a 0.2% variation in gain. The Shockley model of a Field Effect Transistor [3] found in basic electronics texts [1] predicts that:

$$g = \frac{-2\sqrt{(I_{DSS} I_D)}}{V_P} \tag{8}$$

where I_{DSS} is the drain-source current at zero gate-source voltage and the V_P is the "pinch-off voltage" at which the channel first becomes depleted over (nearly) its full width. For this analysis, one only needs to know that both are constants. Differentiating equation 8,

$$\frac{dg}{dI} = \frac{-\sqrt{(I_{DSS}/I_D)}}{V_P} \tag{9}$$

and dividing by equation 8 and re-arranging,

$$\frac{dg}{g} = +\frac{dI_D}{2I_D} \tag{10}$$

Substituting this result into equation 7,

$$\frac{dA}{A} = \frac{1/2\left(dI_D / I_D\right)}{\left(1 + gR\right)} \tag{11}$$

The current variation induced at the output by a full scale signal is,

$$dI_D = \frac{-SAQ_{MAX}}{R} \tag{12}$$

where S is the output node sensitivity, A is source follower gain, and Q_{MAX} is full scale signal. For the 24 μm pixel SITe 2K CCD whose linearity is plotted in figure 6,

$S=2.6\ \mu V/e^-$; $A=0.7$ and $Q_{MAX} = 200\ Ke^-$ so $dI_D = -18\ \mu A$.

and from Ohm's Law,

$$I_D = \frac{V_{OS}}{R} = \frac{(20V)}{(20K\Omega)} = 1mA \tag{13}$$

Substituting for dI_D and I_D, we arrive at the desired expression for percentage gain change in terms of maximum signal:

$$\frac{dA}{A} = -\frac{SAQ_{MAX}}{2V_{OS}\left(1 + gR\right)} = -\frac{\left(2.6\,\mu V/e^-\right)\left(0.7\right)\left(200Ke^-\right)}{2\left(20V\right)\left(1 + 20K\Omega/8.6K\Omega\right)} = -0.27\% \tag{14}$$

The non-linearity predicted by this simple analysis has the right sign and approximately the right magnitude as that observed in the central part of the linearity curves in figure 6, suggesting that FET transconductance variation is a significant component of CCD non-linearity. The contribution from the output FET can be reduced to an undetectable level (25 fold) by replacing the load resistor with a current source. This increases R by a factor of 50 and increases A to unity (equation 5).

8. Proposal for Faster Linearity Measurements

A method is being investigated to obtain a linearity curve from a single exposure after having first obtained two bias frames and a flat field. Re-measuring linearity after parameter adjustments will not in general require new calibration frames, and thus will require only one frame to be acquired.

Flash:
- The LED is pulsed very briefly before each row is shifted.

Flush:
- Light detected by the serial register is clocked out, before parallel transfer.
- The line is then read normally. Rows and/or columns may be binned to reduce readout time.
- The image is overscan subtracted.
- A bias frame acquired the same way (identical timing but without flashing the LED) is subtracted to remove spurious charge and any dark current.

Flat:
- A flat field is taken normally and overscan subtracted.
- A normally acquired bias frame is subtracted.
- The expected ramp image is predicted from the processed flat field by adding the first N lines of the flat field to obtain the Nth line of the synthetic ramp image. This accounts for sensitivity and illumination differences from one pixel to the next. Extrapolation is required to calculate the first line in case it was contaminated by a frame start transient.
- The measured ramp is divided by the synthetic ramp and the average of good columns is plotted. Any deviation from flatness represents a departure from linear.

A pair of ramp mode images can also be used to obtain a rapid variance curve. The difference of two ramp images is calculated to rapidly dispense with sensitivity variations. The variance of each line in the difference frame is plotted against the mean of the sum frame. The slope of the linear part of the curve is the gain in ADU/e$^-$. The point at which it deviates from linear is the full well capacity or saturation level. The Y-intercept is the readout noise squared, though this can be measured more accurately from the statistics of a bias frame.

9. Signal Chain Linearity

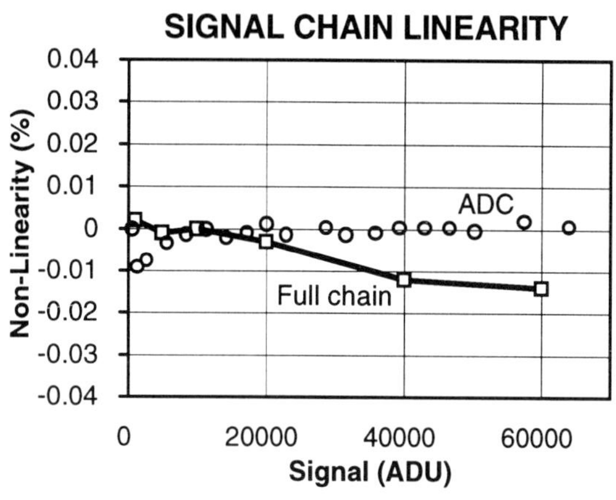

Figure 8: Non-linearity for entire video chain (solid line) and for the ADC & buffer alone (circles). Note the expanded vertical scale

Non-linearity can originate in the controller as well as the CCD. To verify that this was not a significant effect, we first performed a conventional linearity measurement on a sequence of DC voltages injected into the ADC buffer. The mean value in the image was divided by and plotted against the input value measured with a 6 digit voltmeter. The results are consistent with the Crystal Semiconductor CS5101A ADC's worst case non-linearity which is specified to be 0.004% of full scale (1.3 ADU at half scale). The error at low signals in figure 8 (circles) can be modeled by a zero point error of only 0.15 ADU. This is unlikely to be due to drift in the bias level, since the lowest signal point shows no error (For real CCD data, the zero point error is even less likely since it is removed by the overscan, which is measured at the same time as the image data.). The error could be due to Differential Non-Linearity of the ADC (DNL). All ADCs suffer from this to some extent. It is the variation in the voltage step size represented by successive output codes. The Crystal CS5101A's have very low DNL, typically 0.25 ADU RMS. Nonetheless, this can cause an slight offset between the bias level and the true zero point of the linearity curve: subtracting the bias frame from areas with no signal will (by definition) still yield zero in the presence of DNL errors, but subtracting it from a small signal will produce a slight error since the DNL error at that signal level will be uncorrelated. Low level linearity errors as great as 0.45 ADU have been measured in several of our ADC, but this only results in 0.03% non-linearity which is still negligible.

The contribution from the full signal chain was measured by injecting a four level waveform into the preamplifier to simulate the true video signal as much as possible. This Artificial Video signal was generated with an analog multiplexer constructed from fast DG271 analog switches and an OPA627 to buffer the output.

Multiplexer input levels were chosen to simulate CCD Reset Feedthrough, Reset Level (analog ground), Summing Well Feedthrough, and Signal Level. The Signal Level was generated by the same precision voltage source and 6 digit DVM used to measure the ADC linearity. The 0.015% non-linearity measured for the entire signal chain is surprisingly good and is five times less than the lowest non-linearity measured for a CCD at CTIO.

A severe intermittent signal chain non-linearity was discovered and eliminated in 1993. Originally, we adhered to the common practice of placing diodes across the gain setting resistors in the preamplifier to clamp the excursion of the CCD Reset Feedthrough pulse. They are reverse biased by the signal and should have no effect on linearity, however if the reference voltage subtracted from $\mathrm{V_{OS}}$ does not remain accurately matched, the clamping diodes can become slightly forward biased at low signals moving into reverse bias again for larger signals, modulating gain by up to 5%. This is more likely in a DC coupled signal chain, such as that used in Arcon, than for AC coupling. The approximately 20 volt DC component of the video signal must is removed by subtracting a stable reference voltage, which can drift slightly and will not track drift in V_{OS}. We solved this problem simply by removing the diodes and providing high enough preamplifier supply rails to ensure that the CCD Reset feedthrough pulse never saturates.

10. CONCLUSIONS

CCDs typically exhibit an approximately linear 0.2% gain droop with increasing signal before reaching full well, at least in the case of flat illumination. This could plausibly be due to FET transconductance variation and be remedied by using a current source as the load. Further work is needed to quantify and explain an apparent signal deficit of a few electrons at low light levels, and to investigate differences between linearity behavior of point sources and flat illumination.

Intensity stabilized, pulsed LEDs have been shown to be useful tools for obtaining precise linearity measurements rapidly in the laboratory, as well as for performing a variety of other diagnostics. Being compact and operable between the shutter and the CCD, they are appropriate for performing diagnostics on the telescope at times when other work may be in progress which requires the dome lights to be on. Their stability is adequate for long term system gain and QE verification.

Acknowledgements: I am indebted to my colleagues for their collaboration on this long running investigation. Jorge Bravo performed most of the CCD and LED linearity measurements and refined the scripts, which were written by Pedro Gigoux to automate the whole process. Sergio Cathalifaud took the LED stabilization circuit board from a conceptual design to a working printed circuit board. Ramon Galvez and Rolando Rogers developed the Artificial Video Generator used for testing the linearity of the signal chain.

11. References

[1] Millman, J, Halkias, C.C, 1972, Integrated Electronics: Analog and Digital Circuits and Systems, McGraw-Hill, **319**

[2] Motorola Corporation, 1988, Motorola Optoelectronic Device Data, DL118 Rev2, **4-6 4-9 5-43**

[3] Shockley,W. 1952, Proc. IRE, vol. 40, "A Unipolar Field-Effect Transistor", **1365-1376**

APPLICATION OF THE ARCON CCD CONTROLLER TO THE NOAO 8K MOSAIC IMAGER.

ROGER M. SMITH [1]

ABSTRACT

The Arcon CCD controller used at NOAO employs a network of transputers for its control and communication backbone. This distributed parallel processing technology presents two packaging possibilities for extension to large mosaics. Rather than assemble a 16 channel controller in a single large enclosure, we chose to operate each quadrant of the CCD mosaic with a nearly standard quad readout Arcon. This paper describes the advantages and requirements of this approach.

1. INTRODUCTION

Developers of controllers for CCD mosaics gain economies of scale by replicating a basic board set to read from the large number of output amplifiers. We generally have choices for packaging these boards, ranging from a single large enclosure to running a collection of smaller standard controllers. The basic board real estate and thus the board production cost is not very different for either approach since nearly all components have to be replicated or enlarged proportionally to the number of output amplifiers. The single enclosure has so far been the most common choice, though the University of Hawaii has chosen an intermediate path by running two 8 channel controllers. The Arcon board set can be packaged either way and we chose multiple enclosures by preference, not because of architectural constraints. This paper describes our reasoning and recent experience building the controller for the NOAO 8k × 8k Mosaic Imager

1.1 One Big Controller

The clearest advantage of having a single enclosure is that all CCDs can be driven from a single sequencer. Aside from the obvious cost saving, having only one sequencer ensures that the CCDs are read in exact synchronism. This is essential to prevent interference between clocks and video outputs from creating beat frequencies in the data. It is far easier to read the CCDs synchronously than to achieve the $>10^7$ rejection ratio required to avoid seeing the effects of $\approx$10V clock swings on CCD outputs with noise floors on the order of a few microvolts.

The clocks and biases for each CCD need to be driven separately to allow independent optimization of their levels even though the timing is identical. Isolating the CCDs this way also eliminates a potential crosstalk path, and divides the large total load capacitance between separate clock drivers. Thus, the output of a single sequencer must be bussed to multiple clock switching banks. Signal chains must be replicated, while the

[1] Cerro Tololo Inter-American Observatory, Casilla 603, La Serena, Chile . Email:**rsmith@noao.edu**

J. W. Beletic and P. Amico (eds.), Optical Detectors for Astronomy, 265–274.

data can be multiplexed into a single higher speed datalink back to the host. Clearly, the backplane, power supply, and enclosure must also be enlarged.

1.2 Multiple Standard Controllers

We decided that the effort to synchronize the sequencers was clearly outweighed by the following advantages of using standard controllers to operate mosaics:

- No redesign of data links, enclosure, backplane or supply is required.
- The inventory of all boards, backplanes and power supplies is standard. Existing stock and spares may be utilized.
- Controllers can be tested separately on a standard lab dewar. This was crucial to meeting our schedule since the project was a collaboration between geographically separated divisions of NOAO. The controllers built in Chile and the mosaic dewar built in the USA were united only shortly before the first trials on Kitt Peak.
- It was equally important to our schedule to enable various engineers and programmers to work independently on software testing, board commissioning, the synchronization mechanism, and the three major board upgrades being developed concurrently. Our approach allowed the 16 channel mosaic controller to be quickly separated for use in 4 different labs, or as several half size mosaic controllers.
- Data flow parallelism is retained as until as far downstream as possible (the workstation and disk), relieving the need to improve the throughput of existing data transfer mechanisms.
- The controller hardware is scaleable without redesign. There is a threshold at 4, where another Sbus interface card would be added and one might even consider adding an entire workstation!
- A failure in one Arcon does not affect the others. The remainder of the mosaic can be operated with any of the standard Arcons removed (or turned off).
- The complete Arcon which served as a spare does double duty as a fully functional lab system.
- The use of four separate sequencers allowed for the possibility that the focal plane might have to be populated with detectors from several different manufacturers, or be operated with different waveforms for some other reason. Esoteric modes of operation where some CCDs are readout while others are being exposed could also be supported.

Since separate imaging and frame storage areas are not required, each 2k × 4k buttable CCD requires the same number of clocks and biases as half a quad readout CCD. If vertical pairs of CCDs had been connected to each Arcon, then each 2k × 8k slice of the focal plane would have appeared exactly like a standard quad readout CCD from the point of view of clock naming and pixel sorting, however we chose to connect 4k square quadrants to each Arcon for wiring convenience, and because this layout is likely to recur in other applications.

2. SOFTWARE MODIFICATIONS FOR THE MOSAIC

Each column in Fig.1 is a standard Arcon, minus its real time display controller. The "Nexus" in Fig. 1 is physically close to the host workstation where it serves as primary control node and as a large data FIFO to accommodate Unix latency. The transputer networks in each of the four Arcons were easily connected to allow the entire mosaic to operate as a single instrument: all "Nexus" transputer modules were housed in a common enclosure where their unused lateral transputer links were connected in a ring to facilitate command distribution and response collection, without disturbing the function of the links to the host workstation and lower level transputers. For control purposes, the entire Mosaic transputer network appears as a single controller: commands and status flow across a single transputer serial link to the host. Code is developed for the entire multiprocessor network using integrated tools and is loaded via a single link from the workstation. At the same time, any column(s) of the multiprocessor matrix can be downloaded and operated independently for diagnostic purposes

Minor extensions were made to the command parsers and communication packet format to support broadcast of commands and collection of responses to/from all "Arcons", while also allowing a specific Arcon to be targeted. These changes fitted naturally into the existing design, and most of the code was left untouched. A supervisory task was added to coordinate picture taking and in particular to present the full mosaic as a single multi-amplifier detector, with the ability to describe CCD positions and define Regions-of-Interest in global coordinates.

Whereas a "single controller model" is followed for commands and status, the data are carried by 4 separate links as though coming from 4 separate controllers to avoid bottlenecks. Raw data are buffered as a stream of unmodified message packets in Unix swap space (/tmp) by 4 separate daemons handling the transputer to Sbus communication. Each packet contains pixels in temporal order from a given controller, and thus interleaves data from various CCD outputs. A single (modified) picture formatting task interprets the packets, sorts the pixels and writes an IRAF image per CCD amplifier, applying the appropriate flips to ensure that all sub-images have the correct spatial orientation irrespective of the orientation of the CCD or position of the readout amplifier(s) selected. All reasonable amplifier combinations are supported.

We have traditionally written a single image to disk for an entire quad-readout CCD, however it was decided to write a single image *per amplifier* for the mosaic, so that it would be possible to process the images using standard IRAF commands such as ccdproc. Sub-images from a given exposure are given the same name with a suffix denoting the readout amplifier (e.g.: *object1234_<ampID>*) so that they can be referenced collectively (e.g.: *object1234**). A new hierarchical image format is under development for IRAF, which will provide a mechanism by which the image can be referenced collectively without seeing the underlying structure. In anticipation of this, the Arcon header generation mechanism and format has been enhanced to partition information per observing session, per exposure, per controller, and/or per CCD. For the time being, each sub-image receives all header information.

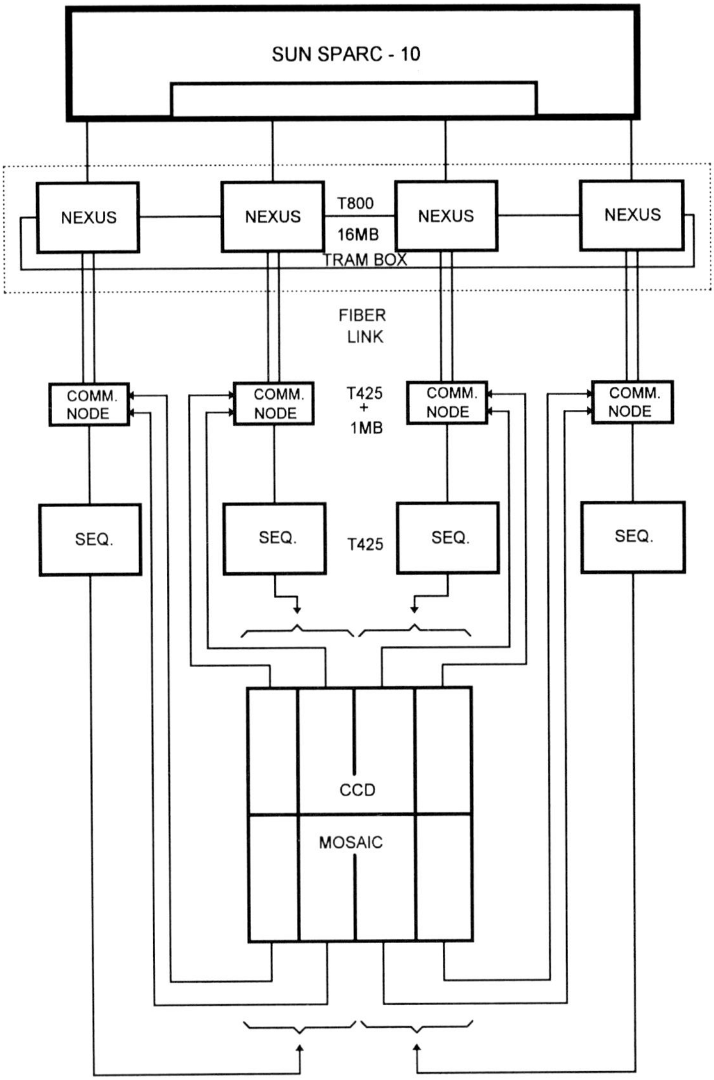

Figure 1: Simplified block diagram of the Mosaic CCD controller showing only the multiprocessor network. Each column is equivalent to a standard quad readout Arcon. For control purposes, the network looks like a single controller, while data returns from 4 standard controllers in parallel.

An X-Windows based real time display is being developed which will receive a copy of the data en route to the disk and format this onto a modified ximtool display window, providing immediate access to the image before the readout completes, and avoiding unnecessary disk I/O overheads. Since this is also still under development, scripts have been written to retrieve the image from disk and display it in a standard IRAF ximtool window *after readout completes*. The sub-images are located on the display to reflect their true relative positions (including gaps) using CCD coordinate information placed in the headers by the controller. Overscans are trimmed off and each image is mapped independently to approximately compensate for bias level and gain differences from one amplifier to another. A modified version of the IRAF imexamine task has been created to provide an interim quick look data inspection tool, which is familiar to the observer and has proven to be extremely useful during telescope tests.

3. Hardware Modifications for the Mosaic

3.1 Grounding

The goal of any grounding system is to minimize voltage differentials. This can be achieved by minimizing ground impedances, breaking loops in order to block the circulating currents (connecting grounds in a star configuration), carrying high currents and voltage sense connections on separate conductors, and by transmitting signals differentially to render harmless those differentials which are unavoidable. Placing the ground star point for a given Arcon or the connection between Arcons at the focal plane is avoided since it would drive unnecessary ground return currents through the high resistance of the dewar wiring. (The high electrical resistance and inductance is dictated by the requirement for high thermal resistance between the detector mount and the warm electronics.) The ground star-point in all Arcons is located on the backplane, which is a small multilayer planar PCB guaranteeing very low ground impedance over a wide frequency range. This wideband low impedance is crucial for the connection of analog and digital grounds, due to the flow of significant high frequency return currents induced by clock distribution and pixel timing signals connecting digital and analog circuitry.

The Arcon enclosure is connected to the ground star point and to the dewar. Both serve as the electrostatic shield. Obviously, there is a common shield for all Arcons they are all connected to the same dewar. Various ground loops are formed via the separate power supply cables, each of which must have their own earth lead for safety when the Arcons are separated. However ground hum is well below the noise floor since the current flow induced by flux coupling is minimal, the impedance through the dewar is very low, and most importantly the detector wiring is not in this current path. CCD substrates are wired to the corresponding Arcon's ground star point, are independent from each other and are isolated from the detector mount and dewar. Small currents, which could flow between Arcons via stray capacitance or inductance, are minimized by operating the controllers in lock step so that the clock transients are nearly identical minimizing the differentials which drive the currents.

The isolation of CCDs from the dewar at the detector mount is a standard feature of Arcon systems which has proven to be very effective at rejecting interference. In the case of the Loral 2k × 4k CCDs currently in use and the SITe CCDs on order, isolation is provided intrinsically by the CCD packaging method. (One must request that SITe omits the bond wire connecting the base plate to the CCD substrate.)

The CCD substrate potential is used as the reference side of the twisted pair carrying the signal. It is coupled by a 1 μF capacitor to the differential preamplifier so that in-band interference picked up by the signal wire is rejected, while a 20 KΩ resistor to the same input adds a precision voltage to cancel the DC component of V_{OS}. The capacitor not only couples common mode noise, but it filters the noise on the precision

DC voltage and thermal noise produced by the 20 KΣΩ resistor. Since the voltage across the common-mode-noise coupling capacitor is very nearly constant, there are no capacitor time constant or dielectric absorption artifacts added to the signal as occurs when the signal path is AC coupled. CCD substrate voltage fluctuations are dominated by small transients induced by clock return currents flowing through the resistive dewar wiring. These are synchronous with pixel timing and thus not an error source provided that they are stable. The differential preamplifier input stage also rejects crosstalk due to capacitive or inductive coupling between signals.

My original intention was for the dewar to be isolated from the rest of the instrument and thus the telescope environment, but this request was lost when a project staffing change occurred. Fortunately, experience with the mosaic and other Arcon applications shows that performance is not compromised, provided that the detector is isolated from the dewar.

3.2 Synchronization of readout waveforms

The interference to the signal in one Arcon from the unsynchronized clocks of its neighbours was observed to be surprisingly low (about 10 e^- per source Arcon, at 8 e^-/ADU), but nonetheless still undesirable. Synchronizing the sequencers is only necessary during readout. We achieved this by distributing a common 5 MHZ master clock to all transputers from which they each generated their 25 MHz processor clocks using PLL based frequency multipliers which are internal to the transputers. All sequencer transputers execute identical code. Waveform tables are identical at present, although the requirement is only that the total pixel (and parallel shift, etc.) times be identical. Each readout "waveform table" begins with a *wait for interrupt*. After the software has requested a readout from all sequencers and they are all waiting, a common trigger pulse is generated by writing to a general purpose output bit which is distributed to all Arcons.

The PLL inside the transputer multiplies the 5 MHz master clock to produce a higher frequency which it then divides to generate 25 MHz. This means the relative phase of the various processor clocks depends on the way the PLLs and dividers power up. Although the requirement for the phase to be fixed is satisfied, the indeterminate phase of the various PLLs does not allow the sequencers to be synchronized to better than one clock cycle. We measure a 57 ns maximum delay between corresponding waveforms which is slightly longer than the 40 ns clock cycle due to propagation delays. The delays between waveforms remain stable to within our measurement capability (1 ns) for the duration of the readout. Several transputers had to be replaced since they failed to synchronize, presumably due to some problem holding lock in the internal frequency multiplier.

Our original plan for distributing the master clock was to use a common clock for the Nexus transputers, which are housed in the same enclosure at the workstation end of the fibers. Data and idle packets in the fiberlink are encoded synchronously using this clock which is recovered by the decoder in the receiver in each Arcon box. This was a very elegant solution, requiring no extra cabling or additional hardware, however the recovered clock suffered from too much timing jitter. Such jitter generates noise in the double correlated sampler which is proportional to signal, since changes in slope time modulate gain. Jitter is most easily detected by plotting noise versus input offset:

differences in the duration of the up-slope and down-slope cause the DC level into the dual slope integrator to be incompletely rejected.

We attempted to filter the jitter by regenerating the master clock with various combinations of phase-locked loops, multipliers and dividers to average the phase, but with no success. We reverted to a simple differential driver/receiver to transmit the clock from Arcon to Arcon over a RG-59 (75 ohm) coaxial cable. To avoid an extra cable and connectors the "readout trigger" was impressed on the same coax as the master clock by mixing in a negative going pulse during the low time of the clock cycle. The disadvantages of using driver/receivers are that the ground isolation is reduced to about 3 kΩ and each Arcon must be designated as master or slave with a jumper.

4. DRINKING FROM THE FIRE-HOSE

While it was always obvious that there would be challenges dealing with the volume of data and with testing the large number boards and CCDs, the scale of the task is still impressive when experienced directly.

Raw images require 140 MB of storage space or twice this when processed, unless care is taken to scale the data to stay within 16 bits. Readout completes in as little as 90 seconds at the current minimum pixel time of 10.5 μs . We expect to achieve 45 second readout with the new CCDs having two functional output amplifiers. In practice we have been reading at 13 μs/pixel/channel while we are limited by the disk write time.

With a Sun Sparc-10 processor, 300 MB of RAM and 6 GB of disk space available for image storage, the disk write took 5 to 6 minutes from start of readout, some 4 minutes more than the readout time! While the pipelining of the data allows the next exposure to be started as soon as readout completes, the duration of that exposure must be sufficient to allow the disk to catch up so that the RAM cache does not overflow. The engineer generally wants to see the image before proceeding and has to wait for disk write to complete and then fetch the image again to display it. On our first attempt, this took 15 minutes to complete, for a total acquisition time of 21 minutes! This was clearly intolerable even as an interim measure.

By simply reducing the ximtool display buffer (the *stdimage* parameter in IRAF) from 4096^2 to 512^2 to reflect the number of screen pixels actually in use, we reduced the number of pixels to be scaled and displayed by a factor of 64, cutting the display time from 15 minutes to about one. To see full resolution one displays the sub-image of interest in a separate display buffer making it possible to flip quickly between full area and full resolution.

Upgrading the CPU to a *HyperSparc* processor (made by Ross, Texas) increased processor clock frequency from 25 to 125 MHz resulted in disk write taking only about two minutes longer than the readout. This is still an order of magnitude longer than the write time implied by the 10 MB/s disk write rate. This time has been shown to be mostly consumed by the IRAF image I/O routines which will be replaced by the new code to explicitly support the Mosaic image format. Furthermore, disk write will be less time critical when the real-time display tool updates the display directly. Meanwhile images can be acquired and displayed with the HyperSparc in under 5 minutes, of which only 2 are dead time between exposures.

Our ultimate goal is to be able to perform instrumental calibration and archiving for each night's observations during the next day. A second workstation has been purchased to handle the planned real-time display, disk I/O, and data reduction leaving the present machine to control the instrument and buffer the data without competition. The new data reduction workstation is a Sun Ultra 2 Model 2200 with two 200 MHz UltraSPARC-I cpu chips. Each CPU has 32 KB of on-chip cache plus 1 MB of external cache, 512 MB RAM, three Fast-Wide SCSI channels (16 bits running at 10 MHz for a potential data rate of 20 MByte/sec), a Creator graphics card with 24-bit 3-D graphics accelerator imaging at 1280x1024 resolution, a Turbo GX+ graphics card (8-bit, 1280x1024) and a Fast (100 Mbit/s) Ethernet connection. Five Micropolis 9GB, 7200 RPM Fast-Wide SCSI disks are striped together into one 45 GB file system providing space for 150 processed images (32 bit) or twice as many raw images. DLT or Exorbyte Mammoth tape drives are being considered for purchase.

Even routine tasks such as jotting down notes on the noise, gain and cosmetic defects for all 8 CCDs can be surprisingly time consuming. Additional impediments such as very slow response time from the tools makes system testing a major obstacle to timely deployment. Till now the development work has been done without the benefit of the huge increase in computing power promised by the new workstation, and the detailed optimization required be the current problematic Loral CCDs has been given low priority while the tools are being improved.

We have made much more extensive use than we normally do of methods to limit the number of pixels read out. Rather than use conventional regions-of-interest, we devised ways to quickly reconfigure to run a single controller, or read all controllers but ignore data from some when crosstalk or synchronization was relevant. It has proven useful to be able to run subsets of the controllers for diagnostic purposes also.

In Arcon, regions-of-interest are automatically expanded beyond the coordinates given by the user, such that pixels on all output nodes are read or are all skipped to maintain synchronization between CCD outputs. We are developing software to support global regions-of-interest for the mosaic, in which output from entire CCDs is automatically suppressed for if they lie outside the region to cut down on the amount of unwanted data.

We frequently employ *test* exposures, for which it is preferable to bin on the CCD to match the number of pixels to the display resolution rather than to allow the display software to sample the data. This not only preserves more information but also decreases the readout time dramatically. The disk write time becomes hidden within the readout time, and the time to display the image from disk is similarly shortened, so that a full field test image can be obtained in less than 30 seconds. This mode is so useful that we plan to support the *test exposure* command with a separate detector setup menu to make switching back and forth between scientific and test configurations rapid and failsafe.

Even the task of checking that the various sub-images had been correctly flipped and arranged on the display presented unexpected difficulties. For example, a different method had to be devised to project a test pattern since our standard projectors fell far short of illuminating the full area. We finally settled on a LED shining down a meter long blackened tube in front of a mask lying on the upturned dewar window.

Some additional work on the software was required to improve debug output and error messages to clearly identify in which subsystem messages originated!

With so much data flowing on parallel links, we found it was necessary to automate our diagnostics for data integrity. Cyclic Redundancy Checking is performed on packet headers to avoid protocol violations but checking the data itself is too slow. To verify the error rate for data, we utilized the artificial data generation hardware on all Arcon ADC cards to provide a fixed data pattern (8 bit row counter, 6 bit column counter, 2 bit amplifier tag). This exercises all codes and detects the position in the image of lost data which can otherwise be hard to determine when transferring data by DMA. We found that the Bit Error Rate rose to one part in 10^7 with the failures occurring in the TRAMbox or link to the Sun always on the 4th least significant bit in any byte. This bit was set erroneously but never cleared erroneously, and no bytes were lost or added. Lowering the supply voltage to 4.8V (!) and providing higher cooling airflow reduced the Bit Error Rate to $2x10^8$, corresponding to typically 5 single bit errors per full mosaic readout. These figures are for the most problematic data patterns, whereas with artificial patterns more closely resembling real data one saw far fewer errors. We do not see errors in data from the CCD since these are difficult to distinguish from the normal cosmic ray background and noise.

5. PERFORMANCE SUMMARY

Mosaics are excellent tools for studying crosstalk mechanisms since crosstalk can occur in many combinations: between signals from the same CCD, different CCDs controlled by the same Arcon, or between different Arcons. We have made only enough measurements to determine that further work is required in this area. By measuring the effect of single bright columns on line averages in other CCDs, we found that crosstalk between two Arcons can varies from as great as one part in 2000 to as little as one part in 160,000 simply by looking at the response in the CCD that was the "source" of the high crosstalk to a bright column in the CCD that was the recipient of the crosstalk in the first case!

Apart from this, the mosaic controller is performing as well as other applications of Arcon. Dead time between exposures is limited to just the readout time, which is currently under two minutes and will shrink to less than one minute in 1997 when the new video card or new CCDs are installed. Image storage and display requires an additional 2-3 minutes, which is passable but will drop radically when the new reduction workstation is used. Real-time display software is under development. The total time to read and display a test exposure is already under 30 seconds for a full field image when binned 4 by 4.

Synchronization is maintained for the entire readout, with waveform skew less than 57 ns, so that the interference between Arcons is converted to a small bias level shift which is completely removed during overscan subtraction. The readout noise is limited by the CCDs and not the controllers.

To date (Oct 1996), the system has been tested for about 10 nights on each of the Kitt Peak 36" and 4m telescopes in preparation for release to observers early in 1997 as originally scheduled. The controller has been handed over to the team in Tucson who are testing the instrument as a whole and optimizing the controller to operate the current front illuminated and cosmetically poor CCDs. They will install superior thinned CCDs which are due to be delivered early next year. Meanwhile the controller team at CTIO is finishing the video upgrade and beginning construction of controllers for a clone so that wide field imaging can be supported at the US national centers in both hemispheres simultaneously.

Acknowledgments: This project and the associated board upgrades involved most of the technical staff at CTIO, who I have pleasure in thanking for their cooperation, dedication and cheerfulness under pressure: Electronics Engineers - Ron Lambert, Eduardo Mondaca, Gustavo Rahmer, Rolando Rogers, Javier Rojas, & Tom Wolfe (Tucson); Electronics technicians - Sergio Cathalifaud, Rodolfo Diaz, Pedro Ojeda & Fernando Saleh; Electronics Draftsman - Rodrigo Leiva; Instrument Makers - Fabian Collao and his crew, Mechanical Design/Drafting - Alfonso Cisternas; PCB layout - Dee Stover (Tucson). Programming - Pedro Gigoux, Jim Hughes, Dan Smith & Gary Lee Webb; Project scientists - Steve Heathcote (software) & Alistair Walker (hardware). This has been a major intercontinental collaboration with pressing schedules and would not have been successful without the support and team spirit of our counterparts and particularly Rich Reed who managed the engineering effort in Tucson. We also thank Neil Gaughan for travelling to Chile to give his advice and initiate the detailed resource scheduling for the controller portion of the mosaic project.

CLEANING CONTAMINATED CCDs

ANTON NORUP SØRENSEN[1]

ABSTRACT

On a few occasions, CCDs operated by the Copenhagen University Observatory have been contaminated to an extent easily visible by naked eye inspection. Attempts to clean the devices by pumping and baking have generally not been successful, as the treatment in most cases proved to be insufficient, still leaving the CCD contaminated, and in some cases resulted in a fatal malfunction due to thermal stress. Here, an alternative approach of chemically cleaning CCDs is described. The method has so far been applied to two chips strongly contaminated by oil, and in both cases the devices emerged clean with performance as before the contamination. Our first attempt to wash a CCD is described in detail below.

1. INTRODUCTION

The DFOSC spectrograph/focal reducer mounted at the Danish 1.54m telescope at La Silla is currently equipped with a Loral $2k^2$ CCD detector, serial number W11-4. In order to achieve the required sensitivity in the 300 nm - 500 nm wavelength range, the chip needs UV-flooding.

Due to problems with maintaining a stable UV-flooded quantum efficiency and vacuum, in June 1996 the dewar was sent to ESO-Garching for leak testing, baking, cleaning and mechanical modifications, to be performed by Sebastian Deiries (ESO) and Niels Michaelsen (CUO). The CCD was removed and sent to Copenhagen University Observatory in order to measure the performance before re-installation in the DFOSC dewar.

2. BAKING

Upon arrival, visual inspection of the CCD revealed it to be contaminated by a considerable layer, possibly oil, covering the entire surface.

In an attempt to remove the contamination, the CCD was baked in vacuum. During four hours, the temperature was gradually increased to +116 °C, the pressure decreasing to 10^{-3} mbar. The temperature was kept constant for 18 hours, during which the pressure decreased to $2.6 \cdot 10^{-4}$ mbar. The heating was stopped, and after two hours, dry synthetic air at room temperature was slowly let in to reach ambient pressure.

Visual inspection after baking revealed that little, if any, change had occurred. The CCD was still severely contaminated.

The CCD was installed in a dewar and the camera was cooled to -100 °C in order to measure the unflooded quantum efficiency in the passbands of eleven filters ranging from 334 nm to 1060 nm. The measurement is plotted in the left part of Fig. 1.

[1] Astronomical Observatory, Niels Bohr Institute for Astronomy, Physics and Geophysics, Copenhagen. Email: **norup@astro.ku.dk**

J. W. Beletic and P. Amico (eds.), Optical Detectors for Astronomy, 275–278.

The dewar was then allowed to warm up to room temperature and filled to atmospheric pressure with dry synthetic air. The chip was UV-flooded for 15 minutes, pumped, and cooled immediately after.

The QE measured immediately after cooling, also shown in Fig. 1 is in the 300 to 700 nm range about 10% lower than what was measured by Mike Lesser (Steward Observatory CCD lab).

The sensitivity was monitored at regular intervals in the following days. A rapid decrease was found: About 8% was lost during the first day after flooding, whereafter the decrease slowed somewhat, apparently changing exponentially towards the unflooded QE. The change in time can be seen in Fig. 1.

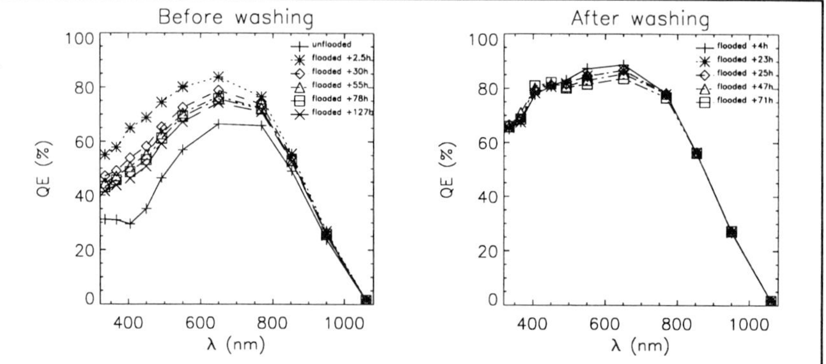

***Figure 1**: Plots of QE of W11-4 at different wavelengths, measured during a few days. First measurement is made before UV-flooding, the second immediately after cooling down after flooding. The development of the QE is then monitored, displaying a severe lack of stability. QE of the washed W11-4 immediately after UV-flooding and during the three following days. The QE has increased and the stability has improved, but still shows a considerable chromatic change.*

The structure of the flat field can be examined in Fig. 2 for exposures at 405 nm. Before flooding, the sensitivity dropped rapidly near the edges. Immediately after flooding, the QE is higher at the edges, indicating a more effective flooding there. All over the CCD, small specks with high sensitivity are seen. In time, the non-uniformity grows stronger, as the central area looses sensitivity while the edges and specks decay at a slower rate. The current pattern of decay seems to be caused by the oil contamination, inhibiting UV-flooding and speeding up the decay in the central region.

3. WASHING

As the oil coating seemed the most probable cause of the lack of QE-stability, this had to be removed by washing the CCD. A procedure for washing was provided by Mike Lesser. Practical considerations led to some deviations from the procedure, though. A set-up grade CCD was selected to be the first subject to the washing process, in order to prevent damage from improper handling to the W11-4.

Acetone and isopropylalcohol (IPA) was purchased from a pharmacy. This quality seems to be sufficient, although we now use higher purity solvents. Bottles of each liquid were heated in a bath of 50 °C warm water, corks slightly loosened in order to avoid a pressure buildup. The temperature of the liquids were not measured directly, but seemed after an hour still to be significantly cooler than the bath due to vaporization cooling. Two 60 milliliter injection syringes without needles were filled with each of the liquids. Before, the plastic of the syringe was confirmed not to react with either acetone or IPA.

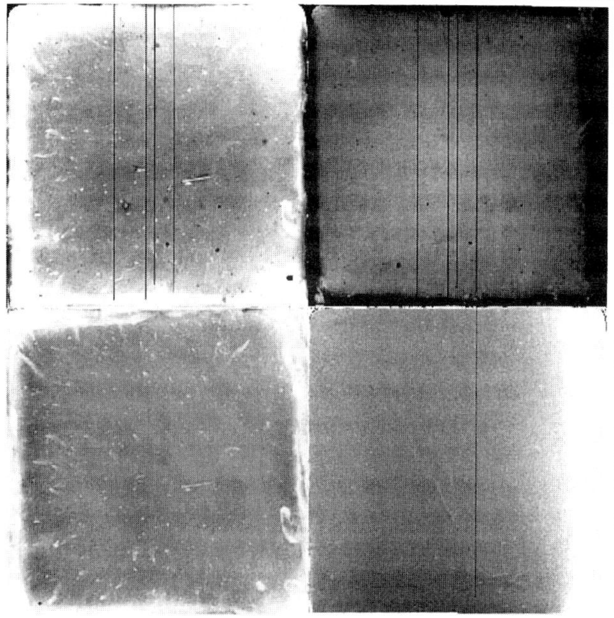

Figure 2: Upper left: Before washing: Full frame flat field at 405 nm immediately after UV flooding. Lower left: Before washing: Ratio to a flat field obtained one day later. The white and black corresponds to deviations of 20% from the median level. The flat field is non-uniform with higher sensitivity at the edges and in specks of the size of a few pixels. Flat fields acquired later are showing even stronger irregularities with the sensitivity falling off most rapidly in the central region. Upper right: After washing: Full frame flat field at 405 nm immediately after UV flooding. Lower right: After washing: Ratio to a flat field obtained three days later. The white and black corresponds to deviations of 3% from the median level. Clearly, the flat field properties have improved after washing.

Holding the CCD with the connector at the bottom, pointing approx. 45° downwards, the CCD was first flushed with acetone applied to the upper edge in a sweeping motion. Most of the syringe was emptied in this way during about ten seconds. Immediately afterwards, before the acetone dries, flushing was done with IPA in the same manner. Then, the CCD was dried with a jet of dry air, initially directed from above to force the majority of the remaining liquid off the chip before drying.

The chip was then baked for at least an hour at +45 °C in vacuum before installing in a dewar. Measurement of QE after UV-flooding showed that the CCD had not sustained any damage from the solvents.

The W11-4 chip was then washed following the same procedure. The majority of the visible oil layer was removed, but a few faint streaks were left. As nothing similar was seen on the first CCD, the streaks probably consists of oil that was dissolved, but not completely flushed off.

We now pre-heat the CCD before washing, to about +50 °C by placing it on a metal bar that has been soaked in the warm water for some time. A piece of paper in between ensures a slow increase in temperature. In this way, the thermal shock from the warm liquids is reduced and the efficiency of the solvents is increased.

4. PERFORMANCE AFTER WASHING

The CCD was UV-flooded for 20 minutes in dry air at atmospheric pressure immediately after installation in the dewar. After evacuating the dewar for one hour, cooling to -100 °C was started.

QE measurement was made immediately after operating temperature was reached and during the following three days. The results are displayed in the right part of Fig. 1. The initial QE is higher than measured before washing and is within a few percent of the measurements by Lesser. Change in sensitivity is apparent during time, but is much smaller than seen before washing. The change is now chromatic. No decay longwards of 800 nm is seen. In the 500 nm to 700-nm range, a decay of approx. 2% per day is occurring, with no convergence towards a stable level seen during the short monitoring time. Around 400 nm, a tendency towards increasing sensitivity is seen. The measurement errors are relatively large here, but the change seems significant.

Due to the rather different color dependence, we may be observing a phenomenon unrelated to the surface charge. While a significant decay of the sensitivity at wavelengths longer than 800 nm was observed before washing, no decay is seen now. From this it is concluded that the UV-flooding is stable, but the QE is now changing due do other factors. The color dependent change in QE has also been seen on chips that were never washed and a long time after washing, so it does not appear to be related to damage from the solvents.

In Fig. 2, the structure of the flat field after washing is displayed. The uniformity of the flat field has improved, leaving only a small fraction of the pixels outside 3% of the median level. A modest drop in sensitivity is seen at the edges, and the high-sensitivity spots are difficult to detect. During the next three days, some small-scale structure appears, but at an amplitude much smaller than before, apparently converging towards a stable high-QE state.

5. CONCLUSION

At the re-commissioning at La Silla we were not able to reproduce the high stability demonstrated in the CUO lab, forcing us to use the CCD unflooded. Nevertheless, the improvement in performance after washing is very encouraging, as the washing procedure seems effective in de-contaminating the CCD, without damaging it. By the time of writing, the method has been applied to another contaminated CCD with equally satisfactory results. Therefore, washing as described is suggested to be a standard procedure for cleaning contaminated CCDs.

CCD FLATNESS MEASURING SYSTEM AT UCO/LICK OBSERVATORY

R. J. STOVER[1], *M. WEI*[1], *Y. LI*[1]

In the detector development laboratory of the University of California Observatories, Lick Observatory, we have set up a very simple but effective system for measuring CCD flatness. In the very large spectrographs being built for the Keck II telescope, as well as in many other instruments, the cameras are between f/1 and f/2. These high speeds make it imperative that our detector deviate from flatness by no more than about 10 micrometers. The figure shows, in a very schematic way, the system we are using.

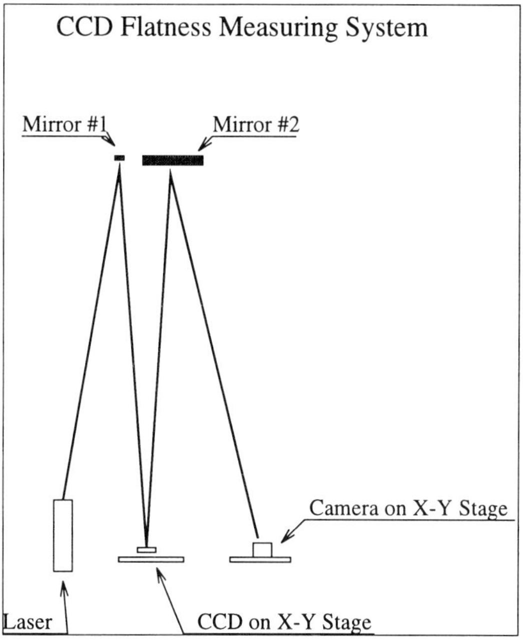

It consists of nothing more than a laser, which produces a narrow, focused beam, an X-Y translation table upon which our CCD can be placed, and some mirrors to direct the laser beam first to the CCD and then to a position recording system. Our position recording system consists of a miniature CCD camera mounted on the pen carriage of an X-Y pen plotter. A computer is used to capture the images from the miniature CCD camera and drive the pen plotter to follow any motion of the laser beam. In an even simpler system the position recording system can be replaced with a piece of graph paper upon which laser spot positions are marked manually.

[1] University of California Santa Cruz, CA 95064 USA. Email: **richard@ucolick.org**

J. W. Beletic and P. Amico (eds.), Optical Detectors for Astronomy, 279–280.

Our X-Y translation table is the bed of an old milling machine. This gives us enough carrying capacity to test our CCDs while they are in the dewar, and cooled to operating temperature. To make a measurement we simply scan the CCD across the laser beam and note where the laser beam is reflected. A perfectly flat CCD would result in no movement of the reflected beam. Any movement in the reflected beam indicates a change in the slope of the reflecting CCD surface, and therefore reveals a non-flat surface. The sensitivity of the system depends on the accuracy with which the reflected spot position can be measured. In our system, we can easily measure typical 1-micrometer variations in CCD flatness.

THE ORTHOGONAL TRANSFER CCD

JOHN TONRY[1] *AND BARRY E. BURKE*[2]

ABSTRACT

The Orthogonal Transfer CCD (OTCCD) is a device to carry out image motion compensation within a CCD. This paper presents three main points about the OTCCD: 1) Science grade OTCCD devices have been produced and work better than expected. 2) The big concern we had about the OTCCD was that pocket pumping would make it useless, but this appears to be a fairly minor problem. 3) Finally, the atmosphere is more benign than we might have feared, and significant gains in sharpening the point spread function (PSF) can be had simply with image motion compensation.

The structure of the OTCCD is slightly non-conventional to permit parallel charge shifting in all four directions instead of merely up and down. This is achieved by replacing the usual vertical channel stop with a fourth gate. In addition, two of the usual three-phase gates are changed slightly in topology into two triangles and the third three-phase gate is symmetric with this new fourth gate, as illustrated in Figure 1. The overlap between all four gates is the site of a reduced channel stop.

We initially built a 64 × 64 test device to demonstrate that this would work, and documented the results in [1]. We were fortunate to be able to piggy-back a larger device on the chords of the wafers being devoted to the UH-MIT/LL 2k×4k CCD development consortium. This new device has a 512 × 512 area of 15μm pixels with OT gates and an adjoining 512 × 512 3-phase frame-store area. On the MDM 2.4-m telescope these pixels subtend 0.17arcsec and the entire chip is therefore 1.5' × 3'.

The operation of the OTCCD is illustrated in Figure 2. The frame store area is left uncovered and is used to track a guide star. When the telescope has been pointed to place the science field in the OT region, we then rotate the instrument to locate a good guide star in the frame-store region. With current, unoptimized algorithms, a backside OTCCD can track a V=16 star at 10 Hz, and the probability of finding such a star within 2' of any given location is nearly unity. Once the guide star is selected, the shutter is opened and a patch containing the guide star is clocked out at a rate between 5 and 100 Hz, depending on the brightness of the guide star. The guide star image is centroided in a SPARCstation, and if it has moved from the previous image, the guide star patch is updated and the electrons in the OT region are shifted accordingly. Depending on the speed at which we track the guide star and the duration of the exposure, we may execute 1000-10,000 such iterations before the shutter is closed and

[1] Institute for Astronomy, University of Hawaii, Honolulu, HI. Email: **jt@hokupa.ifa.hawaii.edu**
[2] Lincoln Laboratory, Massachusetts Institute of Technology, Lexington, MA 02173

J. W. Beletic and P. Amico (eds.), Optical Detectors for Astronomy, 281–291.

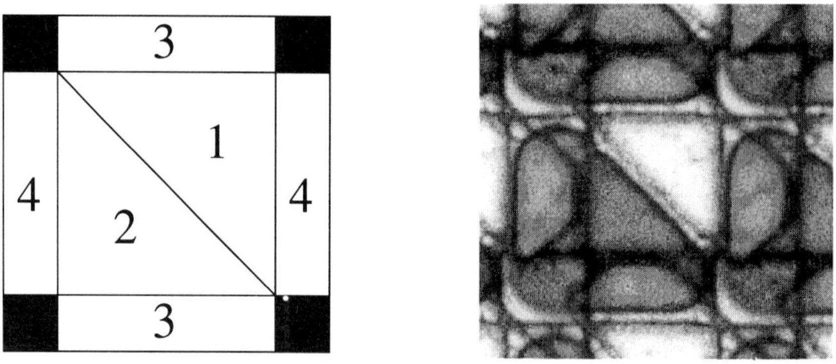

Figure 1: the four gates in the OTCCD are shown schematically on the left and in a photomicrograph on the right. In this implementation the gate #1 is a layer of aluminum and shows up as bright in the photomicrograph. By setting gate #4 low, we can carry out normal clocking in a downwards direction by shifting 1--2--3--1, or upwards with 1--3--2--1. However, the structure is rotation symmetric so that we can also set gate #3 low and shift left with 1--2--4--1, or right with 1--4--2--1.

the image in the OT region is read out. The readout time for the guide star patch is a few msec, depending on the size, and the time to do the OT shift is about 100 μsec.

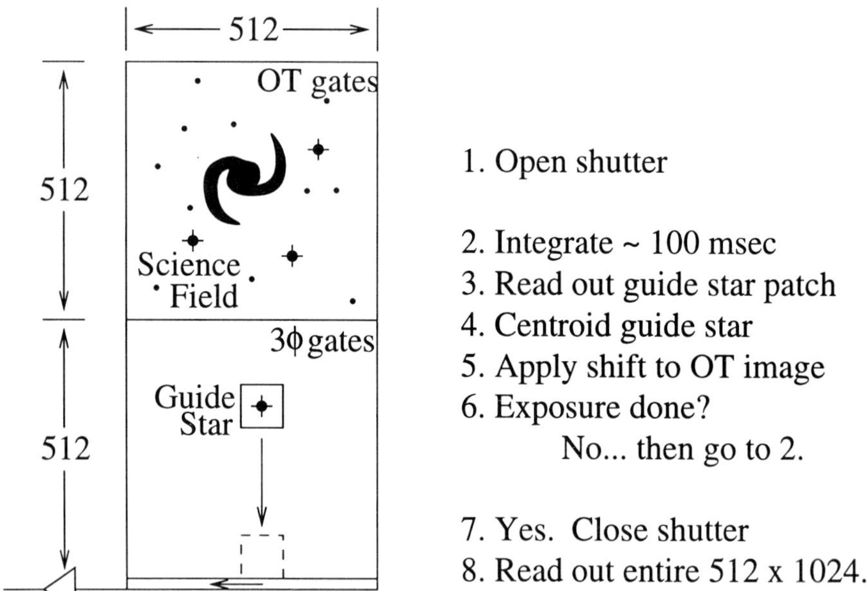

Figure 2: The use of the OTCCD is illustrated on the left: a science field is placed in the OT region and a guide star is located in the frame-store region. The sequence listed on the right is the basic loop for tracking the guide star and shifting the image collecting in the OT region to follow it.

Figure 3 illustrates the basic effect of this operation. A dewar containing an OTCCD was pivoted in the front and supported by a spring in back. An artificial guide star (a glint off of a ball bearing) was imaged onto the frame-store region, and a poster was imaged onto the OT region. The shutter was opened and the dewar shaken at about

10 Hz during the exposure. Without OT tracking the image is very blurred; when OT tracking is enabled the image is sharp despite violent gyrations of the scene. The dust shadows (e.g. upper left) are sharp in the unshifted image and they are blurred out in the OT-shifted image. However, note the pattern left behind by pockets in the OT-shifted image. The severity of this problem was our most serious concern from the outset of this project.

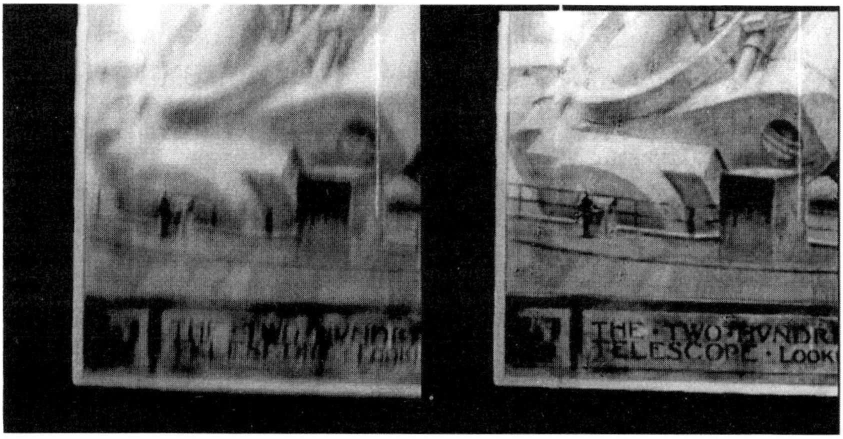

Figure 3:*The left panel shows a blurred image caused by shaking the dewar during an exposure. The right panel shows the improvement gained by OT tracking while the dewar is shaken.*

The software for using the OTCCD is rather pleasant to use. When you start an exposure, a video window pops up which shows the incoming guide star images, the centroid, and a leaky average of the guide star. A strip chart shows you the current FWHM of the guide star, and if you think the telescope is going out of focus you can tweak it in or out. (An advantage of the OT shifting is that any image motion caused by the focus mechanism will be nullified.) When the exposure is finished, the video window disappears and the CCD reads out and writes to disk.

It is important to realize that for some projects a small improvement in PSF can lead to huge improvements in science. A surface brightness fluctuation (SBF) survey is a case in point. Another example is photometry of gravitational lenses to look for time delays. Image motion compensation is not full adaptive optics, nor will it ever rival the resolution of HST, but it can lead to significantly better science. Figure 5 demonstrates a typical improvement achieved with OT tracking. These are two 150 second images of a globular cluster (M71). The first, with only normal guiding, reveals the very good seeing (0.73arcsec) which we have achieved at the MDM 2.4-m telescope on Kitt Peak. The second image had OT tracking enabled on a rather faint guide star (at 7.4 Hz) and has markedly sharper images (0.50arcsec). Although 0.73" seeing is not constant at MDM observatory, the degree of improvement, about 0.2arcsec, did occur again and again, over the course of four nights, in seeing ranging from 0.7 to 1.1arcsec.

The background in Figure 5 is dark enough that it is possible to see the holes left behind by the pockets.

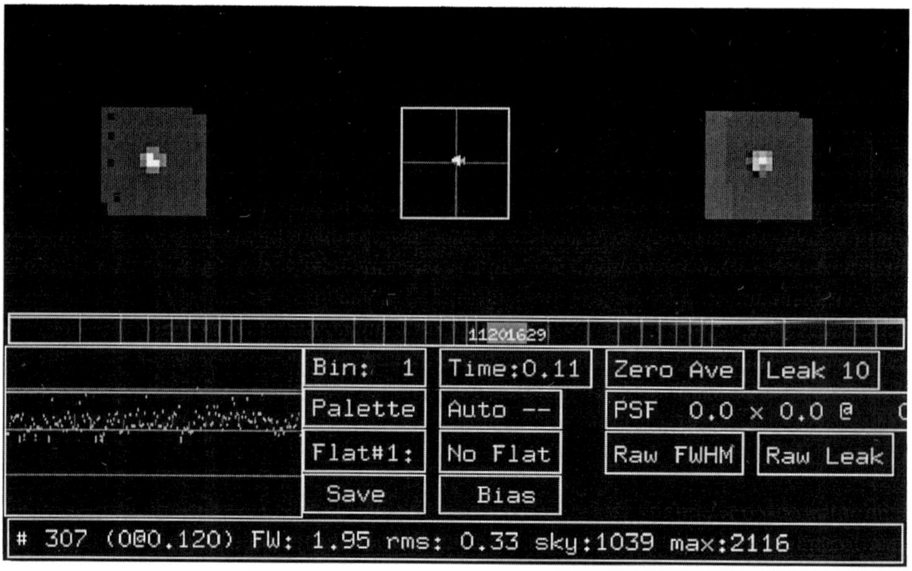

Figure 4:*The video display window seen at the telescope when running the OTCCD.*

Figure 6 is another pair of images with the stretch adjusted to make the background quite dark and the pockets more apparent. This exposure was 500 seconds, during which there were 4650 OT shifts. Even a very small pocket of a fraction of an electron could pocket-pump a very noticeable divot in the background, which is only a few hundred electrons per pixel. Since these chips have *many* pockets which are that deep, we anticipated serious problems. As a matter of course, we recorded the history of

Figure 5: *Two images of the globular cluster M71 are taken with normal guiding only (left; 0.73arcsec) and with OT tracking (right; 0.50arcsec).*

the OT shifts which occur during an exposure (necessary in order to flatfield an image), and wrote software to replay these shifts while exposing on a flatfield, thinking that this would help remove the myriad of divots I expected to see.

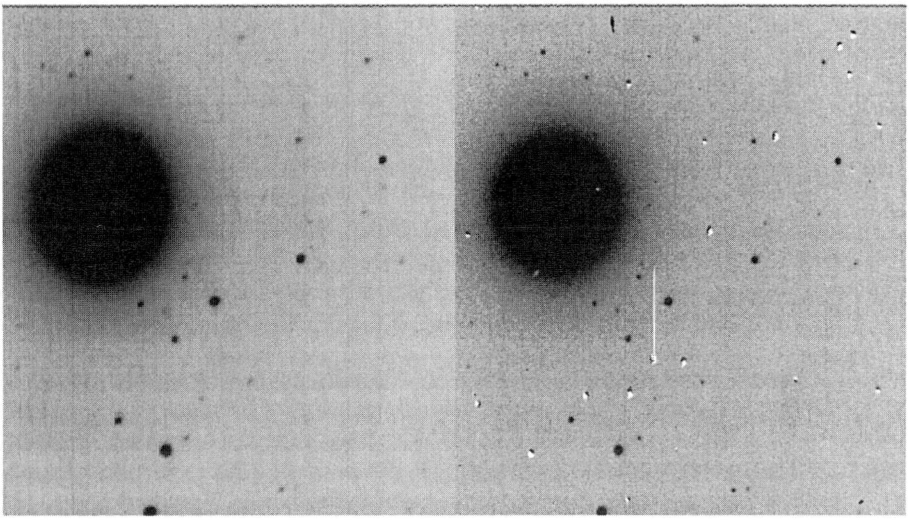

Figure 6:Two images of the galaxy NGC6703 are taken with normal guiding only (left; 0.82arcsec) and with OT tracking (right; 0.58arcsec).

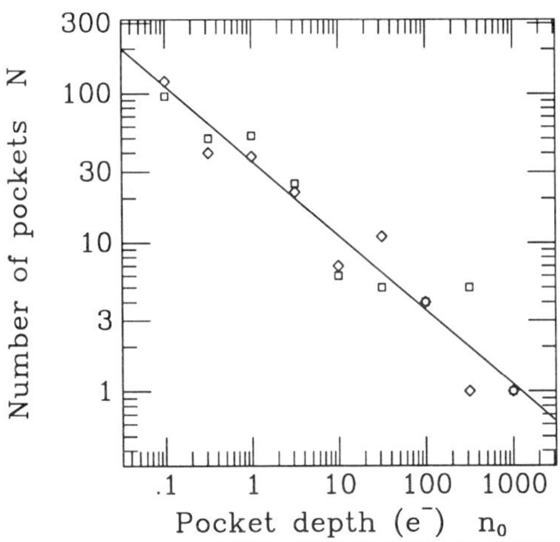

***Figure 7:**The number of pockets as a function of pocket depth.*

We were very pleasantly surprised to find that there are only about two dozen divots visible in the images we took. They are very deep - the pocket pumping is right down to the bias level, but they comprise less than 1 percent of the area of the chip (and could be even less if the OT guide star signal was used to guide the telescope as well as shift the OT region). Pocket pumping in the lab revealed that these divots correspond to the pockets which are around 20 electrons or deeper, and that shallower pockets simply do not create visible divots in the image.

The reason for this can be understood from the theory of diffusive draining of pockets. Figure 7 shows the number of pockets in this chip as a function of pocket depth. The number N goes roughly as the inverse square root of the pocket depth: $N \approx n_0^{-1/2}$. The pockets are about three times as numerous under the OT region as in the 3-phase region, they have low sensitivity to temperature, they are somewhat sensitive to clock speed, particularly faster than 10μsec

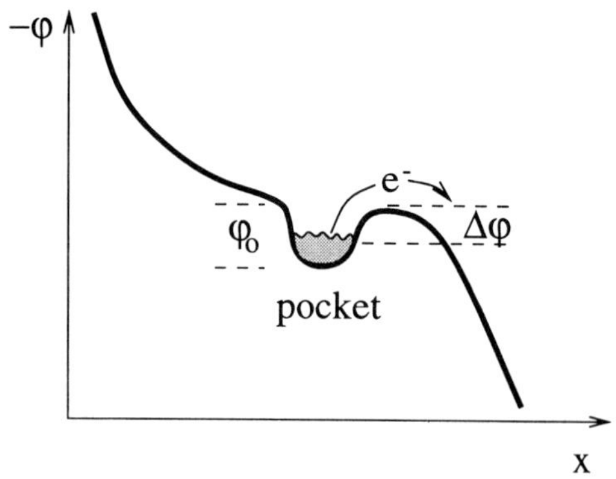

Figure 8: A partially filled pocket in the process of draining.

per shift, they are very sensitive to clock voltage, and they are almost all unidirectional: most trap charge when clocking up or right although some of the larger ones trap charge when clocking down or left. This probably means that the pockets are electrostatic potential minima.

There is a diffusion equation discussed by [2] which describes the draining of a pocket illustrated in Figure 8:

$$\frac{dn(t)}{dt} = -k_1 n(t) = \exp(-\frac{q\Delta\varphi(t)}{kT})$$

where n is the number of electrons in the pocket, the full depth of the pocket is φ_0, the potential barrier that electrons in the partially filled pocket must overcome is $\Delta\varphi$, and k_1 is a diffusion constant. We estimate that k_1^{-1} is very fast, of order nanoseconds, and for typical sizes and capacitances that $\varphi_0 \approx 0.02$V (n_0 / 50 e^-), where n_0 is the full depth of the pocket in electrons. At typical operating temperature kT ≈ 0.02 eV.

The solution of this diffusion equation is an exponential integral, but the behavior is simple to describe: if the pocket is less deep than kT it drains very fast, in nsec, and if it is deeper than kT it drains very slowly, with a characteristic time scale of $\tau \approx k_1^{-1} \exp(q\varphi_0/kT)$. This, we think, is the reason that we see few pockets in our images. The pause of 10-100msec between OT shifts permits all the small pockets ($n_0 \ll 50e^-$) to drain completely, but the large pockets ($n_0 \approx 50e^-$) are still present.

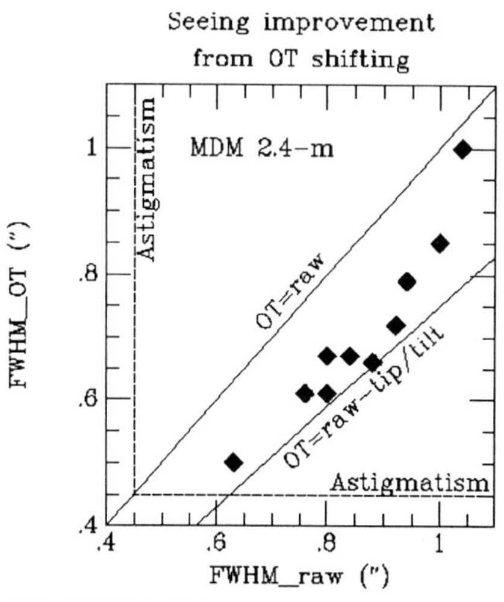

Figure 9: The image FWHM with OT tracking is plotted against the image FWHM with normal guiding. These are derived from pairs

Figure 9 shows the improvement in image quality which we achieved over the course of four nights by OT tracking. These points are derived from pairs of images, taken in quick succession, where one used normal guiding only and the other had OT tracking applied. Generally speaking, the image FWHM improved by about 0.2", and we found that the RMS image motion was a decent predictor of this improvement. This image motion did vary slightly with the overall seeing, and the lower diagonal line is a fit to raw FWHM minus image motion RMS. The 2.4-m telescope has about 0.4" of astigmatism, and the image, which came out at 0.50" with OT tracking, required very delicate focusing. (This astigmatism arises from the mirror support, not the optics, since this is a very thin mirror.)

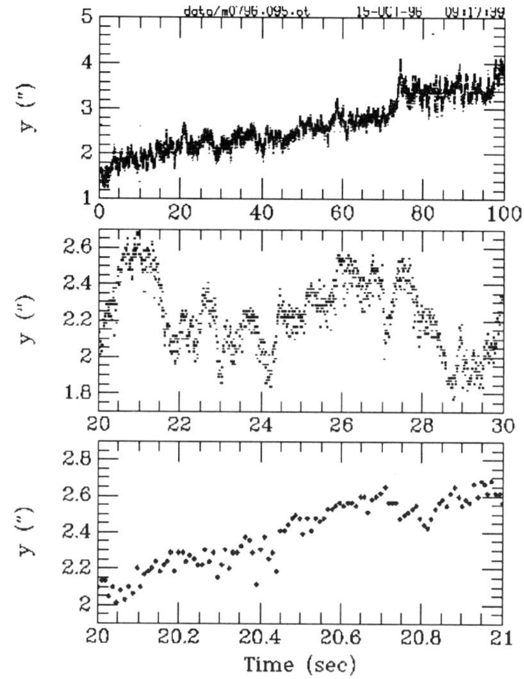

Figure 10: A typical time series of (unguided) image motion in one dimension is plotted against time. Both the panel spanning 100 seconds (hence indicative of ≈0.1 Hz) and the panel spanning 1 second (hence indicative of ≈10 Hz) show rather little power, other than an overall drift. The central panel, by contrast, spanning 10 seconds and indicative of ≈1 Hz, has a lot of power.

We were also surprised to find that the image motion was slower than what I had expected. Current theories of atmospheric distortions based on Kolmogorov turbulence would suggest that the motion should have maximum power at about 10 Hz for a 2.4-m aperture, so I was prepared to have to track much faster than that. However, as Figure 10 demonstrates, most of the power occurred at much lower frequencies, perhaps 0.5--3 Hz.

This is displayed as a power spectrum in Figure 11, along with the predictions by [3] for the image motion from Kolmogorov turbulence which would be expected for images with 1.0" FWHM and 0.25 " FWHM. The characteristic knee of these curves is where the maximum power occurs because the rise to lower frequencies is slower than the f^{-1} needed to overcome the shrinking df in log coordinates. We show two different power spectra to extend the frequency range. The open symbols were obtained from a bright star in 0.85" seeing, and the solid points were from the M71 observation in 0.50" seeing. (It's amusing to note the spikes at 0.4 Hz and 0.6 Hz. These are the errors caused by the normal autoguider when the seeing gets too good for the autoguider's 1" pixels.) The main difference between these two power spectra occurs at low frequencies.

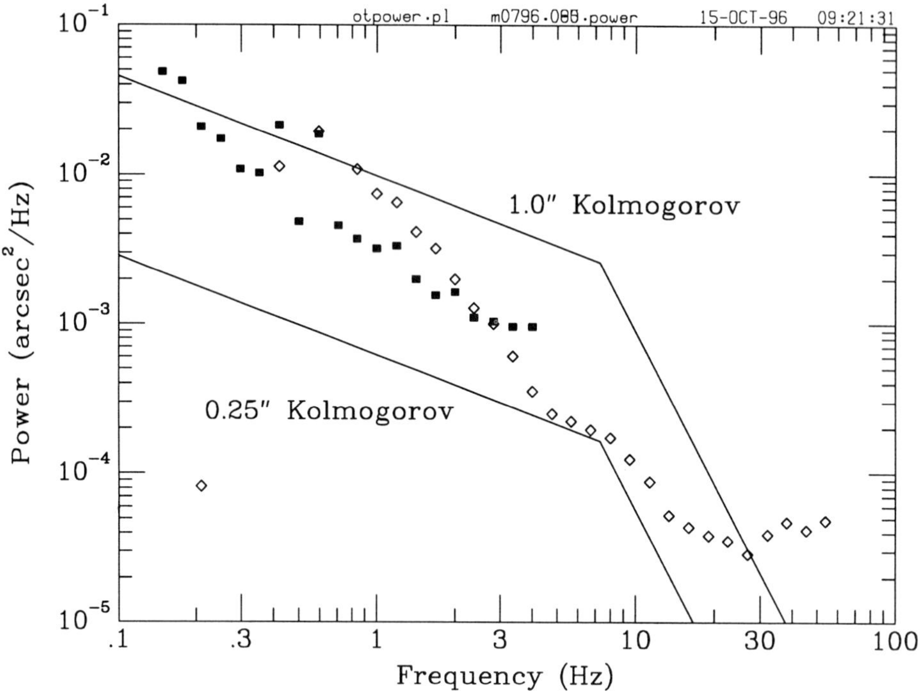

Figure 11: *A typical power spectrum of image motion. Two different observations are plotted to extend the frequency coverage, and the power at negative frequencies has been added to the positive frequency bins. The flattening out of the power spectrum beyond 20 Hz is the noise floor of our centroiding routine. The two lines are predictions for image motion due to Kolmogorov turbulence.*

It is apparent that the image motion power does not closely resemble Kolmogorov turbulence. As stated above, the power is more heavily weighted to low frequencies. We certainly expect that there is Kolmogorov turbulence being generated in the upper atmosphere, but it can have an amplitude no greater than 0.25" to be consistent with these observations. Where then does this extra power arise? We do not have the data to distinguish telescope shake from atmospheric turbulence, but much (if not most) of this power must come from the atmosphere because the OT tracking entirely removes telescope shake, and we are not seeing the 0.25" images which Kolmogorov theory suggests should remain.

We believe this image motion is generated low - either warm air around the telescope or rising out of the dome, or a boundary layer within a few hundred meters altitude. This sort of turbulence might not have developed a full Kolmogorov spectrum and could be heavily weighted to large eddies and slow turnover times. Support for this view comes from Figure 12, which shows that there is virtually no variation in the FWHM (or shape) of stars as we move away from the guide star which is being centroided. The size of the OTCCD is only 3' along its longest dimension, but out to about 2.5' we see none of the characteristic elongation of the PSF which is a normal consequence of the decorrelation of image motion with angular distance. This indicates that the isoplanatic angle (more precisely the isokinetic angle) is much bigger than one would expect for turbulence at an altitude of many kilometers.

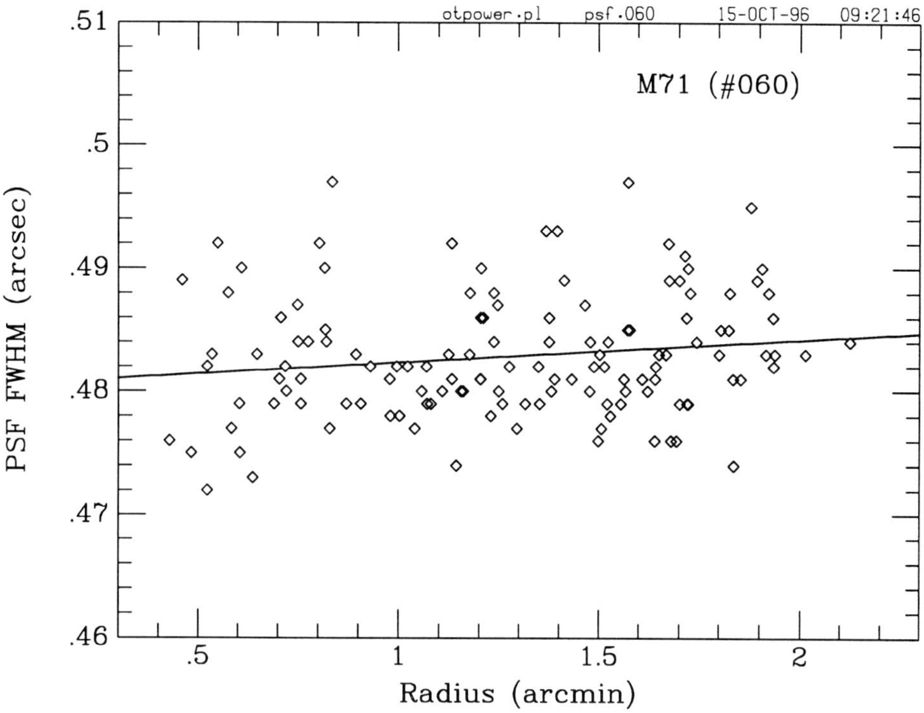

Figure 12: *The variation in image FWHM as a function of distance from the guide star.*

Regardless of whether this image motion comes from telescope shake or from non-Kolmogorov, low level turbulence, it is good news for people who need sharp images. The best possibility would be that Kitt Peak, a good but not outstanding site, commonly enjoys 0.25" seeing and telescope shake (which the OTCCD can remove over arbitrary field sizes) is a major contributor to image FWHM. The possibility which we believe is correct is that the very fast, Kolmogorov turbulence is *not* usually the major contributor to image FWHM, and the main culprit is slower, more benign turbulence generated near the telescope, and therefore having a coherence over a relatively large angle. Of course, this turbulence will be very site, season, and telescope specific, but for the MDM 2.4-m telescope it seems to contribute 0.2" of image size.

To summarize, we have built an OTCCD which is large enough to be a useful scientific device, and it works as expected. The OTCCD is a good way to carry out image motion compensation. It is cheap and simple (we designed the current OTCCD so that only one SDSU/Leach analog board is required), and it offers photon limited guiding, since the Lincoln Laboratory amplifier has only 1.5e$^-$ read noise. Obviously, this is much simpler and cheaper than a mechanical system involving a tip-tilt mirror, and we hope that we have pointed the way to software which makes this system easy and pleasant to use.

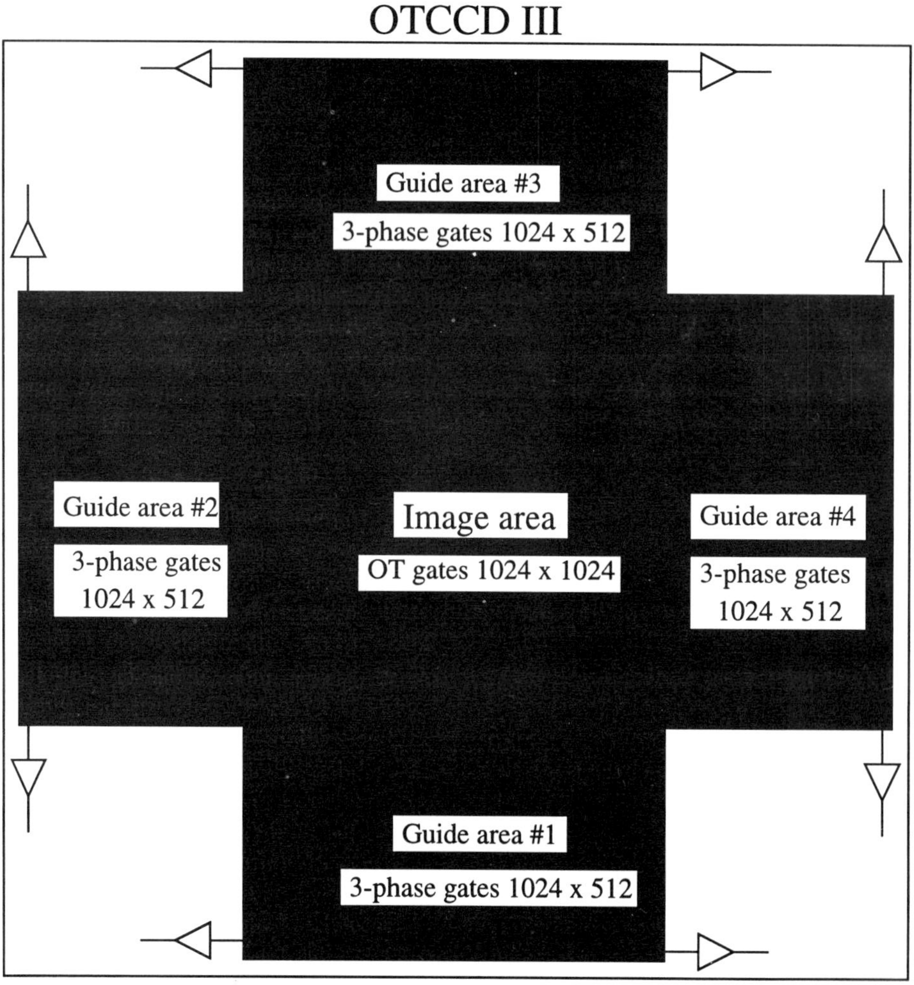

***Figure 13**: Possible layout for a third generation OTCCD with a central 1024×1024 of OT pixels surrounded by four1024×512 arrays of 3-phase pixels to act as guide star regions.*

We have found that the atmosphere is much nicer than we might have been led to believe. The image motion is generally slower than one would expect from fully developed Kolmogorov turbulence (0.5-3 Hz), and we see a very wide isokinetic angle (many arcminutes). We believe that this means that the turbulence occurs low, and has not yet generated a full Kolmogorov spectrum. For our telescope and our site, we found that there was about 0.2arcsec RMS of image motion, and the image FWHM could be improved by just about that much.

The pockets and traps in these chips have proven not to be as serious a problem as we had feared. Only the large pockets are effective at pumping charge, and they occupy a small fraction of the chip. We do not know why the OT region has many more pockets than the 3-phase region, but we suspect the aluminum gate may have something to do with it. (We chose to use aluminum for the gate for reasons of safety and expediency. We did not have time or money to make a mask for a poly gate, and we did not want to introduce any extra processing of the wafers which might jeopardize the UH/LL consortium chips.) We hope to carry out experiments with poly gates and deeper implants to try to reduce the number of pockets.

We would like to raise interest in building a larger format OTCCD. The present device seemed that it would cover the isokinetic patch, but we are seeing that this is not true. Also, it would be helpful to have larger guide areas, arranged all around the chip, so as to improve the probability of finding a bright guide star and to avoid the nuisance of having to rotate the instrument to acquire a guide star. Figure 13 illustrates a possible geometry for a 1024 × 1024 device, but of course an even larger OTCCD would be straightforward to build.

This work was sponsored by the Department of the Air Force, the Lincoln Laboratory Advanced Concepts Committee, and by NSF grant AST-9314665. Paul Schechter was the ever capable observer who coaxed the July weather into relenting long enough to permit him to collect the data presented here.

REFERENCES

[1] Burke, B.E., Reich, R.K., Savoye, E.D., & Tonry, J.L. (1994) IEEE J. Electron Devices 41 2482.

[2] Chatterjee, P.K., Taylor, G.W., & Tasch, A.F. (1979) IEEE Trans. on Electron Devices, ED-26, 6, 871.

[3] Parenti, R. (1993) Lincoln Labs Technical Report #966.

A TECHNIQUE FOR THE DETERMINATION OF CO-PLANARITY IN MOSAIC CCD CAMERAS.

SIMON TULLOCH[1] AND DAVE KING[1]

ABSTRACT

We have developed an optical non-contact scanner able to measure the detector surface profile of both single chip and mosaic CCD cameras. A precision of 1 μm in vertical displacement has been demonstrated when measuring 30 mm square CCDs. We have achieved similar performance when scanning mosaics with dimensions of up to 65 mm square. The system relies on using the CCD under test to measure the position of a pin-hole image projected onto its surface and so will only work with active operational mosaics that are capable of producing good quality images.

1. INTRODUCTION.

With the development of mosaic CCD cameras and very large area CCDs, it becomes increasingly hard to maintain device co-planarity, which is particularly important in fast optical systems. For example a CCD with a corner to centre bow of 50 μm , in an f3 beam , may defocus stellar images to a diameter of 17 μm at the device corners even if perfectly focused at the centre. It is therefore important to be able to measure chip flatness and mosaic co-planarity before delivery of the camera system to the observatory. The measurement system should obviously be non-contact and should measure devices at their operating temperatures. We have designed and built our own co-planarity measurement system here at RGO. It can measure CCD surface profile to an RMS accuracy of about 1 μm and can accommodate CCD mosaics up to 100 mm square. The measurement technique relies on using the CCD under test to record a spot or line image projected onto its surface by a scanning optical head. This precludes the measurement of poor image quality devices since hot pixels, poor CTE, bright and dark columns will distort the shape of the scan-lines.

2. ALTERNATIVE METHODS OF SURFACE PROFILE MEASUREMENT.

Interferometery is a popular method for measuring individual chips; two beams of plane-coherent light, one reflected from the CCD surface and one from a reference flat are combined on an imaging camera to yield a fringe pattern. Surface profile is determined from analysis of this pattern. This technique will only work for continuous surfaces and is therefore unsuitable for mosaic cameras.

Another popular method for surface profiling is *'Laser Triangulation'*. Here, a tightly focused laser beam is shone at the chip surface producing a small *'beacon'* . This beacon is observed by an off-axis microscope, which records its vertical position using either a linear CCD sensor or an analogue position sensitive detector. Sub-micron

[1] Royal Greenwich Observatory, Madingley Road, Cambridge, CB3 OEZ, UK. Email: **smt@ast.cam.ac.uk**

J. W. Beletic and P. Amico (eds.), Optical Detectors for Astronomy, 293–298.

resolution is claimed by the manufacturers. We have tested 3 of these laser devices but found each unable to meet our accuracy requirements. It is thought that the low reflectivity of the CCDs combined with scattering of the measurement beam from the cryostat window are to blame.

Other groups are considering using microscopes to focus onto the CCD surface for profile measurements. This may be difficult with backside illuminated devices due to the lack of surface features on which to focus.

3. SCANNER CONSTRUCTION.

As can be seen from figures 1 and 2, the main bulk of the system consists of a support box measuring 50 cm square and 36 cm high. The camera to be measured is placed on top of this box with the detector plane horizontal and the optical window facing down. The cameras produced by RGO have three point kinematic mounts for attachment to the telescope. Each mounting point has a screw adjuster for taking out tilt and for coarsely adjusting the focus. By use of a custom made cryostat interface plate it should be possible to mount cameras of any design on this support box.

Inside the box is a pair of orthogonal mounted horizontal translation stages. Each has a displacement range of ± 50 mm and a repeatability of 1 μm, they are computer controlled and their motion is fully programmable. When translating, the stage motion does not depart from a straight line by more than ± 1 μm, this is known as the *'tracking error'*.

Figure 1. *The INT Prime Focus Mosaic Camera mounted on the Flatness Scanner*

On top of the stages is the scan-head itself, it measures 98 mm in diameter and 130 mm high. Care is taken in the positioning of this scan-head with relation to the translation stages to ensure that the center of gravity of the top stage and scan-head does not at any time lie outside the base of the lower stage, a situation that would compromise translation accuracy. The scan-head consists of a symmetrical optical system containing two 40 mm diameter achromatic doublets in combination with two meniscus lenses. One doublet - meniscus pair acts as a f1.78 collimator for an LED -illuminated 50 μm pin-hole, the second focuses this collimated beam back down to a 75 μm FWHM spot image on the CCD.

This second lens pair is mounted in a threaded barrel that allows it to be focused vertically by +/-2 mm. Between the two pairs is fixed an aperture plate with a 14 mm diameter on-axis hole surrounded by a ring of 4 equi-spaced 6 mm diameter holes near the lens periphery. A slot allows a brass aperture selector plate to be inserted to uncover one or more of these holes.

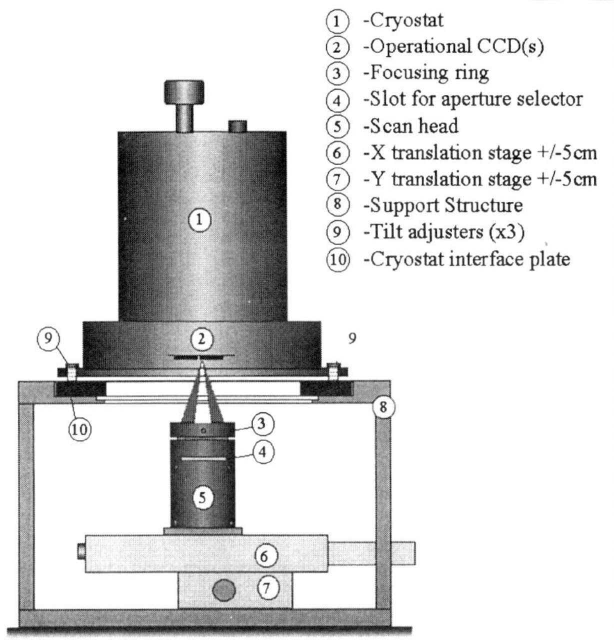

***Figure 2**. Mechanical schematic of scanner*

4. Scanner Operation.

4.1. Measurement Principle.

The principle of operation is derived from the Hartman Shutter technique used for focusing spectrographs. The scan-head can project the spot image onto the CCD from two directions, the angle between the two beams being 28.5 degrees. If the CCD does not lie precisely at the focal plane of the scanner then a double spot image will be recorded, the separation of the spots being proportional to the vertical distance by which the CCD is out of focus. The centroid of each spot image can be determined using the IRAF 'splot' function to an accuracy of better than 0.02 pixels. The optical arrangement is shown in Fig. 3.

In practice, centroiding can be done more accurately by recording two sets of spot images each taken with a separate beam. This is particularly the case if the two spots overlap. Surface maps can be made by recording spot images at many points on the CCD surface; as will be described later this is done by continuously scanning a spot across the chip producing line images. One advantage of this measurement system is that only two sets of line images are required to measure the profile of a whole CCD device.

4.2. Focusing the Scanner

Before measuring the surface profile it is important to correctly focus the scanner and make sure that the plane of the CCD is parallel to that of the XY translation plane. This is done by the iterative adjustment of the scan-head focusing ring and the three kinematic mounts. This is probably the most time consuming part of the measurement process; with practice this can be achieved in less than an hour. Coarse focusing and tilt removal are done by adjusting the three kinematic mounts so that the cryostat face plate is parallel to and 4 mm above the interface plate on top of the scan box. A 4 mm slip gauge can be used for this purpose. The scan-head is then focused fully down and moved using the translation stages to the centre of the CCD or CCD mosaic and an aperture selector plate inserted to uncover two of the peripherally placed apertures. The LED in the scan-head is then pulsed on for a few msec to produce a double spot image on the CCD. As the scan-head is gradually focused up, the separation of these spots should decrease until, when at the exact focal position, only a single spot is visible. All that remains is to remove the tilt in the system by measuring spot images near the edges of the chip and making the necessary adjustments with the three kinematic mounts. After focusing and tilt removal, no part of the CCD or CCD Mosaic surface should be greater than +/-100 μm from the scan-head focal plane. Residual tilt can be removed from the data later using a PC spreadsheet application.

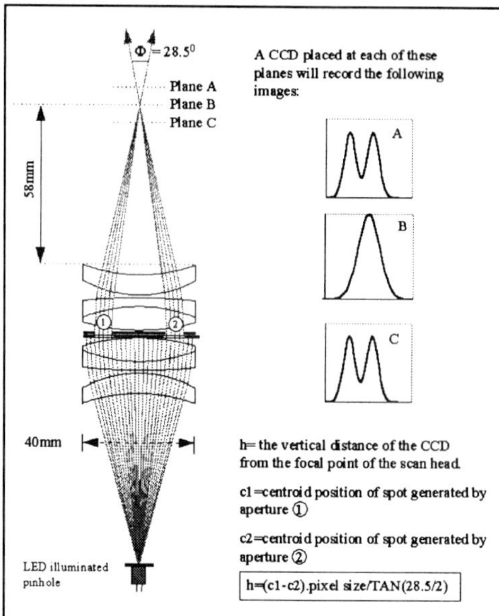

Figure 3. *Optical diagram*

4.3. Performing a Surface Profiling Scan

Once focusing and tilt removal are complete the CCD or CCD Mosaic can then be profiled. The scan-head is moved to a position off the edge of the CCD under measure. An aperture selector plate is inserted to select a single peripheral aperture in the scan-head and the LED is switched on. The translation stage controller is programmed to move the scan-head across the CCD to build up a series of equally spaced line images lying parallel to the CCD columns. The LED is then switched off and the scan-head returned to the starting position once more. After this image is read out, a second diametrically opposed aperture is selected in the scan-head and the translation stage program run again. The second image is then read out and compared with the first. Figure 4 shows an example of two scan line images from a Loral CCD. In the case of a perfectly flat CCD, the two sets of scan-lines will be perfectly parallel to each other. In the case of a bowed device, the separation of the lines in the first image from the corresponding line in the second image will tend to increase at the image centre. A thorough comparison requires the centroid of each scan-line to be calculated on a row by row basis. This can be done easily using IRAF using the ‘splot’ data processing routine which is designed to centroid spectral lines but works equally well on the line images produced by the scanner. It is a little slow, however, so it may be preferable to vertically bin each of the scan-line frames before processing. If the final binned image has as many rows as it does vertical scan-lines the resultant height data can be conveniently handled in a square format data table. ‘Splot’ can be confused by a high image bias, which should be subtracted off before centroid processing. Up to thirty such scan-lines can be fitted on a 30 mm wide CCD before scattered light from one line begins to distort its neighbours. The exposure level of the lines should exceed 10,000 e^- per pixel, which requires quite slow translation speeds. Covering a 30 mm square CCD with 30 scan-lines will take about 11 minutes so lower resolution scans using more widely spaced lines may be preferable for larger or mosaic devices. The first and second apertures should be selected such that a line connecting them lies perpendicular to the scan-lines. It is also important to remember which image was taken with which aperture. An IRAF script, which has been written to analyse the images, prompts the user for the identification of the image taken with the aperture lying closest to column 1 of the CCD i.e. closest to the readout amplifier. If this information is lost, there will be ambiguity as to whether the CCD exhibits a concave or convex bow.

The final processing stage is done by exporting the output of *'splot'* from IRAF to an EXCEL spreadsheet. The exported data takes the form of a table of height values measured in microns relative to the focal plane of the scan-head. Tilt is removed by fitting a reference plane to three corners of the data table and then subtracting this plane from the original data to produce a second table of *'de-tilted'* data.

The co-planarity scanner has also proved useful for checking cross-talk in split frame readout CCDs and evaluating the performance of autoguider cameras. In these applications, the on-axis aperture of the scan-head is selected and the scanner then provides a convenient artificial star.

4.4. Performance

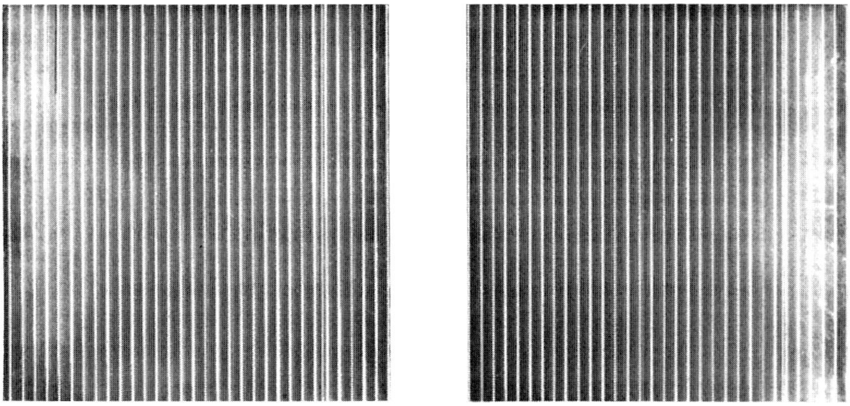

Figure 4a, 4b *. A scan line image pair from a Loral 2K 3EB CCD.*

The accuracy of this system was limited by the performance of the translation stages. The 'centroiding noise' or the vertical measurement error due to the small uncertainty with which the centroid of the scan-lines could be determined was 0.8 μm RMS. Measurement scans that used only one translation stage gave better noise performance than the two dimensional scans used for surface mapping.

5. CCD PROFILE SCAN RESULTS.

A number of surface profile scans have been performed at RGO on various CCDs. The results are presented below.

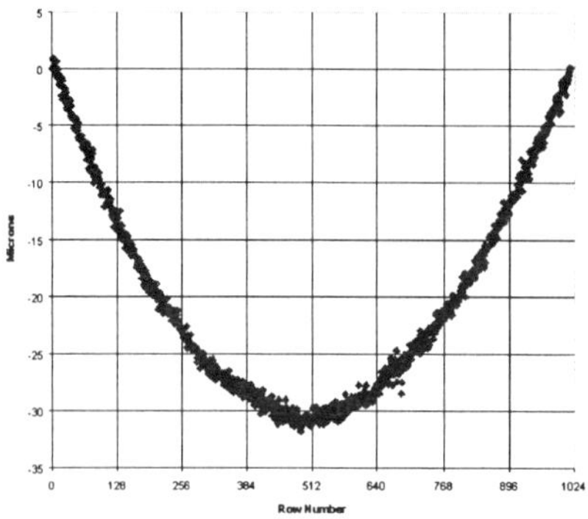

Figure 5

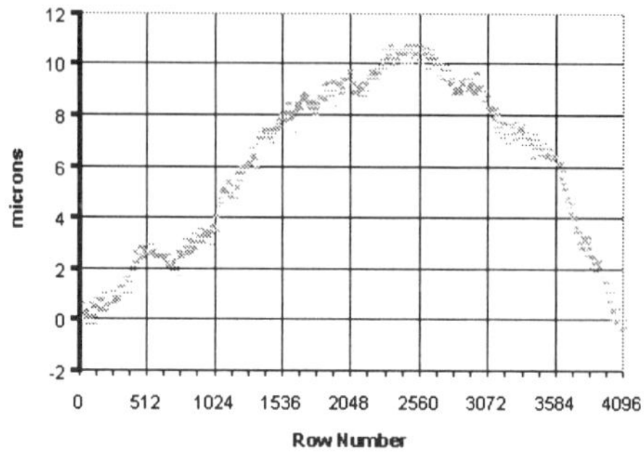

Figure 6

5.1. TEK 1K square, 24 μm pixel, thinned CCD.

This exhibited a pronounced convex bow of about 30 um from edge to centre. (Fig. 5)

5.2. EEV42 2k x 4k, 13.5 μm pixel, thinned CCD.

This device was exceptionally flat considering its large area (55 x 28 mm). (Fig. 6)

5.3. A Three Chip Mosaic

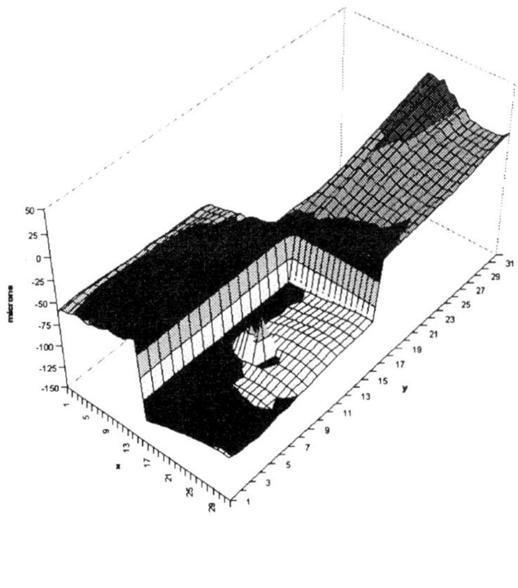

The chip in the foreground is not coplanar with the other two. Shim adjustment of the individual CCD positions is necessary in this case.

Acknowledgements: I am grateful to Sue Worswick for her help in designing and assembling the scan head optics. Thanks also to the RGO Mechanical Workshop, which manufactured the scanner.

CCD CAMERA CALIBRATION AND TESTING USING COMPACT LED LIGHT SOURCES

SIMON TULLOCH[1]

ABSTRACT

Two types of LED based light sources have been developed to assist in CCD camera commissioning and to perform rapid checks of camera operation at the telescope. The first is a compact device able to project flat fields onto a detector for defect mapping and as a check that UV flooded CCD devices have a uniform QE. The second design is a temperature compensated LED lamp that can be used for the determination of CCD Quantum Efficiency in both Imaging Mode and Photo-diode Mode. The lamps can provide narrow band output at various wavelengths between 380 nm and 950 nm depending on the choice of LED.

1. INTRODUCTION

The backside charging of CCDs by UV flooding can result in higher quantum efficiencies (QE) than with other techniques such as boron implantation. The crystal structure of the silicon is undamaged and the mobility of photoelectrons in the surface layers undiminished. A disadvantage, however, is that the chip surface is rendered very sensitive to contamination particularly if the camera cryostat is inadvertently allowed to warm up. Considerable out-gassing of cryo-pumped material can occur within a very short time of cryogen exhaustion and inevitably, some of this finds its way onto the CCD surface. A simple and quick method of QE determination is required, especially at wavelengths shorter than 400 nm where the QE is affected strongly by any contamination. Laboratory based systems for QE measurements are bulky and require the camera to be moved from the telescope environment. If the system contains an incandescent source, there will be the additional problems of long thermal settling times and red-leak in UV filters. We have designed a set of novel light sources based on LEDs that have none of the above disadvantages. This has been made possible by the introduction of high output blue LEDs by Nichia Chemical Industries which have significant output down to 380 nm.

2. A COMPACT FLAT FIELD LIGHT SOURCE

Flat field images can be used very effectively to diagnose when a CCD requires an additional UV flood. The reduction in QE after an UV flooded chip has been allowed to warm up is not constant across the chip surface and deep patterning can be seen in the flat field image. This is shown in Figs. 1 and 2. We have developed a compact flat field source using LEDs measuring only 125 mm in diameter by 70 mm high that can be quickly mounted to the front of our cameras. Its design is based around a geometrical rule that was predicted by a computer spreadsheet model.

---

[1] Royal Greenwich Observatory, Madingley Road, Cambridge, CB3 OEZ, UK. Email: **smt@ast.cam.ac.uk**

J. W. Beletic and P. Amico (eds.), Optical Detectors for Astronomy, 299–302.

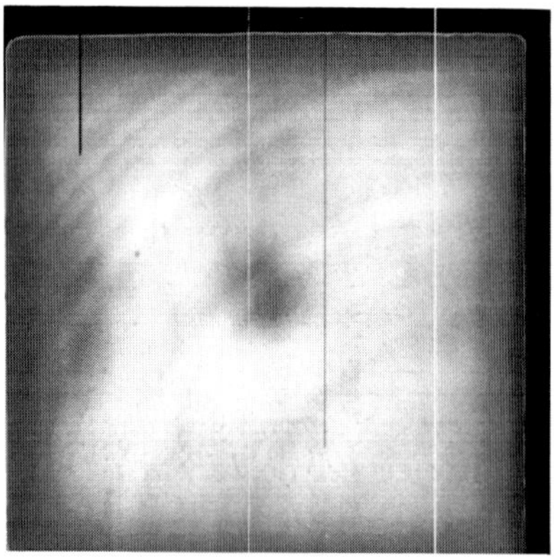

***Figure 1.**: flat field prior to UV flood*

***Figure 2.** Flat field after UV flood*

Both images where obtained using the flat field device described in this article.

The device, shown in Fig. 3 consists of a sand-blasted aluminium disc mounted directly above the cryostat window and parallel to the chip surface. This is illuminated from the camera side by a ring of twelve LEDs arranged in three groups of four. Each group emits light at a different wavelength: 390 nm, 570 nm and 950 nm so that flat fields can be obtained at widely spaced parts of the spectrum. The LEDs in each group are mounted at 90 degree intervals forming a square with sides equal in dimension to the separation of the CCD and sand-blasted disc. Each LED projects a small spot onto the disc near its periphery. Light scattered from these four spots illuminates the CCD with a flat field. If the disc-CCD separation is equal to 3 times the CCD size, the uniformity of this flat field will be better than 0.5%, assuming that the LEDs are of equal brightness. This can only be achieved by device selection or by tuning the drive current to each using a potentiometer. This has not yet been done and the device currently in use produces slightly sloped flat field images that are nonetheless very useful.

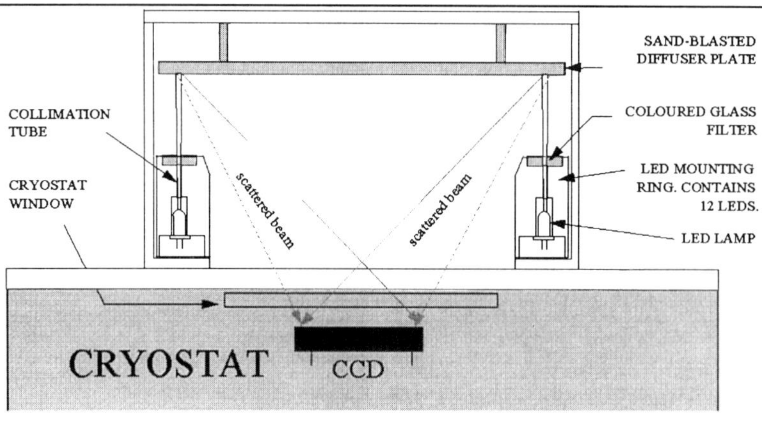

***Figure 3:** The Flat Fielder.*

3. AN LED BASED REFERENCE LIGHT SOURCE

The second design is for a set of compact lamps, also using LED sources that can be attached to the front of a camera for accurate QE measurement at various discrete wavelengths down to 380 nm. They are also useful for determining amplifier gain and linearity. One disadvantage associated with LEDs is their large temperature coefficients; each lamp therefore contains a thermistor sensor and an op-amp compensation circuit to vary the drive current as the temperature varies. Self heating in the LEDs is minimised by operating at low currents, typically 1 mA, wherever possible. In the case of the 380 nm lamp, however, a drive current of 20 mA had to be used to ensure adequate output. This caused a rise in junction temperature of about 10 C. Placing the thermistor actually inside the LED package compensated for this effect and resulted in a self heating error of less than 1% in all cases.

The lamps, shown schematically in Fig. 4, are highly compact, measuring 55 mm in diameter and 70 mm high. They require only a single rail unregulated supply of between 15 and 20V and can be switched on and off in about 2 ms. The lamp optics are very simple: in front of the LED is placed a ground glass diffuser followed by a 5 mm diameter aperture disc. The light then passes through a bandpass interference filter and a 7 mm diameter exit aperture. The bore in which these components are mounted is first sand-blasted and then black anodised to reduce scattered light.

These lamps are best suited for 'diode mode' measurements of CCD QE. For this application, a lamp is attached to the front of the camera with a short spacer barrel and illuminates a spot at the centre of the CCD. An additional aperture at the base of this barrel defines the spot size. Once the lamp is mounted on the camera, a picoammeter connected between RD (Reset Drain) and VSS (Substrate) of the CCD will yield a current proportional to QE. This current can then be compared to that produced in a calibrated photodiode placed in a similar position in front of the lower aperture. Imaging mode measurements are also possible by pulsing the lamps on while the CCD is integrating.

Each lamp is calibrated, using a precision photodiode, in an environmental chamber to check its stability across the typical range of observatory temperatures. The four lamps built so far exhibit maximum errors of +/-1% across the temperature range 0 C to 25 C. Long term stability is currently being investigated.

Acknowledgements: I am grateful to Nathalie Falleur who calibrated the LED lamps used in these instruments and to Terry Dobner who manufactured the light sources.

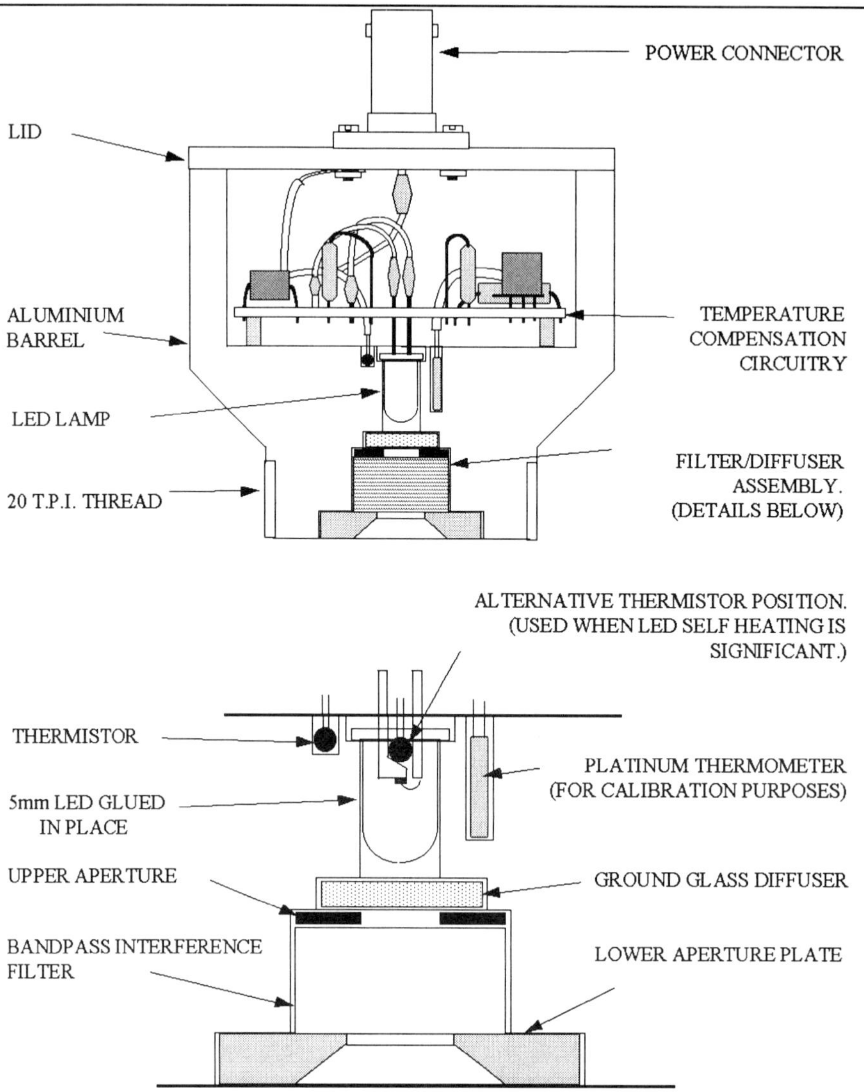

Figure 4: *Reference Light Source*

A FULLY DEPLETED PN-JUNCTION CCD FOR INFRARED-, UV- AND X-RAY DETECTION

C. VON ZANTHIER [1]*

H. BRÄUNINGER, K. DENNERL, R. HARTMANN, G. HARTNER, H. HIPPMANN, A. KALTENBERGER*, E. KASTELIC, W. KINK, N. KRAUSE*, N. MEIDINGER*, G. METZNER, E. PFEFFERMANN, M. POPP*, C. REPPIN, J. RIEDL*, D. STÖTTER*, L. STRÜDER*, J. TRÜMPER, U. WEBER* [2]*

D. CARATHANASSIS, S. ENGELHARD*, TH. GEBHART*, D. HAUFF*, G. LUTZ*, R. H. RICHTER*, H. SEITZ*, P. SOLC [3]*

E. BIEHLER, H. BÖTTCHER, E. KENDZIORRA J. KRÄMER, B. PFLÜGER, R. STAUBERT [4]

P. HOLL, J. KEMMER, P. LECHNER*, B. MAIER*, H. SOLTAU*, R. STÖTTER*, E. BIHLER, H. BÖTTCHER, E. KENDZIORRA, J. KRÄMER, B. PFLÜGER, R. STAUBERT [5]*

ABSTRACT

This paper describes the performance of the Fully Depleted pn-junction CCD (pn-CCD) system, developed for ESA's XMM-satellite mission for soft x-ray imaging and spectroscopy in the single photon counting mode in the 100 eV to 10 keV photon range. The 58 mm x 60 mm large pn-CCD array, designed and fabricated at the Semiconductor Lab (Halbleiterlabor) of the Max-Planck-Institut, uses pn-junctions for registers and as backside structure. This concept naturally enables full depletion of the detector volume independent of the silicon wafer's resistivity and thickness, and as such make it an efficient detector for the x-ray region and the infrared. For high detection efficiency in the soft x-ray region and UV, an ultrathin pn-CCD backside deadlayer has been realized. Each pn-CCD-channel is equipped with its own on-chip JFET amplifier which, in combination with the CAMEX-amplifier and multiplexing chip, facilitates parallel readout and fast data rate: the cooled pn-CCD system can be read out at a data rate up to 3 MHz with an electronic noise floor of ENC < 5 e-.

1. THE PN-CCD CONCEPT

The Fully Depleted pn-Junction CCD (pn-CCD) was first proposed by E. Gatti and P. Rehak in 1983 [1].

[1] KETEK GmbH, Am Isarbach 30, D - 85764 Oberschleißheim, Germany
[2] Max-Planck -Institut für extraterrestische Physik, Gießenbachstr., D-85740 Garching, Germany
[3] Max-Planck -Institut für Physik, Föhringer Ring 6, D - 80805 München, Germany
[4] Institut für Astronomie und Astrophysik der Universität Tübingen, Waldhäuser Str. 64, D - 72076 Tübingen, Germany
[5] Institut für Astronomie und Astrophysik der Universität Tübingen Waldhäuser Straße 64, D - 72076 Tübingen, Germany
* Postal Address: MPI Halbleiterlabor, Paul-Gerhardt-Allee 42, D - 81245 München, Germany

J. W. Beletic and P. Amico (eds.), Optical Detectors for Astronomy, 303–310.

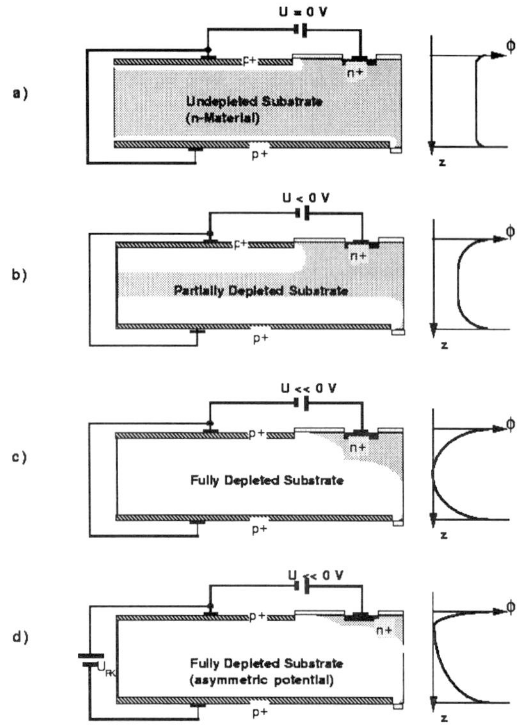

Figure.1.The process of "Sidewards Depleting" the wafer. By increasing the reverse biasing voltages between the two pn-diodes and the n-substrate contact, the non-depleted fieldfree substrate region slowly vanishes. Also shown is the potential across the wafer depth.

It is based on the concept of *sidewards depleting* the semi-conductor, a concept also used for silicon drift detectors [2]. Sidewards depletion works as shown in Fig. 1a) to d). By applying reverse biased voltages between the two pn-junctions of both sides of the wafer and an arbitrarily small substrate contact, it is possible to deplete the full wafer and make it sensitive to ionizing radiation. With this concept full wafer depletion is possible independent of the resistivity and thickness of the silicon wafer material, although high resistivity material is recommended to have the pn-diode voltages not too high.

The electrical potential across the depleted region is also shown in Fig. 1. The potential minimum is given by the valley parallel to the wafer surfaces. In this valley signal electrons in n-material (and holes in p-material) are collected. By increasing the voltage at the backside, the potential minimum can be shifted close to the wafer frontside (Fig. 1d), a necessary condition for a CCD-signal transfer. Note that in any case the surface regions have the strongest electrical fields which is helpful for an efficient charge collection of signal electrons generated by photons near the surfaces.

To contain the signal charge in only a small region within the potential valley the pn-junction on one wafer side must be structured in stripes such that different voltages applied to them form potential wells in transfer direction. Fig. 2 shows this concept for 2.5 kΩ-cm n-type material. The CCD-stripes are called registers. As for conventional CCDs, by applying clocked periodic voltages to the registers, signal charges can be moved to the substrate contact, the most positive region in the wafer. Note that the backside voltage is about 4 times as high as the offset voltage of the transfer registers in order to have the transfer plane closer to the registers; this way the charge transfer is about 10 μm below the registers surface.

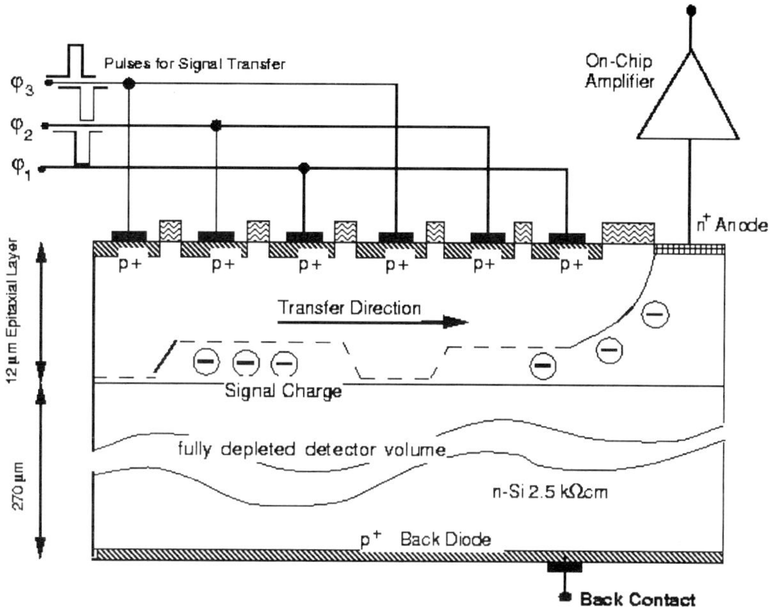

Figure 2. *Cross section through the pn-CCD along one channel. A pixel in the fully depleted pn-CCD volume is formed by the negative voltages applied to the pn-junction transfer registers and the pn-backside diode. By clocking the three independent registers with a periodic pulse scheme the signal charge is transferred to the anode.*

The conception of the low-noise JFET fabrication technology on high resistivity material [3] makes it possible to integrate a low-noise, low current amplifier JFET on the pn-CCD. JFETs inherently have a low 1/f noise performance and can be operated at low current even for low noise applications. Due to the low power consumption, it is possible to integrate an amplifier JFET to each pn-CCD-channel. Reading out the channels in parallel decreases the readout time significantly.

The pn-CCD described above has several advantages:

- It is easy to fully deplete the CCD. The depletion depth depends on the selected wafer thickness only (typically 300 - 500 µm), which is homogeneous and leaves no uncertainty about detection efficiencies. - No wafer thinning is necessary
- With the pn-CCD illuminated from the backside, a very thin and homogeneous pn-junction entrance window can be realized with a deadlayer smaller than 500 Å [11]. A thin entrance window results in a high detection efficiency in the UV and soft x-ray region.
- Due to the parabola like potential between the two CCD surfaces the electric field at the entrance window is strong. As a result, photons converting close at the surface are still fully visible due to the high charge collection efficiency in this region.

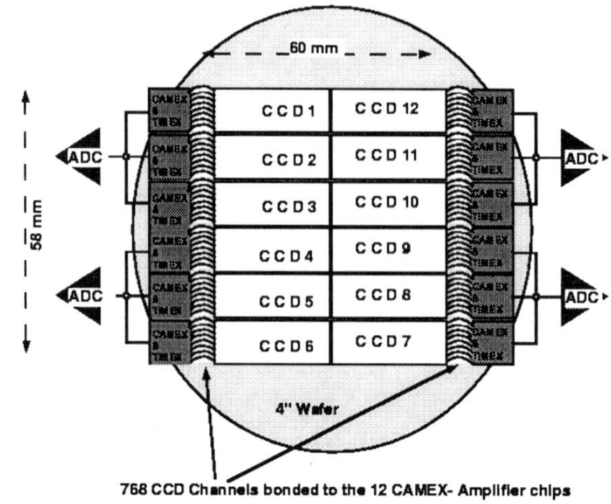

Fig.3. Schematic overall view on the 58 mm x 60 mm large pn-CCD-focal plane for the XMM-satellite. 12 pn-CCD units are integrated monolithically on a 4'' wafer. Each pn-CCD unit is connected to the CAMEX CMOS amplifier and multiplexer chip via wire bonds. Every three outputs of three CAMEX chips are multiplexed to one ADC.

- The absence of active MOS-structures increases the immunity to radiation damage [4]

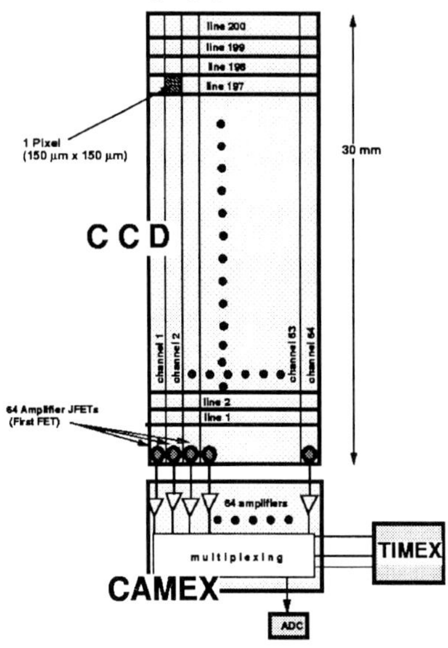

***Figure 4**. Schematic view on a pn-CCD unit in connection with the CAMEX and TIMEX chip. The CAMEX amplifies the pn-CCD signals of the 64 channels in parallel and multiplexes them to an ADC. The TIMEX controls the switching timing for the CAMEX.*

- The pn-CCD fabrication compatibility with JFET fabrication on high-resistivity material allows for the on-chip implementation of low noise, low current CCDs and parallel readout architecture.

2. THE PN-CCD CAMEX SYSTEM FOR XMM

The pn-CCD development is driven by the demands of the x-ray telescopes of ESA's X-ray Multi-Mirror (XMM) satellite to be launched in 1999 [5, 6]. XMM will image and spectroscopically analyse single incoming photons in the 100 eV to 10 keV region coming from the universe. The three telescopes will be equipped with large Wolter II mirrors with an angular resolution of 20'' and a focal length of 7.5m. One of the three telescopes' focal plane cameras will be instrumented with a pn-CCD.

Fig. 3 shows schematically the overall view on the pn-CCD camera for XMM. The sensitive area is 58 mm x 60 mm with a pixel size of 150 μm x 150 μm. The CCD is

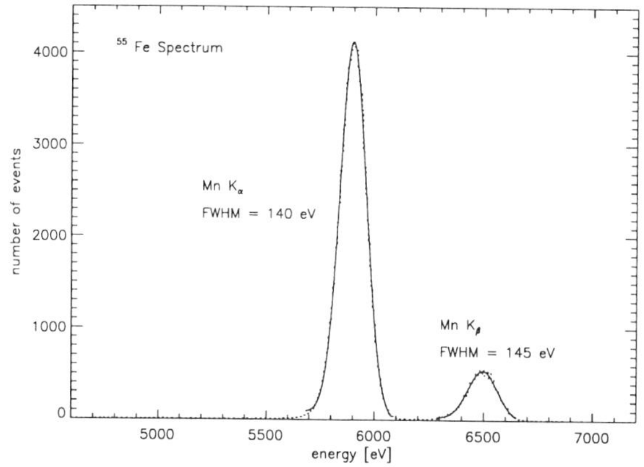

Figure 5. A Mn-K-α spectrum with a FWHM=140 eV, reflecting the high spectroscopic resolution of the pn-CCD-CAMEX system. The electronic noise floor is ENC< 5e-. No events are rejected from this spectrum.

monolithically integrated on a 4'', 2.5 kΩ-cm material wafer. The large pixel size is matched to the telescope's spatial resolution. The large array is subdivided in 12 smaller pn-CCD units each 9.7 mm x 30 mm large (Fig. 4), having 64 channels and 200 lines. Each of the 64 channels is equipped with a low-noise on-chip JFET amplifier, which in turn is wire-bonded to the CAMEX VLSI-chip [7, 8]. The CAMEX is a CMOS 64-channel amplifier and multiplex chip matched to read out the 64 pn-CCD channels in parallel. The parallel pn-CCD readout together with the CAMEX amplifier chip allow for a long low-noise optimized shaping time of typically 5 μs at a high CAMEX data rate of 3 MHz.

For best spectroscopic performance, the pn-CCD is cooled down to about 150 K.

Fig. 5 shows a spectrum taken with 5.9 keV Mn-K-α and 6.49 keV K-β photons. The full width at half maximum at this energy is FWHM = 140 eV. The contributions to the width are the Fano-noise (about 117 eV), the readout noise (40 eV) and noise due to the Charge Transfer Inefficiency (CTI) [9].

3. THE PN-CCD BACKSIDE ENTRANCE WINDOW

A high efficient photon detection in the energy range below 500 eV requires special care for the entrance window because of the short photon absorption length at this energy range. No significant deadlayer is allowed. For a fully depleted CCD, the backside is usually taken as entrance window, which has no absorbing material from the CCD registers and electronics. For the pn-CCD, the backside entrance window is essentially a pn-junction with the deadlayer thickness dominated by the p-implantation depth. With a special implantation technique and a compensating n-implantation step [10], it was possible to keep the p-implantation

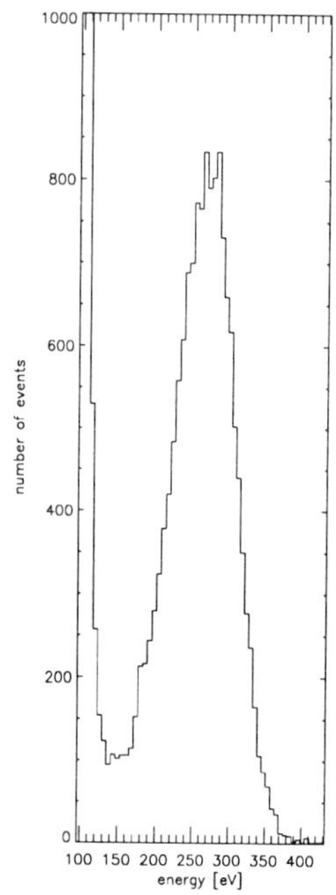

Fig.6. A C-K-α spectrum at 277 eV with a FWHM = 88 eV. No neighboring peaks due to inhomogeneous entrance windows are seen.

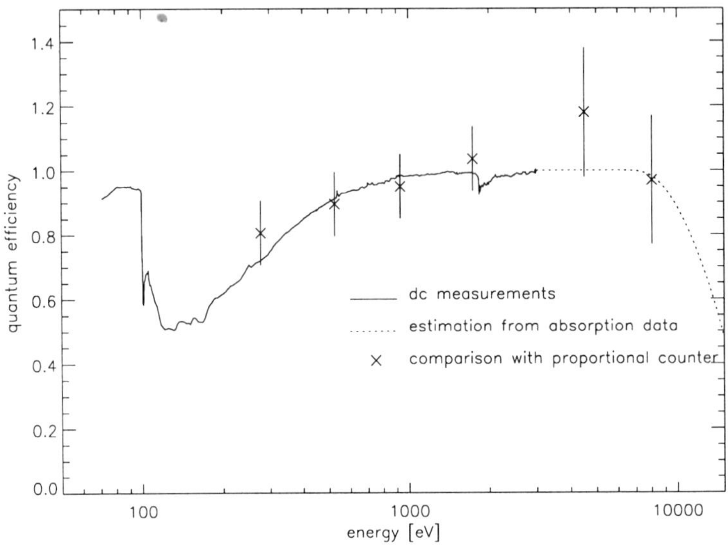

Figure 7: The quantum efficiency curve measured at the BESSY Synchrotron Laboratory in the energy range between 100 and 1000 eV. The high efficiency in the high energy range is due to the 300 μm depletion depth of the detector, the high efficiency at the low energy end is due to the thin backside entrance window.

depth below 40 nm. The strong electric field due to the backside voltage and the compensating n-implantation is able to collect signal electrons at a depth of 70 nm to more than 90%. As a result, the C-line at 277 eV can be fully reproduced with high spectral quality (Fig. 6). No peak distortions can be seen as a consequence of the homogeneous entrance window and good charge collection efficiency.

Fig. 7 shows the detection efficiency of the pn-CCD in the energy region from 100 eV to 10 keV measured in the DC-photocurrent mode and calibrated with a reference diode. The > 90% detection efficiency at 10 keV is due to the large depletion width of 300 um given by the wafer thickness. In the soft x-ray range at above 100 eV, the detection efficiency is still above 60 % due to the thin deadlayer.

4. The pn-CCD for UV and IR Measurements

The work on thin entrance windows generated some interest on how the pn-CCD would perform in the UV region, where the photon absorption length can be below 10 nm. For this purpose, an absolute measurement of the spectral responsivity was performed. Two calibrated diodes from HAMAMATSU (AE-968 for wavelength below 450 nm and Type 1790 for longer wavelengths) served as a reference.

Fig. 8 shows the responsivity and the quantum efficiency extracted from these measurements for two different entrance windows. The lower two curves represent the values of a conventional entrance window with a standard p-implantation. The two curves with the higher values come from devices with the ultrathin entrance window. It turns out that for the ultrathin entrance window the quantum efficiency is well above

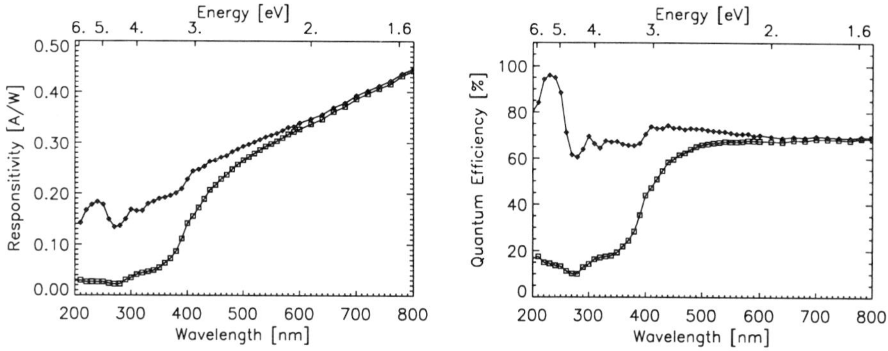

Figure 8.: Spectral responsivity and external quantum efficiency for two detectors with different entrance windows. The SiO_2 thickness is 30 Å for a conventional (hollow squares) and 200 Å for an ultrathin entrance window (filled squares). The detectors were operated at a bias voltage of 1 V

60% from the UV- region up to the IR at 1000 nm. At around 250 nm the quantum efficiency rises even to 95 % due to the onset of the double electron-hole pair creation. Without anti-reflecting coating, this value in the UV is significantly higher than for conventional UV-diodes.

5. Conclusion and Outlook

The pn-CCD is a truly fully depleted CCD, whose depletion width depends on the selected wafer thickness only. In combination with a low-noise readout JFET for each pn-CCD channel and the CAMEX amplifier and multiplexer chip, the pn-CCD system can be read out fully in parallel at low noise (ENC = $5e^-$). The CAMEX multiplexes the 64 analog signals at 3 MHz to an ADC. The pn-CCD is optimized for the XMM telescope requirements for single photon counting, imaging and spectroscopy in the energy range from 100 eV to 10 keV and a pixel size of 150 μm x 150 μm. For other applications however, smaller pixel sizes can be realized. The ultrathin backside entrance window in combination with the large depletion depth of a few 100 μm make the pn-CCD also a fast, highly efficient imaging device in the UV, visible and infrared i.e. at wavelength from 200 nm to 1000 nm.

6. References

[1] E. Gatti, P. Rehak, *Nucl. Instrum. Methods* , **A225** (1984) 608

[2] E. Gatti, P. Rehak, *Nucl. Instrum. Methods*, **A226** (1984) 129

[3] E. Pinotti et al., *Nucl. Instrum. Methods*, **A326** (1993) 85

[4] N. Meidinger et al., *Nucl. Instrum. Methods*, **A377** (1996) 298

[5] Report on ESA's Scientific Satellites, esa SP-1110, (1989) 85

[6] L. Strüder et al., *Nucl. Instrum. Methods*, **A288** (1990) 227

[7] W. Buttler et al., *Nucl. Instrum. Methods*, **A279** (1989) 204

[8] H. Bräuninger et al., *Nucl. Instrum. Methods*, **A326** (1993) 129
[9] H. Soltau et al., *Nucl. Instr Methods*, **A377** (1996) 340
[10] R. Hartmann et al., *Nucl. Instrum. Methods*, **A377** (1996) 191
[11] R. Hartmann et al, *Nucl. Instrum. Methods*, to be published